Modern Industrial Microbiology and Biotechnology

Modern Industrial Microbiology and Biotechnology

Simone Cliff
Marta Cortez

Kruger Brentt
Publishers
2025

Kruger Brentt Publishers UK. LTD.
Company Number 9728962

Regd. Office: 68 St Margarets Road, Edgware, Middlesex HA8 9UU

ISBN: 978-1-78715-442-1

For information on all our publications visit our website at http://krugerbrentt.com/

PREFACE

Modern industrial microbiology and biotechnology represent a convergence of microbiological principles and biotechnological innovations aimed at enhancing production processes across various sectors, including pharmaceuticals, food and beverages, agriculture, and environmental management. This interdisciplinary field leverages the unique properties of microorganisms to develop products and processes that are efficient, sustainable, and environmentally friendly. At the heart of this domain lies the understanding of microbial physiology, genetics, and metabolism, which enables scientists and engineers to harness the capabilities of microbes for industrial applications.

One of the key aspects of modern industrial microbiology is fermentation technology, a process that utilizes microorganisms, such as bacteria, yeast, and fungi, to convert raw materials into valuable products. Fermentation processes have been used for centuries in food and beverage production, including the making of beer, wine, yogurt, and cheese. However, advancements in fermentation technology have led to the development of novel bioprocesses that allow for the production of biofuels, biochemicals, and pharmaceuticals on a large scale. By optimizing conditions such as temperature, pH, and nutrient availability, industries can enhance microbial growth and product yields, making fermentation a cornerstone of modern biomanufacturing.

Another vital component of industrial microbiology is the use of microbial biocatalysts, which are enzymes produced by microorganisms that facilitate biochemical reactions. These enzymes are increasingly being employed in various industrial processes due to their specificity, efficiency, and ability to function under mild conditions. Biocatalysis is used in the synthesis of pharmaceuticals, the production of biodegradable plastics, and the processing of agrochemicals, offering an environmentally friendly alternative to traditional chemical synthesis methods. Additionally, the advent of genetic engineering and synthetic biology has revolutionized industrial microbiology by enabling the modification of microbial genomes for enhanced performance. Techniques such as recombinant DNA technology allow for the introduction of new genes into microorganisms, granting them the ability to produce complex biopharmaceuticals, enzymes, and bioactive compounds that would otherwise be difficult to obtain. This capability not only expands the range of products that can be generated but also improves the efficiency of production processes, making them more economically viable.

Modern industrial microbiology also plays a crucial role in addressing environmental challenges through bioremediation and waste treatment. Microorganisms are employed to degrade pollutants in contaminated soils and water, helping to restore ecosystems affected by industrial activities. This biotechnological approach not only mitigates environmental damage but also promotes the recycling of waste materials into valuable resources, such as biofuels and bioproducts. Furthermore, the integration of biotechnology in agriculture through the use of microbial biofertilizers and biopesticides enhances crop yield and pest resistance while reducing reliance on chemical fertilizers and pesticides, contributing to sustainable farming practices.

As the demand for sustainable and eco-friendly solutions continues to grow, modern industrial microbiology and biotechnology are positioned to play a pivotal role in the future of production processes across various industries. Ongoing research and technological advancements are expected to further enhance the efficiency and applicability of microbial processes, driving innovation in product development and manufacturing. The combination of traditional microbiological techniques with cutting-edge biotechnological approaches creates a dynamic field poised to address global challenges, such as food security, energy sustainability, and environmental protection. Through a commitment to sustainable practices and the responsible use of microbial resources, modern industrial microbiology and biotechnology offer promising pathways toward a more sustainable and resilient future.

Simone Cliff

Marta Cortez

CONTENTS

SECTION 1:
INTRODUCTION

CHAPTER 1

INTRODUCTION OF INDUSTRIAL MICROBIOLOGY AND BIOTECHNOLOGY

1.1 INDUSTRIAL MICROBIOLOGY

Industrial microbiology is a branch of biotechnology that applies microbial sciences to create industrial products in mass quantities, often using microbial cell factories. There are multiple ways to manipulate a microorganism in order to increase maximum product yields. Introduction of mutations into an organism may be accomplished by introducing them to mutagens. Another way to increase production is by gene amplification, this is done by the use of plasmids, and vectors. The plasmids and/ or vectors are used to incorporate multiple copies of a specific gene that would allow more enzymes to be produced that eventually cause more product yield. The manipulation of organisms in order to yield a specific product has many applications to the real world like the production of some antibiotics, vitamins, enzymes, amino acids, solvents, alcohol and daily products. Microorganisms play a big role in the industry, with multiple ways to be used. Medicinally, microbes can be used for creating antibiotics in order to treat infection. Microbes can also be used for the food industry as well. Microbes are very useful in creating some of the mass produced products that are consumed by people. The chemical industry also uses microorganisms in order to synthesize amino acids and organic solvents. Microbes can also be used in an agricultural application for use as a biopesticide instead of using dangerous chemicals and or inoculants to help plant proliferation.

1.1.1 Medical Application

The medical application to industrial microbiology is the production of new drugs synthesized in a specific organism for medical purposes. Production of antibiotics is necessary for the treatment of many bacterial infections. Some natural occurring antibiotics and precursors, are produced through a process called fermentation. The microorganisms grow in a liquid media where the population size is controlled in order to yield the greatest amount of product. In this environment nutrient, pH, temperature, and oxygen are controlled also in order to maximize the amount of cells and cause them not to die before the production of the antibiotic of interest. Once the antibiotic is produced it must be extracted in order to yield an income.

Vitamins also get produced in massive quantities either by fermentation or biotransformation. Vitamin B 2 (riboflavin) for example is produced both ways. Biotransformation is mostly used for the production of riboflavin, and the carbon source starting material for this reaction is glucose. There are a few strains of microorganisms that were engineered to increase the yield of riboflavin produced. The most common organism used for this reaction is *Ashbya gossypii.* The fermentation process is another common way to produce riboflavin. The most common organism used for production of riboflavin through fermentation is *Eremothecium ashbyii.* Once riboflavin is produced it must be extracted from the broth, this is done by heating the cells for a certain amount of time, and then the cells can be filtered out of solution. Riboflavin is later purified and released as final product.

Microbial biotransformation can be used to produce steroid medicaments. Steroids can be consumed either orally or by injection. Steroids play a big role in the control of arthritis. Cortisone is an anti-inflammatory drug that fights against arthritis, as well as several skin diseases. Another steroid used is testosterone, which was produced from dehydroepiandrosterone by using the *Corynebacterium species.*

1.1.2 Food Industry Application

Fermentation

Fermentation is a reaction where sugar can be converted into a gas, alcohols or acids. Fermentation happens anaerobically, which means microorganisms that go through fermentation can function without the presence of oxygen. Yeasts and bacteria are commonly used to mass produce multiple products. Drinking alcohol is a product that is produced by yeasts and bacteria. Alcohol that can be consumed is also known as ethanol, and ethanol is used to power automobiles as a fuel source. Drinking alcohol is produced from natural sugars like glucose. Carbon dioxide is produced as a side product in this reaction and can be used to make bread, and can also be used to carbonate beverages. Fermentation Wine: Alcoholic beverages like beer and wine are fermented by microorganisms when there is no oxygen present.

In this process, once there is enough alcohol and carbon dioxide around in the media the yeast start to die due to the environment becoming toxic to them. There are many strains of yeast and bacteria that can tolerate different amounts of alcohol around in their environment before it becoming toxic, thus one can obtain different alcohol levels in beer and wine, just by selecting a different microbial strain. Most yeast can tolerate between 10 and 15 percent alcohol, but there are some strains that can tolerate up to 21 percent alcohol. Dairy products like cheese and yogurt can also be made through fermentation using microbes. Cheese was produced as a way to preserve the nutrients obtained from milk, through fermentation thus elongating the shelf-life of the product. Microbes are used to convert the lactose sugars into lactic acid through fermentation.

The bacteria used for such fermentation are usually from *Lactococci*, *Lactobacilli*, or *Streptococci* families. Sometimes these microbes are added before or after the acidification step needed for cheese production. Also these microbes are responsible for the different flavors of cheese, since they have enzymes that breakdown milk sugars and fats into multiple building blocks. Some other microbes like mold may be purposely introduced during or before the aging of the cheese, in order to give it a different flavor.

The production of yogurt starts from the pasteurization of milk, where undesired microbes are reduced or eliminated. Once the milk is pasteurized the milk is ready to be processed to reduce fat and liquid content, so what remains is mostly solid content. This can be done by drying the milk so that the liquid evaporates or by adding concentrated milk. Increasing the solid content of the milk also increases the nutritional value since the nutrients are more concentrated. After this step is accomplished, the milk is ready for fermentation where the milk gets inoculated with bacteria in hygienic stainless steel containers and then gets carefully monitored for lactic acid production, temperature and pH.

Enzymes can be produced through fermentation either by submerged fermentation and/ or by solid state fermentation. Submerged fermentation is referred to when the microorganisms are in contact with media. In this process the contact with oxygen is essential. The bioreactors/fermentors that are used to do these mass production of product can store up to 500 cubic meters in volume. Solid state fermentation is less common than submerged fermentation, but has many benefits. There is less need for the environment to be sterile since there is less water, there is a higher stability and concentration for the end product. Insulin synthesis is done through the fermentation process and the use of recombinant *E.coli* or yeast in order to make human insulin also called Humulin.

1.1.3 Agriculture Application

The demand for agricultural products is constantly increasing due to the need of various fertilizers and pesticides. There are long term effects of the overuse of chemical fertilizers and pesticides. Due to the excessive use of chemical fertilizers and pesticides, the soil becomes infertile and a non-sufficient use for growing crops. For that matter, biofertelizers, biopesticides and organic farming come to the rescue.

Biopesticide is a pesticide derived from a living organism or natural occurring substances. Biochemical pesticides can also be produced from naturally occurring substances that can control pest populations in a non-toxic matter. An example of a biochemical pesticide is garlic and pepper based insecticides, these work by repelling insects from the desired location. Microbial pesticides, usually a virus, bacterium, or fungus are used to control pest populations in a more specific manner. The most commonly used microbe for the production of microbial biopesticides is *Bacillus thuringiensis*, also known as Bt. This spore forming bacterium produces a delta-endotoxins in which it causes the insect or pest to stop feeding on the crop or plant because the endotoxin destroys the lining of the digestive system.

1.1.4 Chemical Application

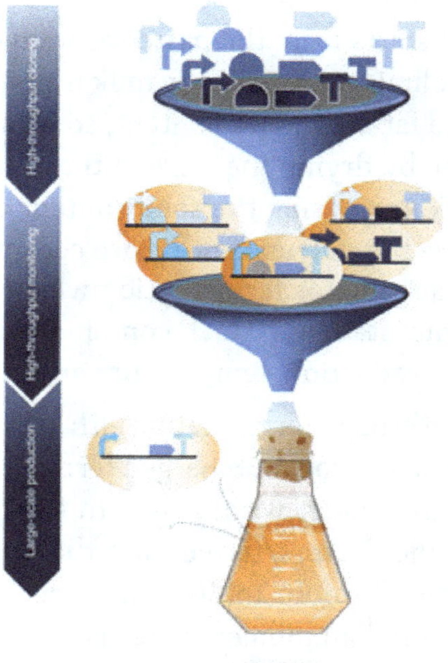

Fig 1: Schematic Workflow for Microbial Factory Optimization

Synthesis of amino acids and organic solvents can also be made using microbes. The synthesis of essential amino acids such as are L-Methionine, L-Lysine, L-Tryptophan and the non-essential amino acid L-Glutamic acid are used today mainly for feed, food, and pharmaceutical industries. The production of these amino acids is due to *Corynebacterium glutamicum* and fermentation. *C.glutamicum* was engineered to be able to produce L-lysine and L-Glutamic acid in large quantities. L-Glutamic acid had a high demand for production because this amino acid is used to produce Monosodium glutamate (MSG) a food flavoring agent. In 2012 the total production of L-Glutamic acid was 2.2 million tons and is produced using a submerged fermentation technique inoculated with *C.glutamicum.* L-Lysine was originally produced from diaminopimelic acid (DAP) by *E.coli*, but once the *C.glutamicum* was discovered for the production of L-Glutamic acid. This organism and other autotrophs were later modified to yield other amino acids such as lysine, aspartate, methionine, isoleucine and threonine. L-Lysine is used for the feeding of pigs and chicken, as well as to treat nutrient deficiency, increase energy in a patient, and sometimes used to treat viral infections. L-Tryptophan is also produced through fermentation and by *Corynebacterium* and *E.coli*, though the production is not as large as the rest of the amino acids it is still produced for pharmaceutical purposes since it can be converted and used to produce neurotransmitters.

The production of organic solvents like acetone, butanol, and isopropanol through fermentation was one of the first things to be produced by using bacteria, since achieving the necessary chirality of the products is easily achieved by using living systems. Solvent fermentation uses a series of *Clostridia* bacterial species. Solvent fermentation at first was not as productive as it is used today.

The amount of bacteria required to yield a product was high, and the actual yield of product was low. Later technological advances were discovered that allowed scientist to genetically alter these strains to achieve a higher yield for these solvents. These Clostridial strains were transformed to have extra gene copies of enzymes necessary for solvent production, as well as being more tolerant to higher concentrations of the solvent being produced, since these bacteria have a range of product in which they can survive in before the environment becomes toxic. Yielding more strains that can use other substrates was also another way to increase the productivity of these bacteria.

1.2 BIOTECHNOLOGY

Biotechnology is a multidisciplinary field that involves the integration of natural sciences and engineering sciences in order to achieve the application of organisms and parts thereof for products and services.

The term *biotechnology* was first used by Károly Ereky in 1919 to refer to the production of products from raw materials with the aid of living organisms. The core principle of biotechnology involves harnessing biological systems and organisms, such as bacteria, yeast, and plants, to perform specific tasks or produce valuable substances.

Biotechnology had a significant impact on many areas of society, from medicine to agriculture to environmental science. One of the key techniques used in biotechnology is genetic engineering, which allows scientists to modify the genetic makeup of organisms to achieve desired outcomes. This can involve inserting genes from one organism into another, and consequently, create new traits or modifying existing ones.

Other important techniques used in biotechnology include tissue culture, which allows researchers to grow cells and tissues in the lab for research and medical purposes, and fermentation, which is used to produce a wide range of products such as beer, wine, and cheese.

The applications of biotechnology are diverse and have led to the development of essential products like life-saving drugs, biofuels, genetically modified crops, and innovative materials. It has also been used to address environmental challenges, such as developing biodegradable plastics and using microorganisms to clean up contaminated sites.

Biotechnology is a rapidly evolving field with significant potential to address pressing global challenges and improve the quality of life for people around the world; however, despite its numerous benefits, it also poses ethical and societal challenges, such as questions around genetic modification and intellectual property rights. As a result, there is ongoing debate and regulation surrounding the use and application of biotechnology in various industries and fields.

1.2.1 Definition

The concept of biotechnology encompasses a wide range of procedures for modifying living organisms for human purposes, going back to domestication of animals, cultivation of plants, and "improvements" to these through breeding programs that employ artificial selection and hybridization. Modern usage also includes genetic engineering, as well as cell and tissue culture technologies. The American Chemical Society defines *biotechnology* as the application of biological organisms, systems, or processes by various industries to learning about the science of life and the improvement of the value of materials and organisms, such as pharmaceuticals, crops, and livestock. As per the European Federation of Biotechnology, biotechnology is the integration of natural science and organisms, cells, parts thereof, and molecular analogues for products and services. Biotechnology is based on the basic biological sciences (e.g., molecular biology, biochemistry, cell biology, embryology, genetics, microbiology) and conversely provides methods to support and perform basic research in biology.

Biotechnology is the research and development in the laboratory using bioinformatics for exploration, extraction, exploitation, and production from any living organisms and any source of biomass by means of biochemical engineering where high value-added products could be planned (reproduced by biosynthesis, for example), forecasted, formulated, developed, manufactured, and marketed for the purpose of sustainable operations (for the return from bottomless initial investment on R & D) and gaining durable patents rights (for exclusives rights for sales, and prior to this to receive national and international approval from the results on animal experiment and human experiment, especially on the pharmaceutical branch of biotechnology to prevent any undetected side-effects or safety concerns by using the products). The utilization of biological processes, organisms or systems to produce products that are anticipated to improve human lives is termed biotechnology.

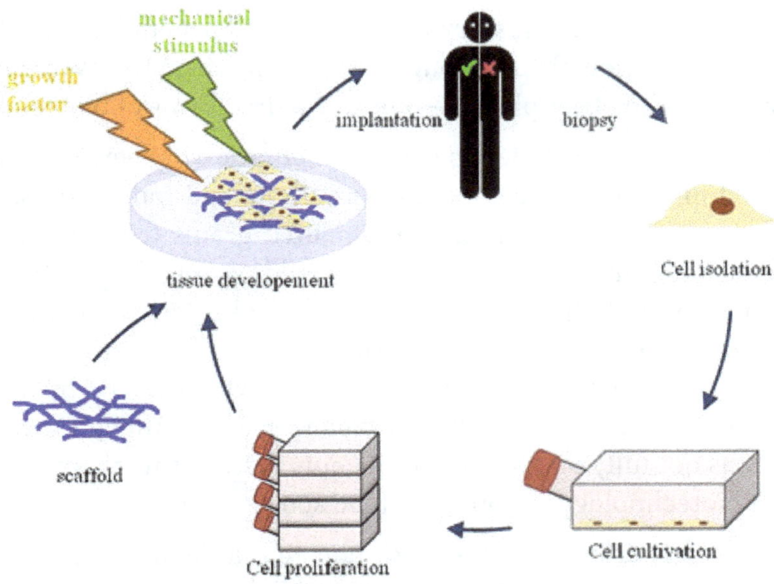

Fig 2: Principles of Tissue Engineering

By contrast, bioengineering is generally thought of as a related field that more heavily emphasizes higher systems approaches (not necessarily the altering or using of biological materials *directly*) for interfacing with and utilizing living things. Bioengineering is the application of the principles of engineering and natural sciences to tissues, cells, and molecules.

This can be considered as the use of knowledge from working with and manipulating biology to achieve a result that can improve functions in plants and animals. Relatedly, biomedical engineering is an overlapping field that often draws upon and applies *biotechnology* (by various definitions), especially in certain sub-fields of biomedical or chemical engineering such as tissue engineering, biopharmaceutical engineering, and genetic engineering.

1.2.2 History

Although not normally what first comes to mind, many forms of human-derived agriculture clearly fit the broad definition of "utilizing a biotechnological system to make products". Indeed, the cultivation of plants may be viewed as the earliest biotechnological enterprise.

Agriculture has been theorized to have become the dominant way of producing food since the Neolithic Revolution. Through early biotechnology, the earliest farmers selected and bred the best-suited crops (e.g., those with the highest yields) to produce enough food to support a growing population. As crops and fields became increasingly large and difficult to maintain, it was discovered that specific organisms and their by-products could effectively fertilize, restore nitrogen, and control pests. Throughout the history of agriculture, farmers have inadvertently altered the genetics of their crops through introducing them to new environments and breeding them with other plants — one of the first forms of biotechnology.

These processes also were included in early fermentation of beer. These processes were introduced in early Mesopotamia, Egypt, China and India, and still use the same basic biological methods. In brewing, malted grains (containing enzymes) convert starch from grains into sugar and then adding specific yeasts to produce beer. In this process, carbohydrates in the grains broke down into alcohols, such as ethanol. Later, other cultures produced the process of lactic acid fermentation, which produced other preserved foods, such as soy sauce. Fermentation was also used in this time period to produce leavened bread. Although the process of fermentation was not fully understood until Louis Pasteur's work in 1857, it is still the first use of biotechnology to convert a food source into another form.

Before the time of Charles Darwin's work and life, animal and plant scientists had already used selective breeding. Darwin added to that body of work with his scientific observations about the ability of science to change species. These accounts contributed to Darwin's theory of natural selection.

For thousands of years, humans have used selective breeding to improve the production of crops and livestock to use them for food. In selective breeding, organisms with desirable characteristics are mated to produce offspring with the same characteristics. For example, this technique was used with corn to produce the largest and sweetest crops.

In the early twentieth century scientists gained a greater understanding of microbiology and explored ways of manufacturing specific products. In 1917, Chaim Weizmann first used a pure microbiological culture in an industrial process, that of manufacturing corn starch using *Clostridium acetobutylicum,* to produce acetone, which the United Kingdom desperately needed to manufacture explosives during World War I.

Biotechnology has also led to the development of antibiotics. In 1928, Alexander Fleming discovered the mold *Penicillium*. His work led to the purification of the antibiotic compound formed by the mold by Howard Florey, Ernst Boris Chain and Norman Heatley - to form what we today know as penicillin. In 1940, penicillin became available for medicinal use to treat bacterial infections in humans.

The field of modern biotechnology is generally thought of as having been born in 1971 when Paul Berg's (Stanford) experiments in gene splicing had early success. Herbert W. Boyer (Univ. Calif. at San Francisco) and Stanley N. Cohen (Stanford) significantly advanced the new technology in 1972 by transferring genetic material into a bacterium, such that the imported material would be reproduced. The commercial viability of a biotechnology industry was significantly expanded on June 16, 1980, when the United States Supreme Court ruled that a genetically modified microorganism could be patented in the case of *Diamond v. Chakrabarty*. Indian-born Ananda Chakrabarty, working for General Electric, had modified a bacterium (of the genus *Pseudomonas*) capable of breaking down crude oil, which he proposed to use in treating oil spills. (Chakrabarty's work did not involve gene manipulation but rather the transfer of entire organelles between strains of the *Pseudomonas* bacterium).

The MOSFET invented at Bell Labs between 1955 and 1960, Two years later, Leland C. Clark and Champ Lyons invented the first biosensor in 1962. Biosensor MOSFETs were later developed, and they have since been widely used to measure physical, chemical, biological and environmental parameters. The first BioFET was the ion-sensitive field-effect transistor (ISFET), invented by Piet Bergveld in 1970. It is a special type of MOSFET, where the metal gate is replaced by an ion-sensitive membrane, electrolyte solution and reference electrode. The ISFET is widely used in biomedical applications, such as the detection of DNA hybridization, biomarker detection from blood, antibody detection, glucose measurement, pH sensing, and genetic technology.

By the mid-1980s, other BioFETs had been developed, including the gas sensor FET (GASFET), pressure sensor FET (PRESSFET), chemical field-effect transistor (ChemFET), reference ISFET (REFET), enzyme-modified FET (ENFET) and immunologically modified FET (IMFET). By the early 2000s, BioFETs such as the DNA field-effect transistor (DNAFET), gene-modified FET (GenFET) and cell-potential BioFET (CPFET) had been developed.

A factor influencing the biotechnology sector's success is improved intellectual property rights legislation—and enforcement—worldwide, as well as strengthened demand for medical and pharmaceutical products.

Rising demand for biofuels is expected to be good news for the biotechnology sector, with the Department of Energy estimating ethanol usage could reduce U.S. petroleum-derived fuel consumption by up to 30% by 2030. The biotechnology sector has allowed the U.S. farming industry to rapidly increase its supply of corn and soybeans—the main inputs into biofuels—by developing genetically modified seeds that resist pests and drought. By increasing farm productivity, biotechnology boosts biofuel production.

1.2.3 Examples

Biotechnology has applications in four major industrial areas, including health care (medical), crop production and agriculture, non-food (industrial) uses of crops and other products (e.g., biodegradable plastics, vegetable oil, biofuels), and environmental uses.

For example, one application of biotechnology is the directed use of microorganisms for the manufacture of organic products (examples include beer and milk products). Another example is using naturally present bacteria by the mining industry in bioleaching. Biotechnology is also used to recycle, treat waste, clean up sites contaminated by industrial activities (bioremediation), and also to produce biological weapons. A series of derived terms have been coined to identify several branches of biotechnology, for example:

- Bioinformatics (or "gold biotechnology") is an interdisciplinary field that addresses biological problems using computational techniques, and makes the rapid organization as well as analysis of biological data possible. The field may also be referred to as *computational biology*, and can be defined as, "conceptualizing biology in terms of molecules and then applying informatics techniques to understand and organize the information associated with these molecules, on a large scale". Bioinformatics plays a key role in various areas, such as functional genomics, structural genomics, and proteomics, and forms a key component in the biotechnology and pharmaceutical sector.
- Blue biotechnology is based on the exploitation of sea resources to create products and industrial applications. This branch of biotechnology is the most used for the industries of refining and combustion principally on the production of bio-oils with photosynthetic micro-algae.
- Green biotechnology is biotechnology applied to agricultural processes. An example would be the selection and domestication of plants via micropropagation. Another example is the designing of transgenic plants to grow under specific environments in the presence (or absence) of chemicals. One hope is that green biotechnology might produce more environmentally friendly solutions than traditional industrial agriculture. An example of this is the engineering of a plant to express a pesticide, thereby ending the need of external application of pesticides. An example of this would be Bt corn. Whether or not green biotechnology products such as this are ultimately more environmentally friendly is a topic of considerable debate. It is commonly considered as the next phase of green revolution, which can be seen as a platform to eradicate world hunger by using technologies which enable the production of

more fertile and resistant, towards biotic and abiotic stress, plants and ensures application of environmentally friendly fertilizers and the use of biopesticides, it is mainly focused on the development of agriculture. On the other hand, some of the uses of green biotechnology involve microorganisms to clean and reduce waste.

- Red biotechnology is the use of biotechnology in the medical and pharmaceutical industries, and health preservation. This branch involves the production of vaccines and antibiotics, regenerative therapies, creation of artificial organs and new diagnostics of diseases. As well as the development of hormones, stem cells, antibodies, siRNA and diagnostic tests.
- White biotechnology, also known as industrial biotechnology, is biotechnology applied to industrial processes. An example is the designing of an organism to produce a useful chemical. Another example is the using of enzymes as industrial catalysts to either produce valuable chemicals or destroy hazardous/polluting chemicals. White biotechnology tends to consume less in resources than traditional processes used to produce industrial goods.
- "Yellow biotechnology" refers to the use of biotechnology in food production (food industry), for example in making wine (winemaking), cheese (cheesemaking), and beer (brewing) by fermentation. It has also been used to refer to biotechnology applied to insects. This includes biotechnology-based approaches for the control of harmful insects, the characterisation and utilisation of active ingredients or genes of insects for research, or application in agriculture and medicine and various other approaches.
- Gray biotechnology is dedicated to environmental applications, and focused on the maintenance of biodiversity and the remotion of pollutants.
- Brown biotechnology is related to the management of arid lands and deserts. One application is the creation of enhanced seeds that resist extreme environmental conditions of arid regions, which is related to the innovation, creation of agriculture techniques and management of resources.
- Violet biotechnology is related to law, ethical and philosophical issues around biotechnology.
- Microbial biotechnology has been proposed for the rapidly emerging area of biotechnology applications in space and microgravity.

- Dark biotechnology is the color associated with bioterrorism or biological weapons and biowarfare which uses microorganisms, and toxins to cause diseases and death in humans, livestock and crops.

Medicine

In medicine, modern biotechnology has many applications in areas such as pharmaceutical drug discoveries and production, pharmacogenomics, and genetic testing (or genetic screening). In 2021, nearly 40% of the total company value of pharmaceutical biotech companies worldwide were active in Oncology with Neurology and Rare Diseases being the other two big applications.

Pharmacogenomics (a combination of pharmacology and genomics) is the technology that analyses how genetic makeup affects an individual's response to drugs. Researchers in the field investigate the influence of genetic variation on drug responses in patients by correlating gene expression or single-nucleotide polymorphisms with a drug's efficacy or toxicity. The purpose of pharmacogenomics is to develop rational means to optimize drug therapy, with respect to the patients' genotype, to ensure maximum efficacy with minimal adverse effects. Such approaches promise the advent of "personalized medicine"; in which drugs and drug combinations are optimized for each individual's unique genetic makeup.

Biotechnology has contributed to the discovery and manufacturing of traditional small molecule pharmaceutical drugs as well as drugs that are the product of biotechnology – biopharmaceutics. Modern biotechnology can be used to manufacture existing medicines relatively easily and cheaply. The first genetically engineered products were medicines designed to treat human diseases. To cite one example, in 1978 Genentech developed synthetic humanized insulin by joining its gene with a plasmid vector inserted into the bacterium *Escherichia coli*. Insulin, widely used for the treatment of diabetes, was previously extracted from the pancreas of abattoir animals (cattle or pigs). The genetically engineered bacteria are able to produce large quantities of synthetic human insulin at relatively low cost. Biotechnology has also enabled emerging therapeutics like gene therapy. The application of biotechnology to basic science (for example through the Human Genome Project) has also dramatically improved our understanding of biology and as our scientific knowledge of normal and disease biology has increased, our ability to develop new medicines to treat previously untreatable diseases has increased as well.

Genetic testing allows the genetic diagnosis of vulnerabilities to inherited diseases, and can also be used to determine a child's parentage (genetic mother and father) or in general a person's ancestry. In addition to studying chromosomes to the level of individual genes, genetic testing in a broader sense includes biochemical tests for the possible presence of genetic diseases, or mutant forms of genes associated with increased risk of developing genetic disorders. Genetic testing identifies changes in chromosomes, genes, or proteins. Most of the time, testing is used to find changes that are associated with inherited disorders. The results of a genetic test can confirm or rule out a suspected genetic condition or help determine a person's chance of developing or passing on a genetic disorder. As of 2011 several hundred genetic tests were in use. Since genetic testing may open up ethical or psychological problems, genetic testing is often accompanied by genetic counseling.

Agriculture

Genetically modified crops ("GM crops", or "biotech crops") are plants used in agriculture, the DNA of which has been modified with genetic engineering techniques. In most cases, the main aim is to introduce a new trait that does not occur naturally in the species. Biotechnology firms can contribute to future food security by improving the nutrition and viability of urban agriculture. Furthermore, the protection of intellectual property rights encourages private sector investment in agrobiotechnology.

Examples in food crops include resistance to certain pests, diseases, stressful environmental conditions, resistance to chemical treatments (e.g. resistance to a herbicide), reduction of spoilage, or improving the nutrient profile of the crop. Examples in non-food crops include production of pharmaceutical agents, biofuels, and other industrially useful goods, as well as for bioremediation.

Farmers have widely adopted GM technology. Between 1996 and 2011, the total surface area of land cultivated with GM crops had increased by a factor of 94, from 17,000 to 1,600,000 square kilometers (4,200,000 to 395,400,000 acres). 10% of the world's crop lands were planted with GM crops in 2010. As of 2011, 11 different transgenic crops were grown commercially on 395 million acres (160 million hectares) in 29 countries such as the US, Brazil, Argentina, India, Canada, China, Paraguay, Pakistan, South Africa, Uruguay, Bolivia, Australia, Philippines, Myanmar, Burkina Faso, Mexico and Spain.

Genetically modified foods are foods produced from organisms that have had specific changes introduced into their DNA with the methods of genetic engineering. These techniques have allowed for the introduction of new crop traits as well as a far greater control over a food's genetic structure than previously afforded by methods such as selective breeding and mutation breeding. Commercial sale of genetically modified foods began in 1994, when Calgene first marketed its Flavr Savr delayed ripening tomato. To date most genetic modification of foods have primarily focused on cash crops in high demand by farmers such as soybean, corn, canola, and cotton seed oil. These have been engineered for resistance to pathogens and herbicides and better nutrient profiles. GM livestock have also been experimentally developed; in November 2013 none were available on the market, but in 2015 the FDA approved the first GM salmon for commercial production and consumption.

There is a scientific consensus that currently available food derived from GM crops poses no greater risk to human health than conventional food, but that each GM food needs to be tested on a case-by-case basis before introduction. Nonetheless, members of the public are much less likely than scientists to perceive GM foods as safe. The legal and regulatory status of GM foods varies by country, with some nations banning or restricting them, and others permitting them with widely differing degrees of regulation.

GM crops also provide a number of ecological benefits, if not used in excess. Insect-resistant crops have proven to lower pesticide usage, therefore reducing the environmental impact of pesticides as a whole. However, opponents have objected to GM crops per se on several grounds, including environmental concerns, whether food produced from GM crops is safe, whether GM crops are needed to address the world's food needs, and economic concerns raised by the fact these organisms are subject to intellectual property law.

Biotechnology has several applications in the realm of food security. Crops like Golden rice are engineered to have higher nutritional content, and there is potential for food products with longer shelf lives. Though not a form of agricultural biotechnology, vaccines can help prevent diseases found in animal agriculture. Additionally, agricultural biotechnology can expedite breeding processes in order to yield faster results and provide greater quantities of food. Transgenic biofortification in cereals has been considered as a promising method to combat malnutrition in India and other countries.

Industrial

Industrial biotechnology (known mainly in Europe as white biotechnology) is the application of biotechnology for industrial purposes, including industrial fermentation. It includes the practice of using cells such as microorganisms, or components of cells like enzymes, to generate industrially useful products in sectors such as chemicals, food and feed, detergents, paper and pulp, textiles and biofuels. In the current decades, significant progress has been done in creating genetically modified organisms (GMOs) that enhance the diversity of applications and economical viability of industrial biotechnology. By using renewable raw materials to produce a variety of chemicals and fuels, industrial biotechnology is actively advancing towards lowering greenhouse gas emissions and moving away from a petrochemical-based economy.

Synthetic biology is considered one of the essential cornerstones in industrial biotechnology due to its financial and sustainable contribution to the manufacturing sector. Jointly biotechnology and synthetic biology play a crucial role in generating cost-effective products with nature-friendly features by using bio-based production instead of fossil-based. Synthetic biology can be used to engineer model microorganisms, such as *Escherichia coli*, by genome editing tools to enhance their ability to produce bio-based products, such as bioproduction of medicines and biofuels. For instance, *E. coli* and *Saccharomyces cerevisiae* in a consortium could be used as industrial microbes to produce precursors of the chemotherapeutic agent paclitaxel by applying the metabolic engineering in a co-culture approach to exploit the benefits from the two microbes.

Another example of synthetic biology applications in industrial biotechnology is the re-engineering of the metabolic pathways of *E. coli* by CRISPR and CRISPRi systems toward the production of a chemical known as 1,4-butanediol, which is used in fiber manufacturing. In order to produce 1,4-butanediol, the authors alter the metabolic regulation of the *Escherichia coli* by CRISPR to induce point mutation in the *glt*A gene, knockout of the *sad* gene, and knock-in six genes (*cat*1, *suc*D, *4hbd*, *cat*2, *bld*, and *bdh*). Whereas CRISPRi system used to knockdown the three competing genes (*gab*D, *ybg*C, and *tes*B) that affect the biosynthesis pathway of 1,4-butanediol. Consequently, the yield of 1,4-butanediol significantly increased from 0.9 to 1.8 g/L.

Environmental

Environmental biotechnology includes various disciplines that play an essential role in reducing environmental waste and providing environmentally safe processes, such as biofiltration and biodegradation. The environment can be affected by biotechnologies, both positively and adversely. Vallero and others have argued that the difference between beneficial biotechnology (e.g., bioremediation is to clean up an oil spill or hazard chemical leak) versus the adverse effects stemming from biotechnological enterprises (e.g., flow of genetic material from transgenic organisms into wild strains) can be seen as applications and implications, respectively. Cleaning up environmental wastes is an example of an application of environmental biotechnology; whereas loss of biodiversity or loss of containment of a harmful microbe are examples of environmental implications of biotechnology. Many cities have installed CityTrees, which use biotechnology to filter pollutants from urban atmospheres.

Regulation

The regulation of genetic engineering concerns approaches taken by governments to assess and manage the risks associated with the use of genetic engineering technology, and the development and release of genetically modified organisms (GMO), including genetically modified crops and genetically modified fish. There are differences in the regulation of GMOs between countries, with some of the most marked differences occurring between the US and Europe. Regulation varies in a given country depending on the intended use of the products of the genetic engineering. For example, a crop not intended for food use is generally not reviewed by authorities responsible for food safety. The European Union differentiates between approval for cultivation within the EU and approval for import and processing. While only a few GMOs have been approved for cultivation in the EU a number of GMOs have been approved for import and processing. The cultivation of GMOs has triggered a debate about the coexistence of GM and non-GM crops. Depending on the coexistence regulations, incentives for the cultivation of GM crops differ.

Database for the GMOs used in the EU

The EUginius (European GMO Initiative for a Unified Database System) database is intended to help companies, interested private users and competent authorities to find precise information on the presence, detection and identification of GMOs used in the European Union. The information is provided in English.

1.2.4 Learning

In 1988, after prompting from the United States Congress, the National Institute of General Medical Sciences (National Institutes of Health) (NIGMS) instituted a funding mechanism for biotechnology training. Universities nationwide compete for these funds to establish Biotechnology Training Programs (BTPs). Each successful application is generally funded for five years then must be competitively renewed. Graduate students in turn compete for acceptance into a BTP; if accepted, then stipend, tuition and health insurance support are provided for two or three years during the course of their PhD thesis work. Nineteen institutions offer NIGMS supported BTPs. Biotechnology training is also offered at the undergraduate level and in community colleges.

SECTION 2:

BIOLOGICAL BASIS OF PRODUCTIVITY IN INDUSTRIAL MICROBIOLOGY AND BIOTECHNOLOGY

CHAPTER 2

MICROORGANISMS IN INDUSTRIAL MICROBIOLOGY AND BIOTECHNOLOGY

2.1 UNDERSTANDING THE MICROORGANISMS

A microorganism, or microbe, is an organism of microscopic size, which may exist in its single-celled form or as a colony of cells. The possible existence of unseen microbial life was suspected from ancient times, such as in Jain scriptures from sixth century BC India. The scientific study of microorganisms began with their observation under the microscope in the 1670s by Anton van Leeuwenhoek. In the 1850s, Louis Pasteur found that microorganisms caused food spoilage, debunking the theory of spontaneous generation. In the 1880s, Robert Koch discovered that microorganisms caused the diseases tuberculosis, cholera, diphtheria, and anthrax.

Because microorganisms include most unicellular organisms from all three domains of life, they can be extremely diverse. Two of the three domains, Archaea and Bacteria, only contain microorganisms. The third domain, Eukaryota, includes all multicellular organisms as well as many unicellular protists and protozoans that are microbes. Some protists are related to animals and some to green plants. Many multicellular organisms are also microscopic, namely micro-animals, some fungi, and some algae, but these are generally not considered microorganisms.

Microorganisms can have very different habitats, and live everywhere from the poles to the equator, in deserts, geysers, rocks, and the deep sea. Some are adapted to extremes such as very hot or very cold conditions, others to high pressure, and a few, such as *Deinococcus radiodurans*, to high radiation environments. Microorganisms also make up the microbiota found in and on all multicellular organisms. There is evidence that 3.45-billion-year-old Australian rocks once contained microorganisms, the earliest direct evidence of life on Earth. Microbes are important in human culture and health in many ways, serving to ferment foods and treat sewage, and to produce fuel, enzymes, and other bioactive compounds. Microbes are essential tools in biology as model organisms and have been put to use in biological warfare and bioterrorism. Microbes are a vital component of fertile soil. In the human body, microorganisms make up the human microbiota, including the essential gut flora. The pathogens responsible for many infectious diseases are microbes and, as such, are the target of hygiene measures.

2.1.1 Discovery

Ancient Precursors

The possible existence of microscopic organisms was discussed for many centuries before their discovery in the seventeenth century. By the 6th century BC, the Jains of present-day India postulated the existence of tiny organisms called nigodas.

These nigodas are said to be born in clusters; they live everywhere, including the bodies of plants, animals, and people; and their life lasts only for a fraction of a second. According to Mahavira, the 24th preacher of Jainism, the humans destroy these nigodas on a massive scale, when they eat, breathe, sit, and move. Many modern Jains assert that Mahavira's teachings presage the existence of microorganisms as discovered by modern science.

The earliest known idea to indicate the possibility of diseases spreading by yet unseen organisms was that of the Roman scholar Marcus Terentius Varro in a first-century BC book entitled *On Agriculture* in which he called the unseen creatures animalia minuta, and warns against locating a homestead near a swamp:

- ... and because there are bred certain minute creatures that cannot be seen by the eyes, which float in the air and enter the body through the mouth and nose and they cause serious diseases.

In *The Canon of Medicine* (1020), Avicenna suggested that tuberculosis and other diseases might be contagious.

Early Modern

Turkish scientist Akshamsaddin mentioned the microbe in his work *Maddat ul-Hayat* (The Material of Life) about two centuries prior to Antonie van Leeuwenhoek's discovery through experimentation:

- It is incorrect to assume that diseases appear one by one in humans. Disease infects by spreading from one person to another. This infection occurs through seeds that are so small they cannot be seen but are alive.

In 1546, Girolamo Fracastoro proposed that epidemic diseases were caused by transferable seedlike entities that could transmit infection by direct or indirect contact, or even without contact over long distances. Antonie van Leeuwenhoek is considered to be one of the fathers of microbiology. He was the first in 1673 to discover and conduct scientific experiments with microorganisms, using simple single-lensed microscopes of his own design. Robert Hooke, a contemporary of Leeuwenhoek, also used microscopy to observe microbial life in the form of the fruiting bodies of moulds. In his 1665 book *Micrographia*, he made drawings of studies, and he coined the term *cell*.

19th Century

Louis Pasteur (1822–1895) exposed boiled broths to the air, in vessels that contained a filter to prevent particles from passing through to the growth medium, and also in vessels without a filter, but with air allowed in via a curved tube so dust particles would settle and not come in contact with the broth. By boiling the broth beforehand, Pasteur ensured that no microorganisms survived within the broths at the beginning of his experiment. Nothing grew in the broths in the course of Pasteur's experiment. This meant that the living organisms that grew in such broths came from outside, as spores on dust, rather than spontaneously generated within the broth. Thus, Pasteur refuted the theory of spontaneous generation and supported the germ theory of disease.

In 1876, Robert Koch (1843–1910) established that microorganisms can cause disease. He found that the blood of cattle that were infected with anthrax always had large numbers of *Bacillus anthracis*. Koch found that he could transmit anthrax from one animal to another by taking a small sample of blood from the infected animal and injecting it into a healthy one, and this caused the healthy animal to become sick. He also found that he could grow the bacteria in a nutrient broth, then inject it into a healthy animal, and cause illness. Based on these experiments, he devised criteria for establishing a causal link between a microorganism and a disease and these are now known as Koch's postulates. Although these postulates cannot be applied in all cases, they do retain historical importance to the development of scientific thought and are still being used today.

The discovery of microorganisms such as *Euglena* that did not fit into either the animal or plant kingdoms, since they were photosynthetic like plants, but motile like animals, led to the naming of a third kingdom in the 1860s. In 1860 John Hogg called this the Protoctista, and in 1866 Ernst Haeckel named it the Protista.

The work of Pasteur and Koch did not accurately reflect the true diversity of the microbial world because of their exclusive focus on microorganisms having direct medical relevance. It was not until the work of Martinus Beijerinck and Sergei Winogradsky late in the nineteenth century that the true breadth of microbiology was revealed. Beijerinck made two major contributions to microbiology: the discovery of viruses and the development of enrichment culture techniques. While his work on the tobacco mosaic virus established the basic principles of virology, it was his development of enrichment culturing that had the most immediate impact on microbiology by allowing for the cultivation of a wide range of microbes with wildly different physiologies. Winogradsky was the first to develop the concept of chemolithotrophy and to thereby reveal the essential role played by microorganisms in geochemical processes. He was responsible for the first isolation and description of both nitrifying and nitrogen-fixing bacteria. French-Canadian microbiologist Felix d'Herelle co-discovered bacteriophages and was one of the earliest applied microbiologists.

2.1.2 Classification and Structure

Microorganisms can be found almost anywhere on Earth. Bacteria and archaea are almost always microscopic, while a number of eukaryotes are also microscopic, including most protists, some fungi, as well as some micro-animals and plants. Viruses are generally regarded as not living and therefore not considered to be microorganisms, although a subfield of microbiology is virology, the study of viruses.

Evolution

Single-celled microorganisms were the first forms of life to develop on Earth, approximately 3.5 billion years ago. Further evolution was slow, and for about 3 billion years in the Precambrian eon, (much of the history of life on Earth), all organisms were microorganisms. Bacteria, algae and fungi have been identified in amber that is 220 million years old, which shows that the morphology of microorganisms has changed little since at least the Triassic period. The newly discovered biological role played by nickel, however - especially that brought about by volcanic eruptions from the Siberian Traps - may have accelerated the evolution of methanogens towards the end of the Permian–Triassic extinction event.

Microorganisms tend to have a relatively fast rate of evolution. Most microorganisms can reproduce rapidly, and bacteria are also able to freely exchange genes through conjugation, transformation and transduction, even between widely divergent species. This horizontal gene transfer, coupled with a high mutation rate and other means of transformation, allows microorganisms to swiftly evolve (via natural selection) to survive in new environments and respond to environmental stresses. This rapid evolution is important in medicine, as it has led to the development of multidrug resistant pathogenic bacteria, *superbugs*, that are resistant to antibiotics.

A possible transitional form of microorganism between a prokaryote and a eukaryote was discovered in 2012 by Japanese scientists. *Parakaryon myojinensis* is a unique microorganism larger than a typical prokaryote, but with nuclear material enclosed in a membrane as in a eukaryote, and the presence of endosymbionts. This is seen to be the first plausible evolutionary form of microorganism, showing a stage of development from the prokaryote to the eukaryote.

Archaea

Archaea are prokaryotic unicellular organisms, and form the first domain of life in Carl Woese's three-domain system. A prokaryote is defined as having no cell nucleus or other membrane bound-organelle. Archaea share this defining feature with the bacteria with which they were once grouped. In 1990 the microbiologist Woese proposed the three-domain system that divided living things into bacteria, archaea and eukaryotes, and thereby split the prokaryote domain.

Archaea differ from bacteria in both their genetics and biochemistry. For example, while bacterial cell membranes are made from phosphoglycerides with ester bonds, archaean membranes are made of ether lipids. Archaea were originally described as extremophiles living in extreme environments, such as hot springs, but have since been found in all types of habitats. Only now are scientists beginning to realize how common archaea are in the environment, with Thermoproteota (formerly Crenarchaeota) being the most common form of life in the ocean, dominating ecosystems below 150 metres (490 ft) in depth. These organisms are also common in soil and play a vital role in ammonia oxidation.

The combined domains of archaea and bacteria make up the most diverse and abundant group of organisms on Earth and inhabit practically all environments where the temperature is below +140 °C (284 °F). They are found in water, soil, air, as the microbiome of an organism, hot springs and even deep beneath the Earth's crust in rocks. The number of prokaryotes is estimated to be around five nonillion, or 5×10^{30}, accounting for at least half the biomass on Earth.

The biodiversity of the prokaryotes is unknown, but may be very large. A May 2016 estimate, based on laws of scaling from known numbers of species against the size of organism, gives an estimate of perhaps 1 trillion species on the planet, of which most would be microorganisms.

Currently, only one-thousandth of one percent of that total have been described. Archael cells of some species aggregate and transfer DNA from one cell to another through direct contact, particularly under stressful environmental conditions that cause DNA damage.

Bacteria

Like archaea, bacteria are prokaryotic – unicellular, and having no cell nucleus or other membrane-bound organelle. Bacteria are microscopic, with a few extremely rare exceptions, such as *Thiomargarita namibiensis*. Bacteria function and reproduce as individual cells, but they can often aggregate in multicellular colonies. Some species such as myxobacteria can aggregate into complex swarming structures, operating as multicellular groups as part of their life cycle, or form clusters in bacterial colonies such as *E.coli*.

Their genome is usually a circular bacterial chromosome - a single loop of DNA, although they can also harbor small pieces of DNA called plasmids. These plasmids can be transferred between cells through bacterial conjugation. Bacteria have an enclosing cell wall, which provides strength and rigidity to their cells. They reproduce by binary fission or sometimes by budding, but do not undergo meiotic sexual reproduction. However, many bacterial species can transfer DNA between individual cells by a horizontal gene transfer process referred to as natural transformation. Some species form extraordinarily resilient spores, but for bacteria this is a mechanism for survival, not reproduction. Under optimal conditions bacteria can grow extremely rapidly and their numbers can double as quickly as every 20 minutes.

Eukaryotes

Most living things that are visible to the naked eye in their adult form are eukaryotes, including humans. However, many eukaryotes are also microorganisms. Unlike bacteria and archaea, eukaryotes contain organelles such as the cell nucleus, the Golgi apparatus and mitochondria in their cells. The nucleus is an organelle that houses the DNA that makes up a cell's genome. DNA (Deoxyribonucleic acid) itself is arranged in complex chromosomes. Mitochondria are organelles vital in metabolism as they are the site of the citric acid cycle and oxidative phosphorylation. They evolved from symbiotic bacteria and retain a remnant genome. Like bacteria, plant cells have cell walls, and contain organelles such as chloroplasts in addition to the organelles in other eukaryotes. Chloroplasts produce energy from light by photosynthesis, and were also originally symbiotic bacteria.

Unicellular eukaryotes consist of a single cell throughout their life cycle. This qualification is significant since most multicellular eukaryotes consist of a single cell called a zygote only at the beginning of their life cycles. Microbial eukaryotes can be either haploid or diploid, and some organisms have multiple cell nuclei.

Unicellular eukaryotes usually reproduce asexually by mitosis under favorable conditions. However, under stressful conditions such as nutrient limitations and other conditions associated with DNA damage, they tend to reproduce sexually by meiosis and syngamy.

Protists

Of eukaryotic groups, the protists are most commonly unicellular and microscopic. This is a highly diverse group of organisms that are not easy to classify. Several algae species are multicellular protists, and slime molds have unique life cycles that involve switching between unicellular, colonial, and multicellular forms. The number of species of protists is unknown since only a small proportion has been identified. Protist diversity is high in oceans, deep sea-vents, river sediment and an acidic river, suggesting that many eukaryotic microbial communities may yet be discovered.

Fungi

The fungi have several unicellular species, such as baker's yeast (*Saccharomyces cerevisiae*) and fission yeast (*Schizosaccharomyces pombe*). Some fungi, such as the pathogenic yeast *Candida albicans*, can undergo phenotypic switching and grow as single cells in some environments, and filamentous hyphae in others.

Plants

The green algae are a large group of photosynthetic eukaryotes that include many microscopic organisms. Although some green algae are classified as protists, others such as charophyta are classified with embryophyte plants, which are the most familiar group of land plants. Algae can grow as single cells, or in long chains of cells. The green algae include unicellular and colonial flagellates, usually but not always with two flagella per cell, as well as various colonial, coccoid, and filamentous forms. In the Charales, which are the algae most closely related to higher plants, cells differentiate into several distinct tissues within the organism. There are about 6000 species of green algae.

2.1.3 Ecology

Microorganisms are found in almost every habitat present in nature, including hostile environments such as the North and South poles, deserts, geysers, and rocks. They also include all the marine microorganisms of the oceans and deep sea. Some types of microorganisms have adapted to extreme environments and sustained colonies; these organisms are known as extremophiles. Extremophiles have been isolated from rocks as much as 7 kilometres below the Earth's surface, and it has been suggested that the amount of organisms living below the Earth's surface is comparable with the amount of life on or above the surface. Extremophiles have been known to survive for a prolonged time in a vacuum, and can be highly resistant to radiation, which may even allow them to survive in space. Many types of microorganisms have intimate symbiotic relationships with other larger organisms; some of which are mutually beneficial (mutualism), while others can be damaging to the host organism (parasitism). If microorganisms can cause disease in a host they are known as pathogens and then they are sometimes referred to as *microbes*. Microorganisms play critical roles in Earth's biogeochemical cycles as they are responsible for decomposition and nitrogen fixation.

Bacteria use regulatory networks that allow them to adapt to almost every environmental niche on earth. A network of interactions among diverse types of molecules including DNA, RNA, proteins and metabolites, is utilised by the bacteria to achieve regulation of gene expression. In bacteria, the principal function of regulatory networks is to control the response to environmental changes, for example nutritional status and environmental stress. A complex organization of networks permits the microorganism to coordinate and integrate multiple environmental signals.

Extremophiles

Extremophiles are microorganisms that have adapted so that they can survive and even thrive in extreme environments that are normally fatal to most life-forms. Thermophiles and hyperthermophiles thrive in high temperatures. Psychrophiles thrive in extremely low temperatures. – Temperatures as high as 130 °C (266 °F), as low as −17 °C (1 °F) Halophiles such as *Halobacterium salinarum* (an archaean) thrive in high salt conditions, up to saturation. Alkaliphiles thrive in an alkaline pH of about 8.5–11. Acidophiles can thrive in a pH of 2.0 or less. Piezophiles thrive at very high pressures: up to 1,000–2,000 atm, down to 0 atm as in a vacuum of space. A few extremophiles such as *Deinococcus radiodurans* are radioresistant, resisting radiation exposure of up to 5k Gy. Extremophiles are significant in different ways. They extend terrestrial life into much of the Earth's hydrosphere, crust and atmosphere, their specific evolutionary adaptation mechanisms to their extreme environment can be exploited in biotechnology, and their very existence under such extreme conditions increases the potential for extraterrestrial life.

Plants and Soil

The nitrogen cycle in soils depends on the fixation of atmospheric nitrogen. This is achieved by a number of diazotrophs. One way this can occur is in the root nodules of legumes that contain symbiotic bacteria of the genera *Rhizobium*, *Mesorhizobium*, *Sinorhizobium*, *Bradyrhizobium*, and *Azorhizobium*. The roots of plants create a narrow region known as the rhizosphere that supports many microorganisms known as the root microbiome.

These microorganisms in the root microbiome are able to interact with each other and surrounding plants through signals and cues. For example, mycorrhizal fungi are able to communicate with the root systems of many plants through chemical signals between both the plant and fungi. This results in a mutualistic symbiosis between the two. However, these signals can be eavesdropped by other microorganisms, such as the soil bacteria, *Myxococcus xanthus*, which preys on other bacteria. Eavesdropping, or the interception of signals from unintended receivers, such as plants and microorganisms, can lead to large-scale, evolutionary consequences. For example, signaler-receiver pairs, like plant-microorganism pairs, may lose the ability to communicate with neighboring populations because of variability in eavesdroppers. In adapting to avoid local eavesdroppers, signal divergence could occur and thus, lead to the isolation of plants and microorganisms from the inability to communicate with other populations.

Symbiosis

A lichen is a symbiosis of a macroscopic fungus with photosynthetic microbial algae or cyanobacteria.

2.1.4 Applications

Microorganisms are useful in producing foods, treating waste water, creating biofuels and a wide range of chemicals and enzymes. They are invaluable in research as model organisms. They have been weaponised and sometimes used in warfare and bioterrorism. They are vital to agriculture through their roles in maintaining soil fertility and in decomposing organic matter. They also have applications in aquaculture, such as in biofloc technology.

Food Production

Microorganisms are used in a fermentation process to make yoghurt, cheese, curd, kefir, ayran, xynogala, and other types of food. Fermentation cultures provide flavour and aroma, and inhibit undesirable organisms.

They are used to leaven bread, and to convert sugars to alcohol in wine and beer. Microorganisms are used in brewing, wine making, baking, pickling and other food-making processes.

Table 1: Example Industrial Uses of Microorganisms

Product	Contribution of Microorganisms
Cheese	Growth of microorganisms contributes to ripening and flavor. The flavor and appearance of a particular cheese is due in large part to the microorganisms associated with it. *Lactobacillus Bulgaricus* is one of the microbes used in production of dairy products
Alcoholic beverages	Yeast is used to convert sugar, grape juice, or malt-treated grain into alcohol. Other microorganisms may also be used; a mold converts starch into sugar to make the Japanese rice wine, sake. *Acetobacter Aceti* a kind of bacterium is used in production of alcoholic beverages
Vinegar	Certain bacteria are used to convert alcohol into acetic acid, which gives vinegar its acid taste. *Acetobacter Aceti* is used on production of vinegar, which gives vinegar odor of alcohol and alcoholic taste
Citric acid	Certain fungi are used to make citric acid, a common ingredient of soft drinks and other foods.
Vitamins	Microorganisms are used to make vitamins, including C, B_2 , B_{12}
Antibiotics	With only a few exceptions, microorganisms are used to make antibiotics. *Penicillin, Amoxicillin, Tetracycline, and Erythromycin*

Water Treatment

These depend for their ability to clean up water contaminated with organic material on microorganisms that can respire dissolved substances. Respiration may be aerobic, with a well-oxygenated filter bed such as a slow sand filter. Anaerobic digestion by methanogens generate useful methane gas as a by-product.

Energy

Microorganisms are used in fermentation to produce ethanol, and in biogas reactors to produce methane. Scientists are researching the use of algae to produce liquid fuels, and bacteria to convert various forms of agricultural and urban waste into usable fuels.

Chemicals and Enzymes

Microorganisms are used to produce many commercial and industrial chemicals, enzymes and other bioactive molecules. Organic acids produced on a large industrial scale by microbial fermentation include acetic acid produced by acetic acid bacteria such as *Acetobacter aceti*, butyric acid made by the bacterium *Clostridium butyricum*, lactic acid made by *Lactobacillus* and other lactic acid bacteria, and citric acid produced by the mould fungus *Aspergillus niger*. Microorganisms are used to prepare bioactive molecules such as Streptokinase from the bacterium *Streptococcus*, Cyclosporin A from the ascomycete fungus *Tolypocladium inflatum*, and statins produced by the yeast *Monascus purpureus*.

Science

Microorganisms are essential tools in biotechnology, biochemistry, genetics, and molecular biology. The yeasts *Saccharomyces cerevisiae* and *Schizosaccharomyces pombe* are important model organisms in science, since they are simple eukaryotes that can be grown rapidly in large numbers and are easily manipulated. They are particularly valuable in genetics, genomics and proteomics. Microorganisms can be harnessed for uses such as creating steroids and treating skin diseases. Scientists are also considering using microorganisms for living fuel cells, and as a solution for pollution.

Warfare

In the Middle Ages, as an early example of biological warfare, diseased corpses were thrown into castles during sieges using catapults or other siege engines. Individuals near the corpses were exposed to the pathogen and were likely to spread that pathogen to others. In modern times, bioterrorism has included the 1984 Rajneeshee bioterror attack and the 1993 release of anthrax by Aum Shinrikyo in Tokyo.

Soil

Microbes can make nutrients and minerals in the soil available to plants, produce hormones that spur growth, stimulate the plant immune system and trigger or dampen stress responses. In general a more diverse set of soil microbes results in fewer plant diseases and higher yield.

2.1.5 Human Health

Human Gut Flora

Microorganisms can form an endosymbiotic relationship with other, larger organisms. For example, microbial symbiosis plays a crucial role in the immune system. The microorganisms that make up the gut flora in the gastrointestinal tract contribute to gut immunity, synthesize vitamins such as folic acid and biotin, and ferment complex indigestible carbohydrates. Some microorganisms that are seen to be beneficial to health are termed probiotics and are available as dietary supplements, or food additives.

Disease

Microorganisms are the causative agents (pathogens) in many infectious diseases. The organisms involved include pathogenic bacteria, causing diseases such as plague, tuberculosis and anthrax; protozoan parasites, causing diseases such as malaria, sleeping sickness, dysentery and toxoplasmosis; and also fungi causing diseases such as ringworm, candidiasis or histoplasmosis. However, other diseases such as influenza, yellow fever or AIDS are caused by pathogenic viruses, which are not usually classified as living organisms and are not, therefore, microorganisms by the strict definition. No clear examples of archaean pathogens are known, although a relationship has been proposed between the presence of some archaean methanogens and human periodontal disease. Numerous microbial pathogens are capable of sexual processes that appear to facilitate their survival in their infected host.

Hygiene

Hygiene is a set of practices to avoid infection or food spoilage by eliminating microorganisms from the surroundings. As microorganisms, in particular bacteria, are found virtually everywhere, harmful microorganisms may be reduced to acceptable levels rather than actually eliminated. In food preparation, microorganisms are reduced by preservation methods such as cooking, cleanliness of utensils, short storage periods, or by low temperatures. If complete sterility is needed, as with surgical equipment, an autoclave is used to kill microorganisms with heat and pressure.

2.2 ROLE OF MICROORGANISMS IN INDUSTRIAL MICROBIOLOGY

Industrial microbiology involves the exploitation of microbial processes for the production of economically valuable products on a commercial scale. Microorganisms such as bacteria, yeasts, and molds have been harnessed for centuries to manufacture products like alcoholic beverages, bread, and antibiotics. With advances in microbiology, biochemistry, and genetic engineering, microorganisms are now being employed in even more sophisticated ways to produce a diverse range of products, including enzymes, vitamins, biofuels, and pharmaceuticals.

2.2.1 Microorganisms in Fermentation Processes

One of the most well-known applications of microorganisms in industrial microbiology is fermentation, a process in which microbes convert sugars into alcohol, gases, or organic acids. Fermentation has been used for centuries in food and beverage production, with microorganisms like **Saccharomyces cerevisiae** (yeast) playing a key role in brewing, baking, and wine-making. In addition to alcohol production, bacteria such as **Lactobacillus** species are employed in the fermentation of dairy products like yogurt and cheese, where they produce lactic acid, giving the food its unique flavor and texture.

Fermentation processes are not limited to food production. Industrial microbiology has expanded the use of microbial fermentation to the production of **antibiotics** (e.g., penicillin from the fungus **Penicillium notatum**), **organic acids** (such as citric acid from **Aspergillus niger**), and **enzymes** (such as amylase and protease from **Bacillus** species). These microbial processes are favored in industry because they are cost-effective, scalable, and environmentally friendly.

2.2.2 Enzyme Production

Microorganisms are crucial in the production of enzymes, which are proteins that catalyze biochemical reactions. These enzymes are widely used in industries such as textiles, food, detergents, and pharmaceuticals. For example, **proteases** and **lipases** produced by **Bacillus** and **Aspergillus** species are employed in the detergent industry to break down proteins and fats in laundry, while **amylases** are used in the starch industry to convert starch into sugars for sweeteners and ethanol production.

The large-scale production of enzymes through microbial fermentation offers an efficient and renewable source of biocatalysts. Microorganisms can be genetically engineered to overproduce specific enzymes, making them indispensable tools in biotechnology. Additionally, microbial enzymes are often more stable and active under a wider range of conditions than their plant or animal counterparts, further enhancing their utility in industrial processes.

2.3 BIOTECHNOLOGY AND GENETIC ENGINEERING

With the advent of genetic engineering, microorganisms have become even more valuable in biotechnology. The ability to modify the genetic material of microorganisms allows scientists to create strains with enhanced metabolic capabilities, optimized for industrial production. These genetically modified organisms (GMOs) can be used to produce pharmaceuticals, biofuels, and environmentally friendly chemicals at an unprecedented scale and efficiency.

2.3.1 Production of Pharmaceuticals

One of the most significant contributions of microorganisms in biotechnology is the production of pharmaceuticals, particularly antibiotics and vaccines. In the early 20th century, the discovery of penicillin revolutionized medicine, and the mass production of antibiotics soon followed. Today, microorganisms are engineered to produce not only antibiotics but also **insulin, growth hormones**, and **monoclonal antibodies**, which are used in the treatment of diseases such as diabetes, cancer, and autoimmune disorders.

Escherichia coli (E. coli) is one of the most commonly used bacteria in pharmaceutical production. Through genetic engineering, scientists have modified E. coli to produce human insulin, which is used to treat millions of diabetes patients worldwide. Similarly, yeast and mammalian cells are used to produce vaccines and therapeutic proteins. Microbial biotechnology has thus played a pivotal role in improving human health by enabling the large-scale production of life-saving medicines.

2.3.2 Biofuel Production

Microorganisms are also employed in the production of biofuels, offering a renewable and sustainable alternative to fossil fuels. **Ethanol** production through the fermentation of sugars by yeast (e.g., **Saccharomyces cerevisiae**) is the most well-known example of microbial biofuel production. This process converts biomass into ethanol, which can be used as a cleaner-burning fuel for vehicles.

In addition to ethanol, **biogas** (methane) and **biodiesel** are also produced using microbial processes. For example, **methanogenic archaea** are used in anaerobic digesters to convert organic waste into methane gas, which can be used for heating, electricity generation, and transportation. Furthermore, genetically engineered microorganisms are being developed to produce **algal biofuels**, which have the potential to become a sustainable energy source due to their high productivity and ability to grow in non-arable land.

2.4 ENVIRONMENTAL BIOTECHNOLOGY AND BIOREMEDIATION

Microorganisms also play a crucial role in environmental biotechnology, particularly in bioremediation—the process of using living organisms to degrade or neutralize pollutants from contaminated sites. Bacteria, fungi, and algae have the natural ability to break down hazardous substances into less toxic forms, making them valuable tools for cleaning up oil spills, heavy metals, and organic pollutants.

Pseudomonas species, for example, are known for their ability to degrade hydrocarbons, and they are used in the treatment of oil-contaminated soils and water. Similarly, certain bacteria can metabolize heavy metals such as mercury, lead, and cadmium, reducing their toxicity and helping to restore polluted environments. Bioremediation is a cost-effective and eco-friendly approach to environmental cleanup, and ongoing research is focused on enhancing the efficiency of these microbial processes through genetic engineering and microbial consortia.

2.5 FUTURE PROSPECTS AND CHALLENGES

The potential of microorganisms in industrial microbiology and biotechnology is vast, but challenges remain. One of the key challenges is the optimization of microbial processes for large-scale production. While microorganisms can be genetically engineered to perform specific tasks, ensuring that these tasks are efficient, cost-effective, and environmentally sustainable at an industrial scale is an ongoing area of research.

Additionally, there are regulatory and ethical concerns surrounding the use of genetically modified microorganisms (GMOs) in biotechnology. Public perception, environmental impact, and biosecurity risks need to be carefully managed to ensure that microbial biotechnology continues to advance responsibly.

The future of microorganisms in biotechnology looks promising, particularly with advancements in synthetic biology and metabolic engineering. Scientists are working to design and build new microbial systems from scratch, creating custom microorganisms with tailored metabolic pathways for specific industrial applications. This approach has the potential to revolutionize industries ranging from pharmaceuticals to renewable energy, offering sustainable and efficient solutions to some of the world's most pressing challenges.

2.6 CONCLUSION

Microorganisms are at the heart of industrial microbiology and biotechnology, offering a diverse range of applications that impact food production, medicine, energy, and environmental management. Through microbial fermentation, enzyme production, genetic engineering, and bioremediation, microorganisms have become indispensable tools in modern industry. As scientific understanding of microbial systems continues to advance, their role in industrial applications will only grow, providing innovative solutions for sustainable development, healthcare, and environmental conservation. The rise of microorganisms in biotechnology marks a new era in which tiny organisms play a massive role in shaping the future of industry and technology.

CHAPTER 3

MOLECULAR BIOLOGY AND BIOINFORMATICS IN INDUSTRIAL MICROBIOLOGY AND BIOTECHNOLOGY

The integration of **molecular biology** and **bioinformatics** has transformed the landscape of industrial microbiology and biotechnology. Molecular biology, which focuses on understanding the genetic and biochemical processes that govern cellular function, provides the foundation for manipulating microorganisms to optimize their industrial applications. Meanwhile, bioinformatics—the use of computational tools to analyze biological data—has become indispensable for managing the vast amounts of information generated through molecular biology research. Together, these disciplines drive innovation in the production of pharmaceuticals, biofuels, agricultural products, and bioremediation techniques.

3.1 UNDERSTANDING THE MOLECULAR BIOLOGY

Molecular biology is a branch of biology that seeks to understand the molecular basis of biological activity in and between cells, including biomolecular synthesis, modification, mechanisms, and interactions.

Though cells and other microscopic structures had been observed in living organisms as early as the 18th century, a detailed understanding of the mechanisms and interactions governing their behavior did not emerge until the 20th century, when technologies used in physics and chemistry had advanced sufficiently to permit their application in the biological sciences. The term 'molecular biology' was first used in 1945 by the English physicist William Astbury, who described it as an approach focused on discerning the underpinnings of biological phenomena—i.e. uncovering the physical and chemical structures and properties of biological molecules, as well as their interactions with other molecules and how these interactions explain observations of so-called classical biology, which instead studies biological processes at larger scales and higher levels of organization. In 1953, Francis Crick, James Watson, Rosalind Franklin, and their colleagues at the Medical Research Council Unit, Cavendish Laboratory, were the first to describe the double helix model for the chemical structure of deoxyribonucleic acid (DNA), which is often considered a landmark event for the nascent field because it provided a physico-chemical basis by which to understand the previously nebulous idea of nucleic acids as the primary substance of biological inheritance. They proposed this structure based on previous research done by Franklin, which was conveyed to them by Maurice Wilkins and Max Perutz. Their work led to the discovery of DNA in other microorganisms, plants, and animals.

The field of molecular biology includes techniques which enable scientists to learn about molecular processes. These techniques are used to efficiently target new drugs, diagnose disease, and better understand cell physiology. Some clinical research and medical therapies arising from molecular biology are covered under gene therapy, whereas the use of molecular biology or molecular cell biology in medicine is now referred to as molecular medicine.

3.1.1 History of Molecular Biology

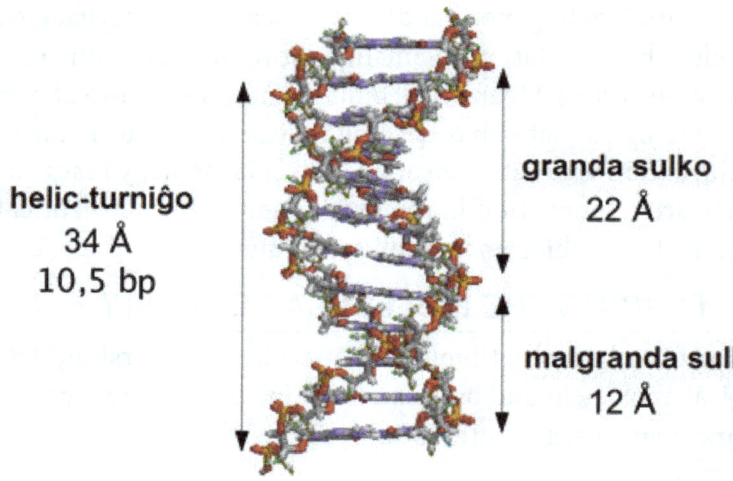

Fig 1: Angle Description in DNA Structure

Molecular biology sits at the intersection of biochemistry and genetics; as these scientific disciplines emerged and evolved in the 20th century, it became clear that they both sought to determine the molecular mechanisms which underlie vital cellular functions. Advances in molecular biology have been closely related to the development of new technologies and their optimization. Molecular biology has been elucidated by the work of many scientists, and thus the history of the field depends on an understanding of these scientists and their experiments.

The field of genetics arose from attempts to understand the set of rules underlying reproduction and heredity, and the nature of the hypothetical units of heredity known as genes. Gregor Mendel pioneered this work in 1866, when he first described the laws of inheritance he observed in his studies of mating crosses in pea plants. One such law of genetic inheritance is the law of segregation, which states that diploid individuals with two alleles for a particular gene will pass one of these alleles to their offspring. Because of his critical work, the study of genetic inheritance is commonly referred to as Mendelian genetics.

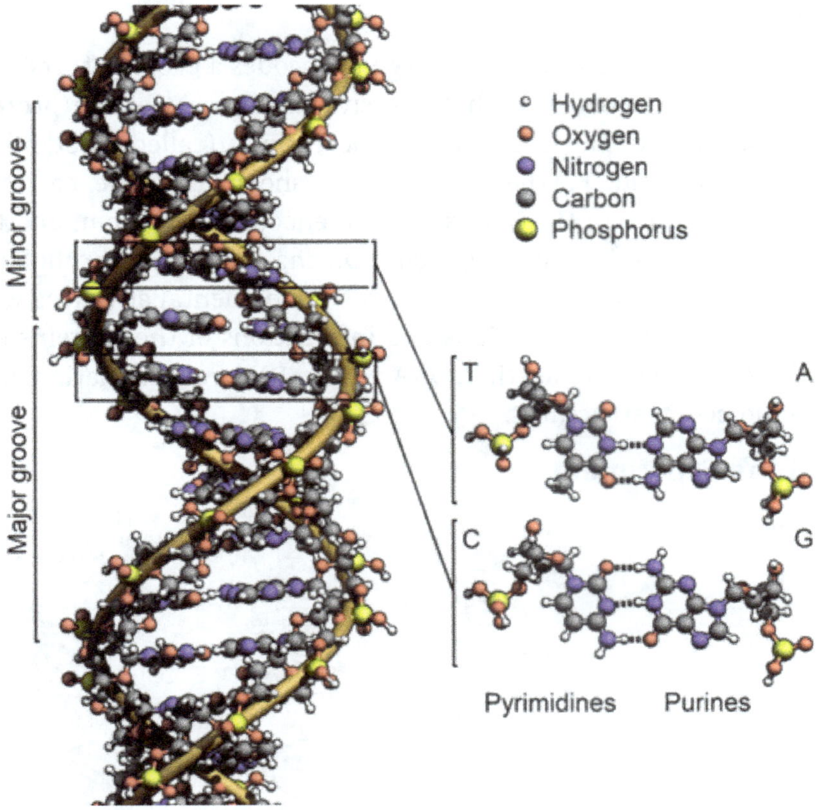

Fig 2: Diagrammatic Representation of Watson and Crick's DNA Structure

A major milestone in molecular biology was the discovery of the structure of DNA. This work began in 1869 by Friedrich Miescher, a Swiss biochemist who first proposed a structure called *nuclein*, which we now know to be (deoxyribonucleic acid), or DNA. He discovered this unique substance by studying the components of pus-filled bandages, and noting the unique properties of the "phosphorus-containing substances". Another notable contributor to the DNA model was Phoebus Levene, who proposed the "polynucleotide model" of DNA in 1919 as a result of his biochemical experiments on yeast. In 1950, Erwin Chargaff expanded on the work of Levene and elucidated a few critical properties of nucleic acids: first, the sequence of nucleic acids varies across species. Second, the total concentration of purines (adenine and guanine) is always equal to the total concentration of pyrimidines (cysteine and thymine). This is now known as Chargaff's rule. In 1953, James Watson and Francis Crick published the double helical structure of DNA, based on the X-ray crystallography work done by Rosalind Franklin which was conveyed to them by Maurice Wilkins and Max Perutz. Watson and Crick described the structure of DNA and conjectured about the implications of this unique structure for possible mechanisms of DNA replication. Watson and Crick were awarded the Nobel Prize in Physiology or Medicine in 1962, along with Wilkins, for proposing a model of the structure of DNA.

In 1961, it was demonstrated that when a gene encodes a protein, three sequential bases of a gene's DNA specify each successive amino acid of the protein. Thus the genetic code is a triplet code, where each triplet (called a codon) specifies a particular amino acid. Furthermore, it was shown that the codons do not overlap with each other in the DNA sequence encoding a protein, and that each sequence is read from a fixed starting point. During 1962–1964, through the use of conditional lethal mutants of a bacterial virus, fundamental advances were made in our understanding of the functions and interactions of the proteins employed in the machinery of DNA replication, DNA repair, DNA recombination, and in the assembly of molecular structures.

3.1.2 Griffith's Experiment

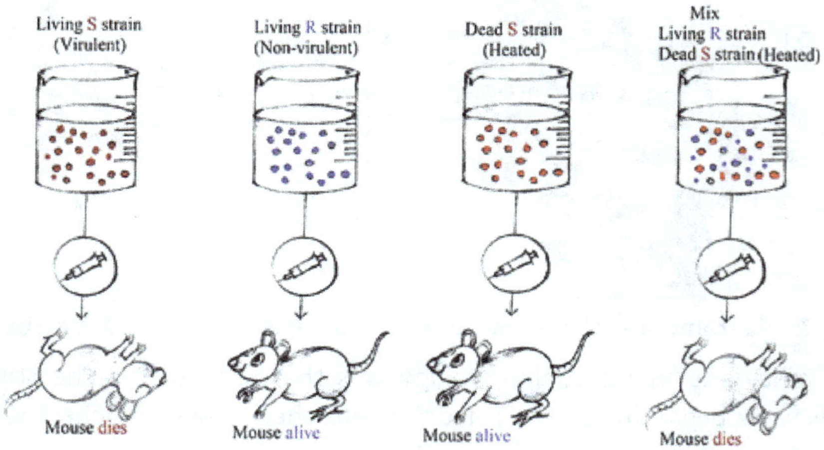

Fig 3: Griffith's Experiment

In 1928, Frederick Griffith, encountered a virulence property in pneumococcus bacteria, which was killing lab rats. According to Mendel, prevalent at that time, gene transfer could occur only from parent to daughter cells. Griffith advanced another theory, stating that gene transfer occurring in member of same generation is known as horizontal gene transfer (HGT). This phenomenon is now referred to as genetic transformation.

Griffith's experiment addressed the pneumococcus bacteria, which had two different strains, one virulent and smooth and one avirulent and rough. The smooth strain had glistering appearance owing to the presence of a type of specific polysaccharide – a polymer of glucose and glucuronic acid capsule. Due to this polysaccharide layer of bacteria, a host's immune system cannot recognize the bacteria and it kills the host. The other, avirulent, rough strain lacks this polysaccharide capsule and has a dull, rough appearance.

Presence or absence of capsule in the strain, is known to be genetically determined. Smooth and rough strains occur in several different type such as S-I, S-II, S-III, etc. and R-I, R-II, R-III, etc. respectively. All this subtypes of S and R bacteria differ with each other in antigen type they produce.

3.1.3 Avery–MacLeod–McCarty Experiment

The Avery–MacLeod–McCarty experiment was a landmark study conducted in 1944 that demonstrated that DNA, not protein as previously thought, carries genetic information in bacteria. Oswald Avery, Colin Munro MacLeod, and Maclyn McCarty used an extract from a strain of pneumococcus that could cause pneumonia in mice. They showed that genetic transformation in the bacteria could be accomplished by injecting them with purified DNA from the extract. They discovered that when they digested the DNA in the extract with DNase, transformation of harmless bacteria into virulent ones was lost. This provided strong evidence that DNA was the genetic material, challenging the prevailing belief that proteins were responsible. It laid the basis for the subsequent discovery of its structure by Watson and Crick.

3.1.4 Hershey–Chase Experiment

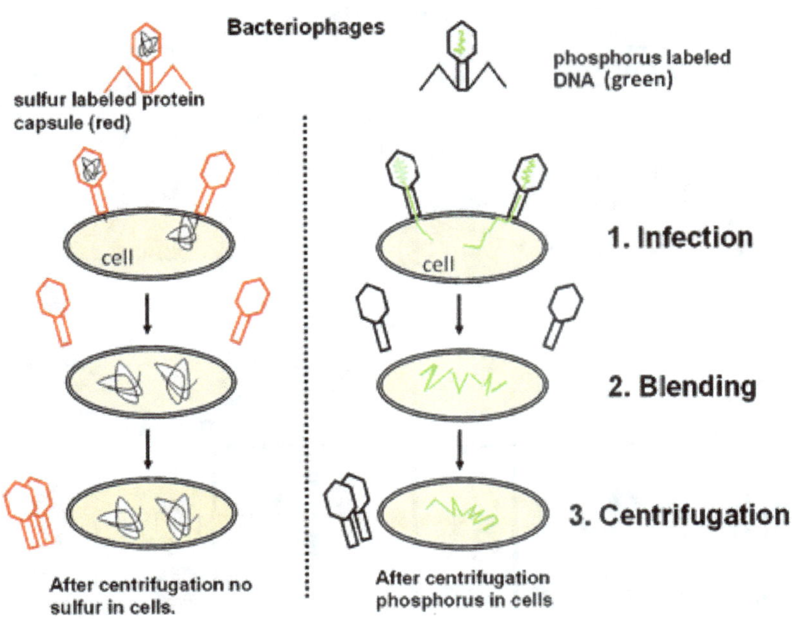

Fig 4: Hershey–Chase Experiment

Confirmation that DNA is the genetic material which is cause of infection came from the Hershey–Chase experiment. They used *E.coli* and bacteriophage for the experiment. This experiment is also known as blender experiment, as kitchen blender was used as a major piece of apparatus. Alfred Hershey and Martha Chase demonstrated that the DNA injected by a phage particle into a bacterium contains all information required to synthesize progeny phage particles. They used radioactivity to tag the bacteriophage's protein coat with radioactive sulphur and DNA with radioactive phosphorus, into two different test tubes respectively. After mixing bacteriophage and *E.coli* into the test tube, the incubation period starts in which phage transforms the genetic material in the *E.coli* cells. Then the mixture is blended or agitated, which separates the phage from *E.coli* cells. The whole mixture is centrifuged and the pellet which contains *E.coli* cells was checked and the supernatant was discarded.

The *E.coli* cells showed radioactive phosphorus, which indicated that the transformed material was DNA not the protein coat.

The transformed DNA gets attached to the DNA of *E.coli* and radioactivity is only seen onto the bacteriophage's DNA. This mutated DNA can be passed to the next generation and the theory of Transduction came into existence. Transduction is a process in which the bacterial DNA carry the fragment of bacteriophages and pass it on the next generation. This is also a type of horizontal gene transfer.

3.1.5 Meselson–Stahl Experiment

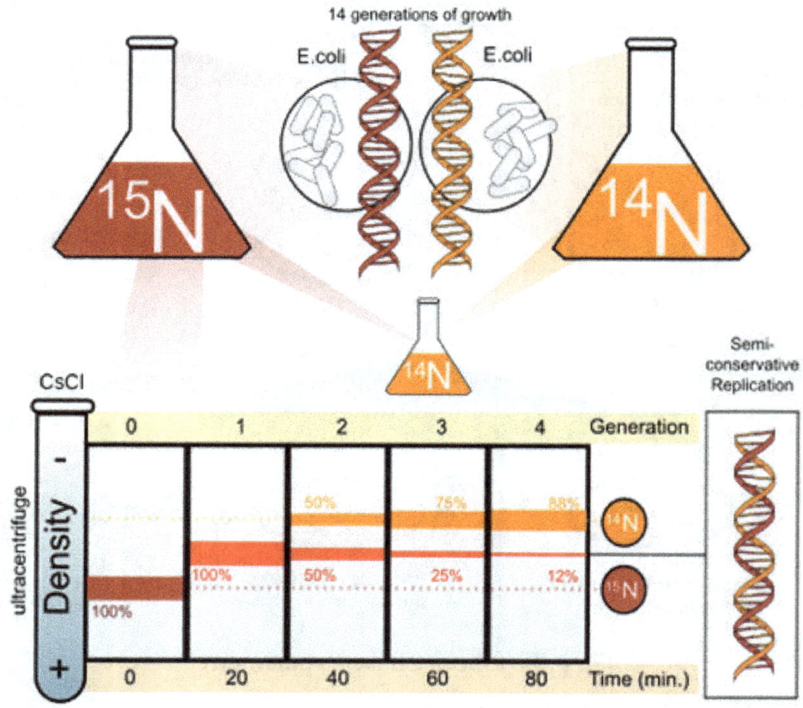

Fig 5: Meselson-Stahl Experiment

The Meselson-Stahl experiment was a landmark experiment in molecular biology that provided evidence for the semiconservative replication of DNA. Conducted in 1958 by Matthew Meselson and Franklin Stahl, the experiment involved growing *E. coli* bacteria in a medium containing heavy isotope of nitrogen (^{15}N) for several generations. This caused all the newly synthesized bacterial DNA to be incorporated with the heavy isotope.

After allowing the bacteria to replicate in a medium containing normal nitrogen (^{14}N), samples were taken at various time points. These samples were then subjected to centrifugation in a density gradient, which separated the DNA molecules based on their density.

The results showed that after one generation of replication in the ^{14}N medium, the DNA formed a band of intermediate density between that of pure ^{15}N DNA and pure ^{14}N DNA. This supported the semiconservative DNA replication proposed by Watson and Crick, where each strand of the parental DNA molecule serves as a template for the synthesis of a new complementary strand, resulting in two daughter DNA molecules, each consisting of one parental and one newly synthesized strand.

The Meselson-Stahl experiment provided compelling evidence for the semiconservative replication of DNA, which is fundamental to the understanding of genetics and molecular biology.

3.1.6 Modern Molecular Biology

In the early 2020s, molecular biology entered a golden age defined by both vertical and horizontal technical development. Vertically, novel technologies are allowing for real-time monitoring of biological processes at the atomic level. Molecular biologists today have access to increasingly affordable sequencing data at increasingly higher depths, facilitating the development of novel genetic manipulation methods in new non-model organisms. Likewise, synthetic molecular biologists will drive the industrial production of small and macro molecules through the introduction of exogenous metabolic pathways in various prokaryotic and eukaryotic cell lines. Horizontally, sequencing data is becoming more affordable and used in many different scientific fields. This will drive the development of industries in developing nations and increase accessibility to individual researchers. Likewise, CRISPR-Cas9 gene editing experiments can now be conceived and implemented by individuals for under $10,000 in novel organisms, which will drive the development of industrial and medical applications.

3.1.7 Relationship to other Biological Sciences

The following list describes a viewpoint on the interdisciplinary relationships between molecular biology and other related fields.

- **Molecular biology** is the study of the molecular underpinnings of the biological phenomena, focusing on molecular synthesis, modification, mechanisms and interactions.
- **Biochemistry** is the study of the chemical substances and vital processes occurring in living organisms. Biochemists focus heavily on the role, function, and structure of biomolecules such as proteins, lipids, carbohydrates and nucleic acids.
- **Genetics** is the study of how genetic differences affect organisms. Genetics attempts to predict how mutations, individual genes and genetic interactions can affect the expression of a phenotype

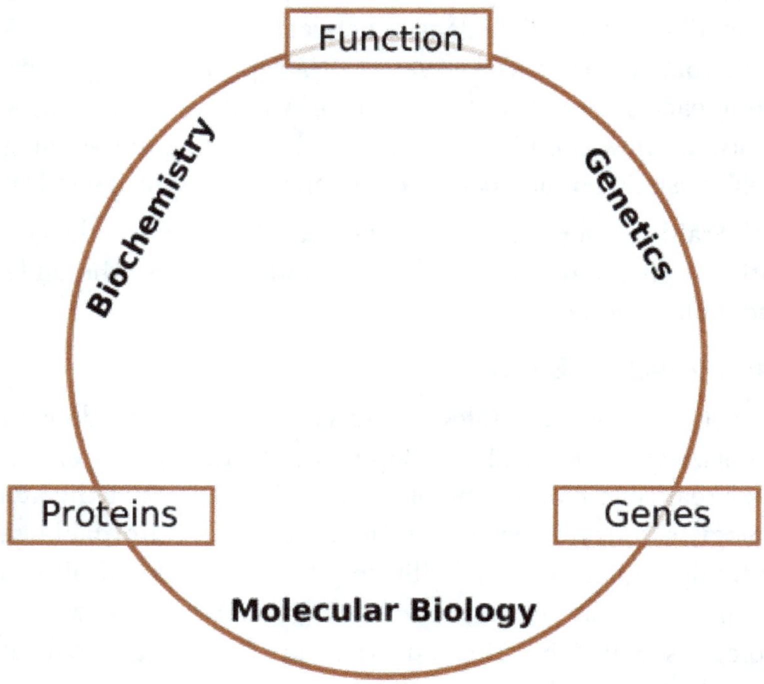

Fig 6: Schematic Relationship between Biochemistry, Genetics and Molecular Biology

While researchers practice techniques specific to molecular biology, it is common to combine these with methods from genetics and biochemistry. Much of molecular biology is quantitative, and recently a significant amount of work has been done using computer science techniques such as bioinformatics and computational biology. Molecular genetics, the study of gene structure and function, has been among the most prominent sub-fields of molecular biology since the early 2000s. Other branches of biology are informed by molecular biology, by either directly studying the interactions of molecules in their own right such as in cell biology and developmental biology, or indirectly, where molecular techniques are used to infer historical attributes of populations or species, as in fields in evolutionary biology such as population genetics and phylogenetics. There is also a long tradition of studying biomolecules "from the ground up", or molecularly, in biophysics.

3.1.8 Techniques of Molecular Biology

Molecular Cloning

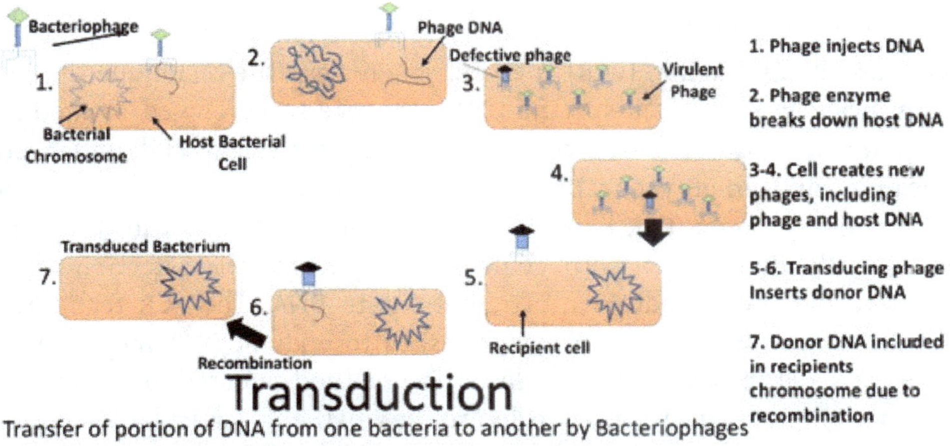

Fig 7: Transduction Image

Molecular cloning is used to isolate and then transfer a DNA sequence of interest into a plasmid vector. This recombinant DNA technology was first developed in the 1960s. In this technique, a DNA sequence coding for a protein of interest is cloned using polymerase chain reaction (PCR), and/or restriction enzymes, into a plasmid (expression vector). The plasmid vector usually has at least 3 distinctive features: an origin of replication, a multiple cloning site (MCS), and a selective marker (usually antibiotic resistance). Additionally, upstream of the MCS are the promoter regions and the transcription start site, which regulate the expression of cloned gene.

This plasmid can be inserted into either bacterial or animal cells. Introducing DNA into bacterial cells can be done by transformation via uptake of naked DNA, conjugation via cell-cell contact or by transduction via viral vector. Introducing DNA into eukaryotic cells, such as animal cells, by physical or chemical means is called transfection. Several different transfection techniques are available, such as calcium phosphate transfection, electroporation, microinjection and liposome transfection. The plasmid may be integrated into the genome, resulting in a stable transfection, or may remain independent of the genome and expressed temporarily, called a transient transfection.

DNA coding for a protein of interest is now inside a cell, and the protein can now be expressed. A variety of systems, such as inducible promoters and specific cell-signaling factors, are available to help express the protein of interest at high levels. Large quantities of a protein can then be extracted from the bacterial or eukaryotic cell. The protein can be tested for enzymatic activity under a variety of situations, the protein may be crystallized so its tertiary structure can be studied, or, in the pharmaceutical industry, the activity of new drugs against the protein can be studied.

Polymerase Chain Reaction

Polymerase chain reaction (PCR) is an extremely versatile technique for copying DNA. In brief, PCR allows a specific DNA sequence to be copied or modified in predetermined ways. The reaction is extremely powerful and under perfect conditions could amplify one DNA molecule to become 1.07 billion molecules in less than two hours. PCR has many applications, including the study of gene expression, the detection of pathogenic microorganisms, the detection of genetic mutations, and the introduction of mutations to DNA. The PCR technique can be used to introduce restriction enzyme sites to ends of DNA molecules, or to mutate particular bases of DNA, the latter is a method referred to as site-directed mutagenesis. PCR can also be used to determine whether a particular DNA fragment is found in a cDNA library. PCR has many variations, like reverse transcription PCR (RT-PCR) for amplification of RNA, and, more recently, quantitative PCR which allow for quantitative measurement of DNA or RNA molecules.

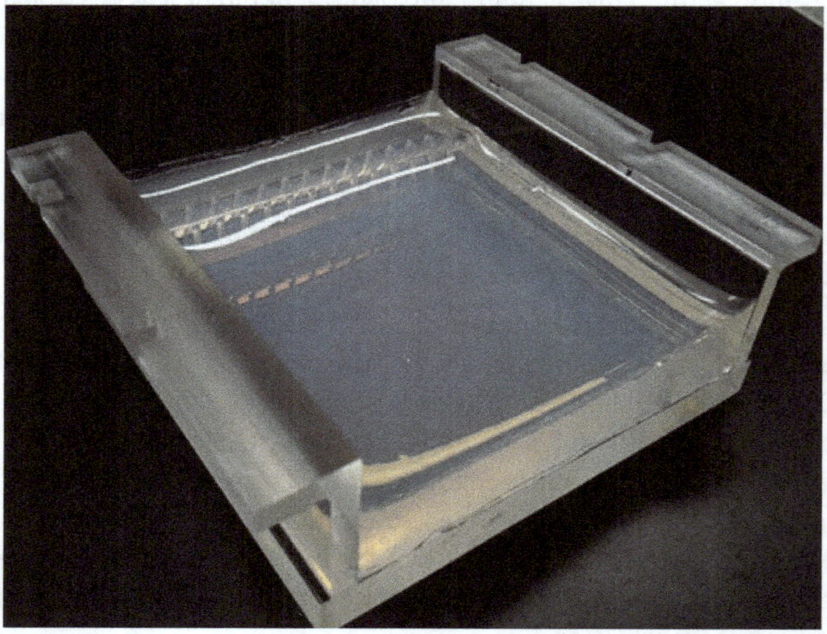

Fig 8: Two Percent Agarose Gel in Borate Buffer Cast in a Gel Tray

Gel Electrophoresis

Gel electrophoresis is a technique which separates molecules by their size using an agarose or polyacrylamide gel. This technique is one of the principal tools of molecular biology. The basic principle is that DNA fragments can be separated by applying an electric current across the gel - because the DNA backbone contains negatively charged phosphate groups, the DNA will migrate through the agarose gel towards the positive end of the current. Proteins can also be separated on the basis of size using an SDS-PAGE gel, or on the basis of size and their electric charge by using what is known as a 2D gel electrophoresis.

The Bradford Protein Assay

The Bradford assay is a molecular biology technique which enables the fast, accurate quantitation of protein molecules utilizing the unique properties of a dye called Coomassie Brilliant Blue G-250. Coomassie Blue undergoes a visible color shift from reddish-brown to bright blue upon binding to protein. In its unstable, cationic state, Coomassie Blue has a background wavelength of 465 nm and gives off a reddish-brown color. When Coomassie Blue binds to protein in an acidic solution, the background wavelength shifts to 595 nm and the dye gives off a bright blue color. Proteins in the assay bind Coomassie blue in about 2 minutes, and the protein-dye complex is stable for about an hour, although it is recommended that absorbance readings are taken within 5 to 20 minutes of reaction initiation. The concentration of protein in the Bradford assay can then be measured using a visible light spectrophotometer, and therefore does not require extensive equipment.

This method was developed in 1975 by Marion M. Bradford, and has enabled significantly faster, more accurate protein quantitation compared to previous methods: the Lowry procedure and the biuret assay. Unlike the previous methods, the Bradford assay is not susceptible to interference by several non-protein molecules, including ethanol, sodium chloride, and magnesium chloride. However, it is susceptible to influence by strong alkaline buffering agents, such as sodium dodecyl sulfate (SDS).

Macromolecule Blotting and Probing

The terms *northern*, *western* and *eastern* blotting are derived from what initially was a molecular biology joke that played on the term *Southern blotting*, after the technique described by Edwin Southern for the hybridisation of blotted DNA. Patricia Thomas, developer of the RNA blot which then became known as the *northern blot*, actually did not use the term.

Southern Blotting

Named after its inventor, biologist Edwin Southern, the Southern blot is a method for probing for the presence of a specific DNA sequence within a DNA sample. DNA samples before or after restriction enzyme (restriction endonuclease) digestion are separated by gel electrophoresis and then transferred to a membrane by blotting via capillary action. The membrane is then exposed to a labeled DNA probe that has a complement base sequence to the sequence on the DNA of interest. Southern blotting is less commonly used in laboratory science due to the capacity of other techniques, such as PCR, to detect specific DNA sequences from DNA samples. These blots are still used for some applications, however, such as measuring transgene copy number in transgenic mice or in the engineering of gene knockout embryonic stem cell lines.

Northern Blotting

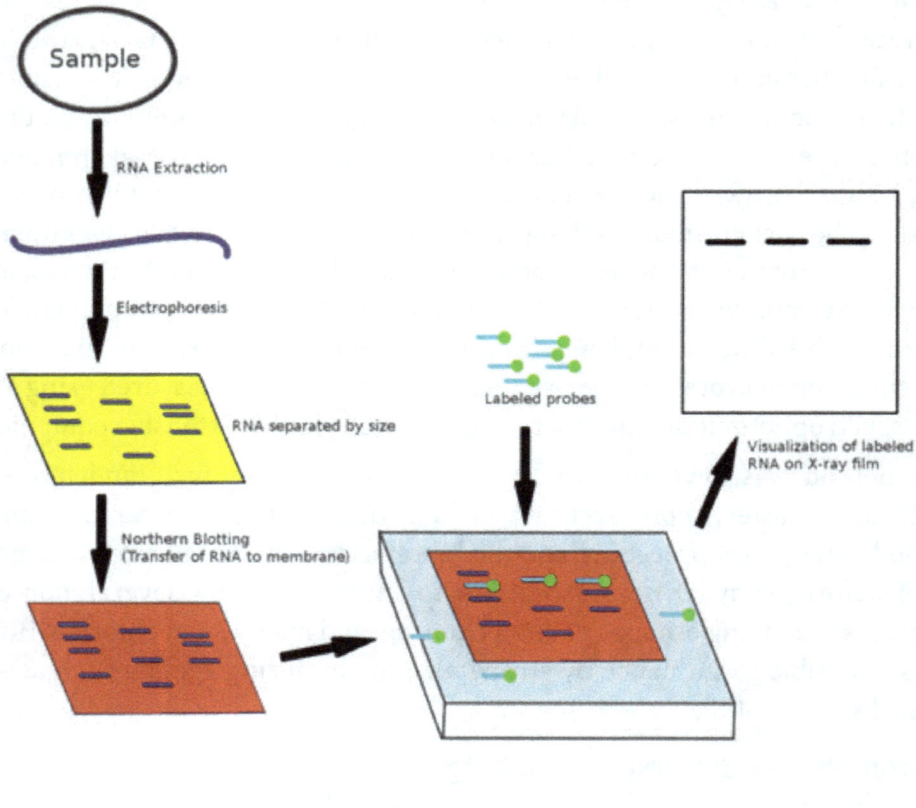

Fig 9: Northern Blot Diagram

The northern blot is used to study the presence of specific RNA molecules as relative comparison among a set of different samples of RNA. It is essentially a combination of denaturing RNA gel electrophoresis, and a blot. In this process RNA is separated based on size and is then transferred to a membrane that is then probed with a labeled complement of a sequence of interest. The results may be visualized through a variety of ways depending on the label used; however, most result in the revelation of bands representing the sizes of the RNA detected in sample. The intensity of these bands is related to the amount of the target RNA in the samples analyzed. The procedure is commonly used to study when and how much gene expression is occurring by measuring how much of that RNA is present in different samples, assuming that no post-transcriptional regulation occurs and that the levels of mRNA reflect proportional levels of the corresponding protein being produced. It is one of the most basic tools for determining at what time, and under what conditions, certain genes are expressed in living tissues.

Western Blotting

A western blot is a technique by which specific proteins can be detected from a mixture of proteins. Western blots can be used to determine the size of isolated proteins, as well as to quantify their expression. In western blotting, proteins are first separated by size, in a thin gel sandwiched between two glass plates in a technique known as SDS-PAGE. The proteins in the gel are then transferred to a polyvinylidene fluoride (PVDF), nitrocellulose, nylon, or other support membrane. This membrane can then be probed with solutions of antibodies. Antibodies that specifically bind to the protein of interest can then be visualized by a variety of techniques, including colored products, chemiluminescence, or autoradiography. Often, the antibodies are labeled with enzymes. When a chemiluminescent substrate is exposed to the enzyme it allows detection. Using western blotting techniques allows not only detection but also quantitative analysis. Analogous methods to western blotting can be used to directly stain specific proteins in live cells or tissue sections.

Eastern Blotting

The eastern blotting technique is used to detect post-translational modification of proteins. Proteins blotted on to the PVDF or nitrocellulose membrane are probed for modifications using specific substrates.

Microarrays

A DNA microarray is a collection of spots attached to a solid support such as a microscope slide where each spot contains one or more single-stranded DNA oligonucleotide fragments. Arrays make it possible to put down large quantities of very small (100 micrometre diameter) spots on a single slide. Each spot has a DNA fragment molecule that is complementary to a single DNA sequence. A variation of this technique allows the gene expression of an organism at a particular stage in development to be qualified (expression profiling). In this technique the RNA in a tissue is isolated and converted to labeled complementary DNA (cDNA). This cDNA is then hybridized to the fragments on the array and visualization of the hybridization can be done. Since multiple arrays can be made with exactly the same position of fragments, they are particularly useful for comparing the gene expression of two different tissues, such as a healthy and cancerous tissue. Also, one can measure what genes are expressed and how that expression changes with time or with other factors. There are many different ways to fabricate microarrays; the most common are silicon chips, microscope slides with spots of ~100 micrometre diameter, custom arrays, and arrays with larger spots on porous membranes (macroarrays). There can be anywhere from 100 spots to more than 10,000 on a given array. Arrays can also be made with molecules other than DNA.

Allele-Specific Oligonucleotide

Allele-specific oligonucleotide (ASO) is a technique that allows detection of single base mutations without the need for PCR or gel electrophoresis. Short (20–25 nucleotides in length), labeled probes are exposed to the non-fragmented target DNA, hybridization occurs with high specificity due to the short length of the probes and even a single base change will hinder hybridization. The target DNA is then washed and the unhybridized probes are removed. The target DNA is then analyzed for the presence of the probe via radioactivity or fluorescence. In this experiment, as in most molecular biology techniques, a control must be used to ensure successful experimentation.

In molecular biology, procedures and technologies are continually being developed and older technologies abandoned. For example, before the advent of DNA gel electrophoresis (agarose or polyacrylamide), the size of DNA molecules was typically determined by rate sedimentation in sucrose gradients, a slow and labor-intensive technique requiring expensive instrumentation; prior to sucrose gradients, viscometry was used. Aside from their historical interest, it is often worth knowing about older technology, as it is occasionally useful to solve another new problem for which the newer technique is inappropriate.

3.2 MOLECULAR BIOLOGY IN INDUSTRIAL MICROBIOLOGY AND BIOTECHNOLOGY

Molecular biology has revolutionized industrial microbiology by enabling the precise manipulation of microorganisms to enhance their productivity and efficiency. Through techniques such as **recombinant DNA technology**, **gene editing**, and **synthetic biology**, scientists can modify the genetic makeup of microorganisms to perform specific tasks that are valuable to industry.

3.2.1 Recombinant DNA Technology

Recombinant DNA (rDNA) technology involves the combination of genetic material from different organisms to create new genetic sequences. This technology is fundamental to modern biotechnology, allowing scientists to insert genes from one organism into another, often using bacteria or yeast as the host. For example, **Escherichia coli** (E. coli) is frequently used to produce **human insulin** by inserting the human insulin gene into the bacterium's genome. This process enables the large-scale production of insulin for diabetes treatment, representing a major success in the field of industrial microbiology. In addition to insulin, rDNA technology has facilitated the production of other pharmaceuticals, such as growth hormones, vaccines, and monoclonal antibodies. The ability to produce these biologically active compounds in microorganisms has reduced production costs, increased scalability, and improved accessibility to essential medicines.

3.2.2 Gene Editing and CRISPR Technology

One of the most groundbreaking developments in molecular biology is the advent of **CRISPR-Cas9** gene-editing technology. This tool allows for precise modifications to the DNA of microorganisms, making it possible to edit or delete specific genes or insert new ones. CRISPR has dramatically accelerated research and development in industrial microbiology, as it enables scientists to optimize microorganisms for desired traits, such as increased productivity, resistance to environmental stress, or enhanced enzymatic activity. For instance, CRISPR has been used to engineer yeast strains that are more efficient at converting sugars into ethanol, improving the production of biofuels. Similarly, bacteria have been genetically edited to enhance their ability to degrade pollutants in bioremediation processes. The versatility of CRISPR makes it a valuable tool for optimizing microorganisms across various industries, from pharmaceuticals to agriculture.

3.2.3 Synthetic Biology

Synthetic biology builds on molecular biology principles by designing and constructing new biological parts, devices, and systems. It goes beyond simply modifying existing organisms, aiming to create entirely new microbial strains that can perform novel functions. In industrial biotechnology, synthetic biology has opened up possibilities for developing microorganisms that can produce bio-based chemicals, pharmaceuticals, and even materials that could replace traditional plastics. For example, synthetic biology is being used to engineer bacteria that produce **bioplastics** from renewable resources. These bioplastics are biodegradable and offer a more sustainable alternative to petroleum-based plastics. By creating microorganisms with customized metabolic pathways, synthetic biology enables the production of a wide range of bio-based products that are environmentally friendly and economically viable.

3.3 BIOINFORMATICS: MANAGING AND ANALYZING BIOLOGICAL DATA

Bioinformatics is an interdisciplinary field of science that develops methods and software tools for understanding biological data, especially when the data sets are large and complex. Bioinformatics uses biology, chemistry, physics, computer science, computer programming, information engineering, mathematics and statistics to analyze and interpret biological data. The subsequent process of analyzing and interpreting data is often referred to as computational biology, though the distinction between the two terms is often disputed.

Computational, statistical, and computer programming techniques have been used for computer simulation analyses of biological queries. They include reused specific analysis "pipelines", particularly in the field of genomics, such as by the identification of genes and single nucleotide polymorphisms (SNPs). These pipelines are used to better understand the genetic basis of disease, unique adaptations, desirable properties (esp. in agricultural species), or differences between populations. Bioinformatics also includes proteomics, which tries to understand the organizational principles within nucleic acid and protein sequences.

Image and signal processing allow extraction of useful results from large amounts of raw data. In the field of genetics, it aids in sequencing and annotating genomes and their observed mutations. Bioinformatics includes text mining of biological literature and the development of biological and gene ontologies to organize and query biological data. It also plays a role in the analysis of gene and protein expression and regulation. Bioinformatics tools aid in comparing, analyzing and interpreting genetic and genomic data and more generally in the understanding of evolutionary aspects of molecular biology. At a more integrative level, it helps analyze and catalogue the biological pathways and networks that are an important part of systems biology. In structural biology, it aids in the simulation and modeling of DNA, RNA, proteins as well as biomolecular interactions.

3.3.1 History

The first definition of the term *bioinformatics* was coined by Paulien Hogeweg and Ben Hesper in 1970, to refer to the study of information processes in biotic systems. This definition placed bioinformatics as a field parallel to biochemistry (the study of chemical processes in biological systems). Bioinformatics and computational biology involved the analysis of biological data, particularly DNA, RNA, and protein sequences. The field of bioinformatics experienced explosive growth starting in the mid-1990s, driven largely by the Human Genome Project and by rapid advances in DNA sequencing technology.

Analyzing biological data to produce meaningful information involves writing and running software programs that use algorithms from graph theory, artificial intelligence, soft computing, data mining, image processing, and computer simulation. The algorithms in turn depend on theoretical foundations such as discrete mathematics, control theory, system theory, information theory, and statistics.

Sequences

5'ATGACGTGGGGA3'

3'TACTGCACCCCT5'

Fig 10: Sequences of Genetic Material are Frequently Used in Bioinformatics and are Easier to Manage Using Computers than Manually

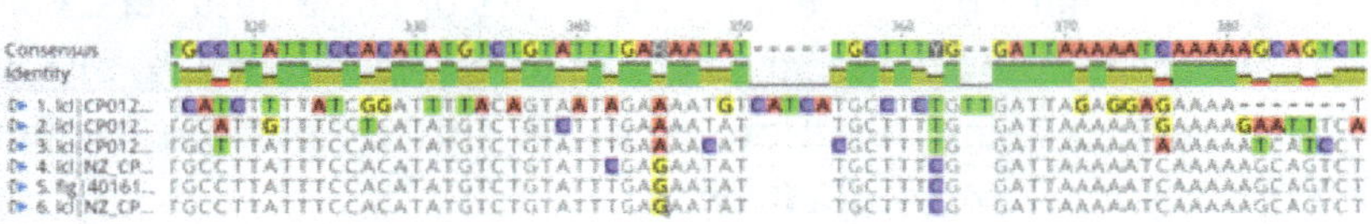

Fig 11: These are Sequences being Compared in a MUSCLE Multiple Sequence Slignment (MSA): Each Sequence Name (Leftmost Column) is from Various Louse Species, while the Sequences Themselves are in the Second Column

There has been a tremendous advance in speed and cost reduction since the completion of the Human Genome Project, with some labs able to sequence over 100,000 billion bases each year, and a full genome can be sequenced for $1,000 or less.

Computers became essential in molecular biology when protein sequences became available after Frederick Sanger determined the sequence of insulin in the early 1950s. Comparing multiple sequences manually turned out to be impractical. Margaret Oakley Dayhoff, a pioneer in the field, compiled one of the first protein sequence databases, initially published as books as well as methods of sequence alignment and molecular evolution. Another early contributor to bioinformatics was Elvin A. Kabat, who pioneered biological sequence analysis in 1970 with his comprehensive volumes of antibody sequences released online with Tai Te Wu between 1980 and 1991.

In the 1970s, new techniques for sequencing DNA were applied to bacteriophage MS2 and øX174, and the extended nucleotide sequences were then parsed with informational and statistical algorithms. These studies illustrated that well known features, such as the coding segments and the triplet code, are revealed in straightforward statistical analyses and were the proof of the concept that bioinformatics would be insightful.

3.3.2 Goals

In order to study how normal cellular activities are altered in different disease states, raw biological data must be combined to form a comprehensive picture of these activities. Therefore, the field of bioinformatics has evolved such that the most pressing task now involves the analysis and interpretation of various types of data. This also includes nucleotide and amino acid sequences, protein domains, and protein structures.

Important sub-disciplines within bioinformatics and computational biology include:

- Development and implementation of computer programs to efficiently access, manage, and use various types of information.
- Development of new mathematical algorithms and statistical measures to assess relationships among members of large data sets. For example, there are methods to locate a gene within a sequence, to predict protein structure and/or function, and to cluster protein sequences into families of related sequences.

The primary goal of bioinformatics is to increase the understanding of biological processes. What sets it apart from other approaches is its focus on developing and applying computationally intensive techniques to achieve this goal. Examples include: pattern recognition, data mining, machine learning algorithms, and visualization. Major research efforts in the field include sequence alignment, gene finding, genome assembly, drug design, drug discovery, protein structure alignment, protein structure prediction, prediction of gene expression and protein–protein interactions, genome-wide association studies, the modeling of evolution and cell division/mitosis.

Bioinformatics entails the creation and advancement of databases, algorithms, computational and statistical techniques, and theory to solve formal and practical problems arising from the management and analysis of biological data.

Over the past few decades, rapid developments in genomic and other molecular research technologies and developments in information technologies have combined to produce a tremendous amount of information related to molecular biology. Bioinformatics is the name given to these mathematical and computing approaches used to glean understanding of biological processes.

Common activities in bioinformatics include mapping and analyzing DNA and protein sequences, aligning DNA and protein sequences to compare them, and creating and viewing 3-D models of protein structures.

3.3.3 Sequence Analysis

Since the bacteriophage Phage Φ-X174 was sequenced in 1977, the DNA sequences of thousands of organisms have been decoded and stored in databases. This sequence information is analyzed to determine genes that encode proteins, RNA genes, regulatory sequences, structural motifs, and repetitive sequences. A comparison of genes within a species or between different species can show similarities between protein functions, or relations between species (the use of molecular systematics to construct phylogenetic trees). With the growing amount of data, it long ago became impractical to analyze DNA sequences manually. Computer programs such as BLAST are used routinely to search sequences—as of 2008, from more than 260,000 organisms, containing over 190 billion nucleotides.

DNA Sequencing

Before sequences can be analyzed, they are obtained from a data storage bank, such as GenBank. DNA sequencing is still a non-trivial problem as the raw data may be noisy or affected by weak signals. Algorithms have been developed for base calling for the various experimental approaches to DNA sequencing.

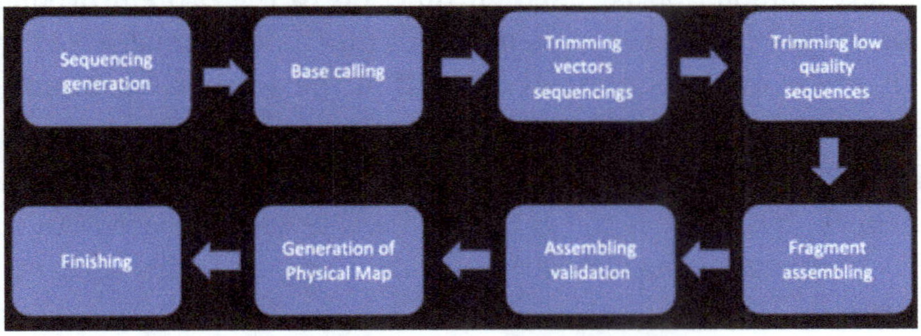

Fig 12: Sequencing Genome Steps

Sequence Assembly

Most DNA sequencing techniques produce short fragments of sequence that need to be assembled to obtain complete gene or genome sequences. The shotgun sequencing technique (used by The Institute for Genomic Research (TIGR) to sequence the first bacterial genome, *Haemophilus influenzae*) generates the sequences of many thousands of small DNA fragments (ranging from 35 to 900 nucleotides long, depending on the sequencing technology). The ends of these fragments overlap and, when aligned properly by a genome assembly program, can be used to reconstruct the complete genome. Shotgun sequencing yields sequence data quickly, but the task of assembling the fragments can be quite complicated for larger genomes. For a genome as large as the human genome, it may take many days of CPU time on large-memory, multiprocessor computers to assemble the fragments, and the resulting assembly usually contains numerous gaps that must be filled in later. Shotgun sequencing is the method of choice for virtually all genomes sequenced (rather than chain-termination or chemical degradation methods), and genome assembly algorithms are a critical area of bioinformatics research.

Genome Annotation

In genomics, annotation refers to the process of marking the stop and start regions of genes and other biological features in a sequenced DNA sequence. Many genomes are too large to be annotated by hand. As the rate of sequencing exceeds the rate of genome annotation, genome annotation has become the new bottleneck in bioinformatics.

Genome annotation can be classified into three levels: the nucleotide, protein, and process levels. Gene finding is a chief aspect of nucleotide-level annotation. For complex genomes, a combination of ab initio gene prediction and sequence comparison with expressed sequence databases and other organisms can be successful. Nucleotide-level annotation also allows the integration of genome sequence with other genetic and physical maps of the genome. The principal aim of protein-level annotation is to assign function to the protein products of the genome. Databases of protein sequences and functional domains and motifs are used for this type of annotation. About half of the predicted proteins in a new genome sequence tend to have no obvious function.

Understanding the function of genes and their products in the context of cellular and organismal physiology is the goal of process-level annotation. An obstacle of process-level annotation has been the inconsistency of terms used by different model systems.

The Gene Ontology Consortium is helping to solve this problem. The first description of a comprehensive annotation system was published in 1995 by The Institute for Genomic Research, which performed the first complete sequencing and analysis of the genome of a free-living (non-symbiotic) organism, the bacterium *Haemophilus influenzae*. The system identifies the genes encoding all proteins, transfer RNAs, ribosomal RNAs, in order to make initial functional assignments. The GeneMark program trained to find protein-coding genes in *Haemophilus influenzae* is constantly changing and improving.

Following the goals that the Human Genome Project left to achieve after its closure in 2003, the ENCODE project was developed by the National Human Genome Research Institute. This project is a collaborative data collection of the functional elements of the human genome that uses next-generation DNA-sequencing technologies and genomic tiling arrays, technologies able to automatically generate large amounts of data at a dramatically reduced per-base cost but with the same accuracy (base call error) and fidelity (assembly error).

Gene Function Prediction

While genome annotation is primarily based on sequence similarity (and thus homology), other properties of sequences can be used to predict the function of genes. In fact, most *gene* function prediction methods focus on *protein* sequences as they are more informative and more feature-rich. For instance, the distribution of hydrophobic amino acids predicts transmembrane segments in proteins. However, protein function prediction can also use external information such as gene (or protein) expression data, protein structure, or protein-protein interactions.

Computational Evolutionary Biology

Evolutionary biology is the study of the origin and descent of species, as well as their change over time. Informatics has assisted evolutionary biologists by enabling researchers to:

- Trace the evolution of a large number of organisms by measuring changes in their DNA, rather than through physical taxonomy or physiological observations alone,
- Compare entire genomes, which permits the study of more complex evolutionary events, such as gene duplication, horizontal gene transfer, and the prediction of factors important in bacterial speciation,
- Build complex computational population genetics models to predict the outcome of the system over time, and
- Track and share information on an increasingly large number of species and organisms.

Comparative Genomics

The core of comparative genome analysis is the establishment of the correspondence between genes (orthology analysis) or other genomic features in different organisms. Intergenomic maps are made to trace the evolutionary processes responsible for the divergence of two genomes. A multitude of evolutionary events acting at various organizational levels shape genome evolution. At the lowest level, point mutations affect individual nucleotides. At a higher level, large chromosomal segments undergo duplication, lateral transfer, inversion, transposition, deletion and insertion. Entire genomes are involved in processes of hybridization, polyploidization and endosymbiosis that lead to rapid speciation. The complexity of genome evolution poses many exciting challenges to developers of mathematical models and algorithms, who have recourse to a spectrum of algorithmic, statistical and mathematical techniques, ranging from exact, heuristics, fixed parameter and approximation algorithms for problems based on parsimony models to Markov chain Monte Carlo algorithms for Bayesian analysis of problems based on probabilistic models.

Pan Genomics

Pan genomics is a concept introduced in 2005 by Tettelin and Medini. Pan genome is the complete gene repertoire of a particular monophyletic taxonomic group. Although initially applied to closely related strains of a species, it can be applied to a larger context like genus, phylum, etc. It is divided in two parts: the Core genome, a set of genes common to all the genomes under study (often housekeeping genes vital for survival), and the Dispensable/Flexible genome: a set of genes not present in all but one or some genomes under study. A bioinformatics tool BPGA can be used to characterize the Pan Genome of bacterial species.

Genetics of Disease

As of 2013, the existence of efficient high-throughput next-generation sequencing technology allows for the identification of cause many different human disorders. Simple Mendelian inheritance has been observed for over 3,000 disorders that have been identified at the Online Mendelian Inheritance in Man database, but complex diseases are more difficult. Association studies have found many individual genetic regions that individually are weakly associated with complex diseases (such as infertility, breast cancer and Alzheimer's disease), rather than a single cause. There are currently many challenges to using genes for diagnosis and treatment, such as how we don't know which genes are important, or how stable the choices an algorithm provides.

Genome-wide association studies have successfully identified thousands of common genetic variants for complex diseases and traits; however, these common variants only explain a small fraction of heritability. Rare variants may account for some of the missing heritability. Large-scale whole genome sequencing studies have rapidly sequenced millions of whole genomes, and such studies have identified hundreds of millions of rare variants. Functional annotations predict the effect or function of a genetic variant and help to prioritize rare functional variants, and incorporating these annotations can effectively boost the power of genetic association of rare variants analysis of whole genome sequencing studies. Some tools have been developed to provide all-in-one rare variant association analysis for whole-genome sequencing data, including integration of genotype data and their functional annotations, association analysis, result summary and visualization. Meta-analysis of whole genome sequencing studies provides an attractive solution to the problem of collecting large sample sizes for discovering rare variants associated with complex phenotypes.

Analysis of Mutations in Cancer

In cancer, the genomes of affected cells are rearranged in complex or unpredictable ways. In addition to single-nucleotide polymorphism arrays identifying point mutations that cause cancer, oligonucleotide microarrays can be used to identify chromosomal gains and losses (called comparative genomic hybridization). These detection methods generate terabytes of data per experiment. The data is often found to contain considerable variability, or noise, and thus Hidden Markov model and change-point analysis methods are being developed to infer real copy number changes. Two important principles can be used to identify cancer by mutations in the exome. First, cancer is a disease of accumulated somatic mutations in genes. Second, cancer contains driver mutations which need to be distinguished from passengers. Further improvements in bioinformatics could allow for classifying types of cancer by analysis of cancer driven mutations in the genome. Furthermore, tracking of patients while the disease progresses may be possible in the future with the sequence of cancer samples. Another type of data that requires novel informatics development is the analysis of lesions found to be recurrent among many tumors.

3.3.4 Gene and Protein Expression

Analysis of Gene Expression

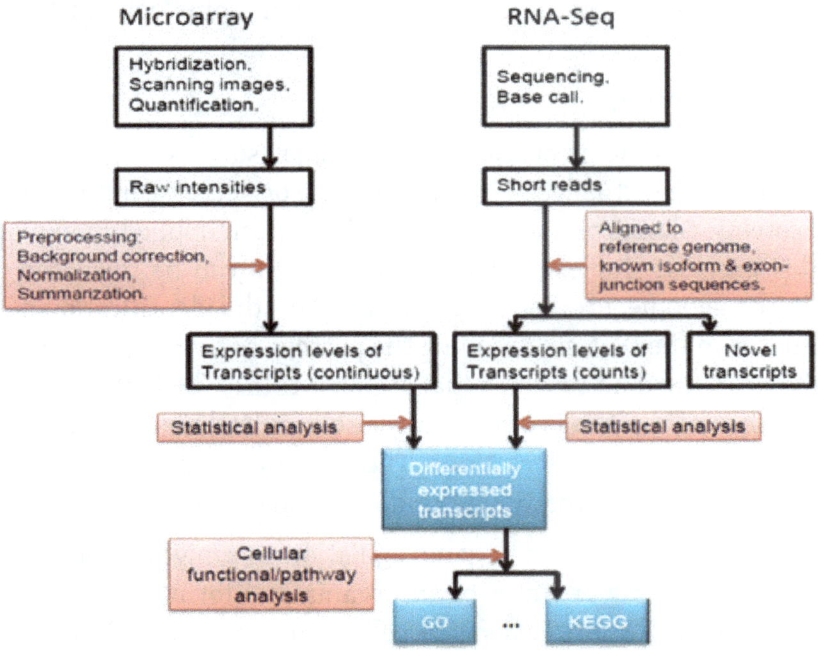

Fig 13: Microarray vs. RNA-Seq

The expression of many genes can be determined by measuring mRNA levels with multiple techniques including microarrays, expressed cDNA sequence tag (EST) sequencing, serial analysis of gene expression (SAGE) tag sequencing, massively parallel signature sequencing (MPSS), RNA-Seq, also known as "Whole Transcriptome Shotgun Sequencing" (WTSS), or various applications of multiplexed in-situ hybridization. All of these techniques are extremely noise-prone and/ or subject to bias in the biological measurement, and a major research area in computational biology involves developing statistical tools to separate signal from noise in high-throughput gene expression studies. Such studies are often used to determine the genes implicated in a disorder: one might compare microarray data from cancerous epithelial cells to data from non-cancerous cells to determine the transcripts that are up-regulated and down-regulated in a particular population of cancer cells.

Analysis of Protein Expression

Protein microarrays and high throughput (HT) mass spectrometry (MS) can provide a snapshot of the proteins present in a biological sample. The former approach faces similar problems as with microarrays targeted at mRNA, the latter involves the problem of matching large amounts of mass data against predicted masses from protein sequence databases, and the complicated statistical analysis of samples when multiple incomplete peptides from each protein are detected. Cellular protein localization in a tissue context can be achieved through affinity proteomics displayed as spatial data based on immunohistochemistry and tissue microarrays.

Analysis of Regulation

Gene regulation is a complex process where a signal, such as an extracellular signal such as a hormone, eventually leads to an increase or decrease in the activity of one or more proteins. Bioinformatics techniques have been applied to explore various steps in this process. For example, gene expression can be regulated by nearby elements in the genome. Promoter analysis involves the identification and study of sequence motifs in the DNA surrounding the protein-coding region of a gene. These motifs influence the extent to which that region is transcribed into mRNA. Enhancer elements far away from the promoter can also regulate gene expression, through three-dimensional looping interactions. These interactions can be determined by bioinformatic analysis of chromosome conformation capture experiments.

Expression data can be used to infer gene regulation: one might compare microarray data from a wide variety of states of an organism to form hypotheses about the genes involved in each state. In a single-cell organism, one might compare stages of the cell cycle, along with various stress conditions (heat shock, starvation, etc.). Clustering algorithms can be then applied to expression data to determine which genes are co-expressed. For example, the upstream regions (promoters) of co-expressed genes can be searched for over-represented regulatory elements. Examples of clustering algorithms applied in gene clustering are k-means clustering, self-organizing maps (SOMs), hierarchical clustering, and consensus clustering methods.

3.3.5 Analysis of Cellular Organization

Several approaches have been developed to analyze the location of organelles, genes, proteins, and other components within cells. A gene ontology category, *cellular component*, has been devised to capture subcellular localization in many biological databases.

Microscopy and Image Analysis

Microscopic pictures allow for the location of organelles as well as molecules, which may be the source of abnormalities in diseases.

Protein Localization

Finding the location of proteins allows us to predict what they do. This is called protein function prediction.

For instance, if a protein is found in the nucleus it may be involved in gene regulation or splicing. By contrast, if a protein is found in mitochondria, it may be involved in respiration or other metabolic processes. There are well developed protein subcellular localization prediction resources available, including protein subcellular location databases, and prediction tools.

Nuclear Organization of Chromatin

Data from high-throughput chromosome conformation capture experiments, such as Hi-C (experiment) and ChIA-PET, can provide information on the three-dimensional structure and nuclear organization of chromatin. Bioinformatic challenges in this field include partitioning the genome into domains, such as Topologically Associating Domains (TADs), that are organised together in three-dimensional space.

3.3.6 Structural Bioinformatics

Finding the structure of proteins is an important application of bioinformatics. The Critical Assessment of Protein Structure Prediction (CASP) is an open competition where worldwide research groups submit protein models for evaluating unknown protein models.

Amino Acid Sequence

The linear amino acid sequence of a protein is called the primary structure. The primary structure can be easily determined from the sequence of codons on the DNA gene that codes for it. In most proteins, the primary structure uniquely determines the 3-dimensional structure of a protein in its native environment. An exception is the misfolded protein involved in bovine spongiform encephalopathy. This structure is linked to the function of the protein. Additional structural information includes the *secondary*, *tertiary* and *quaternary* structure. A viable general solution to the prediction of the function of a protein remains an open problem. Most efforts have so far been directed towards heuristics that work most of the time.

Homology

In the genomic branch of bioinformatics, homology is used to predict the function of a gene: if the sequence of gene A, whose function is known, is homologous to the sequence of gene B, whose function is unknown, one could infer that B may share A's function. In structural bioinformatics, homology is used to determine which parts of a protein are important in structure formation and interaction with other proteins. Homology modeling is used to predict the structure of an unknown protein from existing homologous proteins.

One example of this is hemoglobin in humans and the hemoglobin in legumes (leghemoglobin), which are distant relatives from the same protein superfamily. Both serve the same purpose of transporting oxygen in the organism.

Although both of these proteins have completely different amino acid sequences, their protein structures are virtually identical, which reflects their near identical purposes and shared ancestor.

Other techniques for predicting protein structure include protein threading and *de novo* (from scratch) physics-based modeling. Another aspect of structural bioinformatics include the use of protein structures for Virtual Screening models such as Quantitative Structure-Activity Relationship models and proteochemometric models (PCM).

Furthermore, a protein's crystal structure can be used in simulation of for example ligand-binding studies and *in silico* mutagenesis studies. A 2021 deep-learning algorithms-based software called AlphaFold, developed by Google's DeepMind, greatly outperforms all other prediction software methods, and has released predicted structures for hundreds of millions of proteins in the AlphaFold protein structure database.

3.3.7 Network and Systems Biology

Network analysis seeks to understand the relationships within biological networks such as metabolic or protein–protein interaction networks. Although biological networks can be constructed from a single type of molecule or entity (such as genes), network biology often attempts to integrate many different data types, such as proteins, small molecules, gene expression data, and others, which are all connected physically, functionally, or both.

Systems biology involves the use of computer simulations of cellular subsystems (such as the networks of metabolites and enzymes that comprise metabolism, signal transduction pathways and gene regulatory networks) to both analyze and visualize the complex connections of these cellular processes. Artificial life or virtual evolution attempts to understand evolutionary processes via the computer simulation of simple (artificial) life forms.

Molecular Interaction Networks

Tens of thousands of three-dimensional protein structures have been determined by X-ray crystallography and protein nuclear magnetic resonance spectroscopy (protein NMR) and a central question in structural bioinformatics is whether it is practical to predict possible protein–protein interactions only based on these 3D shapes, without performing protein–protein interaction experiments. A variety of methods have been developed to tackle the protein–protein docking problem, though it seems that there is still much work to be done in this field. Other interactions encountered in the field include Protein–ligand (including drug) and protein–peptide. Molecular dynamic simulation of movement of atoms about rotatable bonds is the fundamental principle behind computational algorithms, termed docking algorithms, for studying molecular interactions.

3.3.8 Biodiversity Informatics

Biodiversity informatics deals with the collection and analysis of biodiversity data, such as taxonomic databases, or microbiome data. Examples of such analyses include phylogenetics, niche modelling, species richness mapping, DNA barcoding, or species identification tools. A growing area is also macro-ecology, i.e. the study of how biodiversity is connected to ecology and human impact, such as climate change.

3.3.9 Others

Literature Analysis

The enormous number of published literature makes it virtually impossible for individuals to read every paper, resulting in disjointed sub-fields of research. Literature analysis aims to employ computational and statistical linguistics to mine this growing library of text resources. For example:

- **Abbreviation Recognition:** Identify the long-form and abbreviation of biological terms.
- **Named-Entity Recognition:** Recognizing biological terms such as gene names.
- **Protein–Protein Interaction:** Identify which proteins interact with which proteins from text.

The area of research draws from statistics and computational linguistics.

High-Throughput Image Analysis

Computational technologies are used to automate the processing, quantification and analysis of large amounts of high-information-content biomedical imagery. Modern image analysis systems can improve an observer's accuracy, objectivity, or speed. Image analysis is important for both diagnostics and research. Some examples are:

- High-throughput and high-fidelity quantification and sub-cellular localization (high-content screening, cytohistopathology, Bioimage informatics).
- Morphometrics.
- Clinical image analysis and visualization.
- Determining the real-time air-flow patterns in breathing lungs of living animals.
- Quantifying occlusion size in real-time imagery from the development of and recovery during arterial injury.
- Making behavioral observations from extended video recordings of laboratory animals.
- Infrared measurements for metabolic activity determination.
- Inferring clone overlaps in DNA mapping, e.g. the Sulston score.

High-Throughput Single Cell Data Analysis

Computational techniques are used to analyse high-throughput, low-measurement single cell data, such as that obtained from flow cytometry. These methods typically involve finding populations of cells that are relevant to a particular disease state or experimental condition.

Ontologies and Data Integration

Biological ontologies are directed acyclic graphs of controlled vocabularies. They create categories for biological concepts and descriptions so they can be easily analyzed with computers. When categorised in this way, it is possible to gain added value from holistic and integrated analysis. The OBO Foundry was an effort to standardise certain ontologies. One of the most widespread is the Gene ontology which describes gene function. There are also ontologies which describe phenotypes.

3.3.10 Databases

Databases are essential for bioinformatics research and applications. Databases exist for many different information types, including DNA and protein sequences, molecular structures, phenotypes and biodiversity. Databases can contain both empirical data (obtained directly from experiments) and predicted data (obtained from analysis of existing data). They may be specific to a particular organism, pathway or molecule of interest. Alternatively, they can incorporate data compiled from multiple other databases. Databases can have different formats, access mechanisms, and be public or private. Some of the most commonly used databases are listed below:

- **Used in Biological Sequence Analysis:** Genbank, UniProt.
- **Used in Structure Analysis:** Protein Data Bank (PDB).
- **Used in Finding Protein Families and Motif Finding:** InterPro, Pfam.
- **Used for Next Generation Sequencing:** Sequence Read Archive.
- **Used in Network Analysis:** Metabolic Pathway Databases (KEGG, BioCyc), Interaction Analysis Databases, Functional Networks.
- **Used in Design of Synthetic Genetic Circuits:** GenoCAD.

3.3.11 Software and Tools

Software tools for bioinformatics include simple command-line tools, more complex graphical programs, and standalone web-services. They are made by bioinformatics companies or by public institutions.

Open-Source Bioinformatics Software

Many free and open-source software tools have existed and continued to grow since the 1980s. The combination of a continued need for new algorithms for the analysis of emerging types of biological readouts, the potential for innovative *in silico* experiments, and freely available open code bases have created opportunities for research groups to contribute to both bioinformatics regardless of funding. The open source tools often act as incubators of ideas, or community-supported plug-ins in commercial applications. They may also provide *de facto* standards and shared object models for assisting with the challenge of bioinformation integration. Open-source bioinformatics software includes Bioconductor, BioPerl, Biopython, BioJava, BioJS, BioRuby, Bioclipse, EMBOSS, .NET Bio, Orange with its bioinformatics add-on, Apache Taverna, UGENE and GenoCAD. The non-profit Open Bioinformatics Foundation and the annual Bioinformatics Open Source Conference promote open-source bioinformatics software.

Web Services in Bioinformatics

SOAP- and REST-based interfaces have been developed to allow client computers to use algorithms, data and computing resources from servers in other parts of the world. The main advantage are that end users do not have to deal with software and database maintenance overheads. Basic bioinformatics services are classified by the EBI into three categories: SSS (Sequence Search Services), MSA (Multiple Sequence Alignment), and BSA (Biological Sequence Analysis). The availability of these service-oriented bioinformatics resources demonstrate the applicability of web-based bioinformatics solutions, and range from a collection of standalone tools with a common data format under a single web-based interface, to integrative, distributed and extensible bioinformatics workflow management systems.

Bioinformatics Workflow Management Systems

A bioinformatics workflow management system is a specialized form of a workflow management system designed specifically to compose and execute a series of computational or data manipulation steps, or a workflow, in a Bioinformatics application. Such systems are designed to:

- Provide an easy-to-use environment for individual application scientists themselves to create their own workflows,
- Provide interactive tools for the scientists enabling them to execute their workflows and view their results in real-time,
- Simplify the process of sharing and reusing workflows between the scientists, and
- Enable scientists to track the provenance of the workflow execution results and the workflow creation steps.

BioCompute and BioCompute Objects

In 2014, the US Food and Drug Administration sponsored a conference held at the National Institutes of Health Bethesda Campus to discuss reproducibility in bioinformatics. Over the next three years, a consortium of stakeholders met regularly to discuss what would become BioCompute paradigm.

These stakeholders included representatives from government, industry, and academic entities. Session leaders represented numerous branches of the FDA and NIH Institutes and Centers, non-profit entities including the Human Variome Project and the European Federation for Medical Informatics, and research institutions including Stanford, the New York Genome Center, and the George Washington University.

It was decided that the BioCompute paradigm would be in the form of digital 'lab notebooks' which allow for the reproducibility, replication, review, and reuse, of bioinformatics protocols. This was proposed to enable greater continuity within a research group over the course of normal personnel flux while furthering the exchange of ideas between groups. The US FDA funded this work so that information on pipelines would be more transparent and accessible to their regulatory staff.

In 2016, the group reconvened at the NIH in Bethesda and discussed the potential for a BioCompute Object, an instance of the BioCompute paradigm. This work was copied as both a "standard trial use" document and a preprint paper uploaded to bioRxiv. The BioCompute object allows for the JSON-ized record to be shared among employees, collaborators, and regulators.

3.3.12 Education Platforms

Bioinformatics is not only taught as in-person masters degree at many universities. The computational nature of bioinformatics lends it to computer-aided and online learning. Software platforms designed to teach bioinformatics concepts and methods include Rosalind and online courses offered through the Swiss Institute of Bioinformatics Training Portal. The Canadian Bioinformatics Workshops provides videos and slides from training workshops on their website under a Creative Commons license. The 4273π project or 4273pi project also offers open source educational materials for free. The course runs on low cost Raspberry Pi computers and has been used to teach adults and school pupils. 4273 is actively developed by a consortium of academics and research staff who have run research level bioinformatics using Raspberry Pi computers and the 4273π operating system. MOOC platforms also provide online certifications in bioinformatics and related disciplines, including Coursera's Bioinformatics Specialization at the University of California, San Diego, Genomic Data Science Specialization at Johns Hopkins University, and EdX's Data Analysis for Life Sciences XSeries at Harvard University.

3.4 APPLICATIONS IN INDUSTRIAL MICROBIOLOGY AND BIOTECHNOLOGY

The integration of molecular biology and bioinformatics has led to significant advancements in various industries, including pharmaceuticals, biofuels, agriculture, and environmental biotechnology.

3.4.1 Pharmaceutical Production

As mentioned earlier, molecular biology techniques such as recombinant DNA technology and gene editing have transformed pharmaceutical production. Bioinformatics aids in identifying potential drug targets, optimizing microbial strains for production, and analyzing the efficacy of therapeutic proteins. The combination of molecular biology and bioinformatics has led to the efficient production of drugs, vaccines, and biologics that improve human health.

3.4.2 Biofuel Production

In the biofuel industry, molecular biology is used to engineer microorganisms that can efficiently convert biomass into biofuels like ethanol, biodiesel, and methane. Bioinformatics tools help optimize these microbial strains by analyzing metabolic pathways and identifying bottlenecks that limit biofuel production. Through metabolic engineering and bioinformatics-guided strain optimization, biofuel production processes have become more efficient and cost-effective.

3.4.3 Agricultural Biotechnology

Molecular biology and bioinformatics are also critical in **agricultural biotechnology**. Genetically modified microorganisms (GMOs) are used to improve crop yields, enhance nutrient content, and protect plants from pests and diseases. Bioinformatics enables the analysis of plant-microbe interactions, helping researchers develop microbial inoculants that promote plant growth or protect crops from pathogens.

3.4.4 Environmental Biotechnology and Bioremediation

In environmental biotechnology, molecular biology and bioinformatics are used to engineer microorganisms for **bioremediation**—the process of using microorganisms to degrade or neutralize environmental pollutants. Bioinformatics helps identify enzymes capable of breaking down hazardous compounds, while molecular biology tools are used to enhance the metabolic pathways of microorganisms to improve their efficiency in cleaning up contaminated environments.

3.5 CONCLUSION: THE FUTURE OF MOLECULAR BIOLOGY AND BIOINFORMATICS IN BIOTECHNOLOGY

The combination of molecular biology and bioinformatics has revolutionized industrial microbiology and biotechnology, offering unprecedented opportunities for innovation. As technology continues to advance, the integration of these disciplines will become even more critical in addressing global challenges such as sustainable energy production, environmental conservation, and healthcare. With ongoing developments in **CRISPR gene editing**, **synthetic biology**, and **metagenomics**, the potential for harnessing microorganisms for industrial applications is limitless. The future of biotechnology lies in our ability to manipulate and optimize microbial systems, and molecular biology and bioinformatics will remain at the forefront of this scientific revolution.

CHAPTER 4

NUTRIENT MEDIA FOR CULTIVATION OF INDUSTRIAL MICROORGANISMS AND PRODUCTION OF MICROBIAL PRODUCTS

In industrial microbiology, the cultivation of microorganisms is fundamental for the production of valuable products such as enzymes, antibiotics, biofuels, and other biochemicals.

To ensure the optimal growth and productivity of these microorganisms, a key factor is the nutrient media used to support their development. Nutrient media supply essential nutrients, energy sources, vitamins, and minerals that microorganisms need for growth and metabolic activity.

Properly designed media not only enhance microbial growth but also increase the efficiency and yield of microbial products.

4.1 TYPES OF NUTRIENT MEDIA

The selection of an appropriate nutrient medium depends on the type of microorganism being cultured and the desired end product. Nutrient media can be classified into several categories based on their composition, preparation, and function.

4.1.1 Complex Media

Complex media, also known as undefined media, contain a mixture of natural products whose exact chemical composition is unknown. These media often include ingredients such as yeast extract, peptone, meat extract, and soybean meal, which provide a wide range of nutrients, including amino acids, vitamins, and minerals.

- **Example:** Yeast extract peptone dextrose (YPD) medium is a widely used complex medium for culturing Saccharomyces cerevisiae in the production of bioethanol and other fermentation products. The yeast extract provides vitamins and amino acids, while peptone supplies nitrogen, and dextrose (glucose) serves as an energy source.

Complex media are particularly useful for cultivating fastidious microorganisms that have complex nutritional requirements. They are also commonly used in industrial applications where the goal is to maximize biomass or product yield without the need for precise control over every nutrient component.

4.1.2 Defined (Synthetic) Media

In contrast to complex media, defined or synthetic media contain known quantities of pure chemicals. Each component is precisely measured, and the chemical composition of the medium is fully understood. Defined media are typically used when researchers or industrial producers need to study the effects of specific nutrients on microbial growth or when consistent and reproducible results are required.

- **Example:** A defined medium for the cultivation of Escherichia coli might include glucose as a carbon source, ammonium chloride as a nitrogen source, and salts such as magnesium sulfate and potassium phosphate to supply essential minerals. The precise control of nutrients allows for optimization of microbial metabolism for the production of desired products, such as recombinant proteins or enzymes.

Defined media are especially important in pharmaceutical and biotechnological industries where consistency, purity, and reproducibility are critical for product quality and regulatory compliance.

4.1.3 Selective Media

Selective media are formulated to support the growth of specific microorganisms while inhibiting the growth of others. This type of medium is often used when a particular microbial strain needs to be isolated or when contaminants need to be suppressed during industrial fermentation processes.

- **Example:** MacConkey agar is a selective medium used to isolate Gram-negative bacteria like E. coli while inhibiting the growth of Gram-positive bacteria. It contains bile salts and crystal violet, which prevent the growth of Gram-positive organisms, while lactose and a pH indicator allow differentiation between lactose fermenters and non-fermenters.

Selective media are valuable in industrial settings where microbial purity is essential, such as in the production of vaccines or probiotics, where contamination by unwanted strains could compromise product safety or efficacy.

4.1.4 Differential Media

Differential media are designed to distinguish between different types of microorganisms based on their metabolic activity or biochemical characteristics. These media often contain indicators that change color or form precipitates in response to specific microbial reactions.

- **Example:** Eosin methylene blue (EMB) agar is a differential medium used to differentiate between lactose fermenting and non-fermenting Gram-negative bacteria. Lactose fermenters, such as **E. coli**, produce a green metallic sheen on the agar, while non-fermenters remain colorless.

Differential media are useful in the initial screening of industrial microbial strains, especially when multiple microorganisms are being tested for their ability to produce a specific metabolite or biochemical product.

4.2 COMPONENTS OF NUTRIENT MEDIA

The success of microbial cultivation in industrial microbiology depends on the proper selection and balance of the components in the nutrient media. Key components of nutrient media include carbon sources, nitrogen sources, salts, growth factors, and trace elements.

4.2.1 Carbon Sources

Carbon is the primary energy source for microorganisms, and the choice of carbon source can significantly impact microbial growth and product formation. Common carbon sources include glucose, sucrose, glycerol, and starch. In some industrial processes, waste products such as molasses or corn steep liquor are used as inexpensive carbon sources.

For example, in the production of citric acid by Aspergillus niger, glucose or sucrose is typically used as the carbon source. The microorganism metabolizes these sugars to produce citric acid, which is widely used in the food and beverage industry.

4.2.2 Nitrogen Sources

Nitrogen is essential for the synthesis of proteins, nucleic acids, and other cellular components. Common nitrogen sources in nutrient media include ammonium salts, nitrates, urea, and peptones. The choice of nitrogen source can influence the metabolic pathways used by the microorganism, affecting both growth rate and product yield. In the production of antibiotics by Streptomyces species, nitrogen sources such as soybean meal or ammonium sulfate are often included in the medium to support high levels of biomass and secondary metabolite production.

4.2.3 Salts and Minerals

In addition to carbon and nitrogen, microorganisms require various salts and minerals to support cellular functions. These include magnesium, potassium, phosphates, calcium, and iron. These ions are involved in enzyme activation, osmotic balance, and metabolic regulation. For instance, magnesium is a cofactor for many enzymes involved in nucleic acid synthesis and energy production. Phosphates are critical for ATP production and serve as buffering agents to maintain the pH of the medium.

4.2.4 Growth Factors and Vitamins

Many microorganisms require additional growth factors or vitamins that they cannot synthesize on their own. These are typically added to the nutrient medium to ensure optimal growth. Common growth factors include biotin, riboflavin, pantothenic acid, and thiamine. In the production of riboflavin (vitamin B2) by the microorganism Ashbya gossypii, the medium must be supplemented with adequate amounts of nitrogen and growth factors to maximize riboflavin synthesis.

4.2.5 Trace Elements

While microorganisms require only small amounts of trace elements such as zinc, copper, manganese, and molybdenum, these elements play critical roles in enzyme function and metabolic processes. Trace elements are often added to nutrient media in the form of salts or chelated compounds to ensure bioavailability. For example, in the industrial production of enzymes by Bacillus subtilis, trace elements such as manganese and calcium are added to stabilize the enzymes and improve their activity during fermentation.

4.3 PRODUCTION OF MICROBIAL PRODUCTS

The production of microbial products in industrial microbiology relies on optimizing both the nutrient media and the environmental conditions under which the microorganisms are grown. Microbial products can be classified into two broad categories: primary metabolites and secondary metabolites.

4.3.1 Primary Metabolites

Primary metabolites are compounds produced during the growth phase of microorganisms and are essential for their normal metabolic processes. These include amino acids, organic acids, alcohols, and enzymes. The production of primary metabolites is closely linked to the composition of the nutrient medium, as the availability of carbon, nitrogen, and other nutrients directly affects the growth rate and metabolic activity of the microorganism.

Example: The production of ethanol by yeast is a primary metabolite process. The yeast converts sugars in the medium into ethanol and carbon dioxide through fermentation. The efficiency of ethanol production depends on the availability of glucose and other nutrients in the medium.

4.3.2 Secondary Metabolites

Secondary metabolites are not essential for microbial growth but are often produced during the stationary phase of growth. These compounds include antibiotics, pigments, and toxins, and they have significant industrial and pharmaceutical applications.

- **Example:** The production of penicillin by the fungus Penicillium chrysogenum is a classic example of secondary metabolite production. The nutrient medium for penicillin production must contain specific carbon and nitrogen sources, as well as precursors such as phenylacetic acid, to maximize the yield of the antibiotic.

4.4 CONCLUSION: THE IMPORTANCE OF OPTIMIZING NUTRIENT MEDIA

In industrial microbiology, the choice and optimization of nutrient media are critical factors that determine the success of microbial cultivation and the production of microbial products. Different microorganisms have unique nutritional requirements, and the composition of the medium must be tailored to meet these needs. By selecting the appropriate nutrient sources, growth factors, and environmental conditions, industries can maximize the yield and efficiency of valuable products such as enzymes, antibiotics, biofuels, and organic acids. As biotechnology advances, the development of new and improved nutrient media will continue to play a vital role in the sustainable and efficient production of microbial products.

CHAPTER 5

BIOSYNTHETIC PATHWAYS FOR METABOLIC PRODUCTS OF MICROORGANISMS

Biosynthetic pathways in microorganisms play a crucial role in converting simple molecules into complex compounds necessary for growth, survival, and interaction with the environment.

These pathways consist of a series of enzyme-mediated reactions that lead to the production of primary and secondary metabolites. The knowledge and manipulation of these pathways have significant applications in industrial microbiology, biotechnology, pharmaceuticals, agriculture, and biofuels.

The production of metabolic products such as amino acids, antibiotics, organic acids, and bioactive molecules is mediated by specific biosynthetic routes. By understanding and engineering these pathways, scientists can enhance the production of valuable compounds, optimize yields, and create new biotechnological applications.

5.1 UNDERSTANDING THE BIOSYNTHESIS

Biosynthesis, i.e., chemical synthesis occurring in biological contexts, is a term most often referring to multi-step, enzyme-catalyzed processes where chemical substances absorbed as nutrients (or previously converted through biosynthesis) serve as enzyme substrates, with conversion by the living organism either into simpler or more complex products. Examples of biosynthetic pathways include those for the production of amino acids, lipid membrane components, and nucleotides, but also for the production of all classes of biological macromolecules, and of acetyl-coenzyme A, adenosine triphosphate, nicotinamide adenine dinucleotide and other key intermediate and transactional molecules needed for metabolism. Thus, in biosynthesis, any of an array of compounds, from simple to complex, are converted into other compounds, and so it includes both the catabolism and anabolism (building up and breaking down) of complex molecules (including macromolecules).

Biosynthetic processes are often represented via charts of metabolic pathways. A particular biosynthetic pathway may be located within a single cellular organelle (e.g., mitochondrial fatty acid synthesis pathways), while others involve enzymes that are located across an array of cellular organelles and structures (e.g., the biosynthesis of glycosylated cell surface proteins).

5.1.1 Elements of Biosynthesis

Elements of biosynthesis include: precursor compounds, chemical energy (e.g. ATP), and catalytic enzymes which may need coenzymes (e.g. NADH, NADPH). These elements create monomers, the building blocks for macromolecules. Some important biological macromolecules include: proteins, which are composed of amino acid monomers joined via peptide bonds, and DNA molecules, which are composed of nucleotides joined via phosphodiester bonds.

5.1.2 Properties of Chemical Reactions

Biosynthesis occurs due to a series of chemical reactions. For these reactions to take place, the following elements are necessary:

- **Precursor Compounds:** These compounds are the starting molecules or substrates in a reaction. These may also be viewed as the reactants in a given chemical process.
- **Chemical Energy:** Chemical energy can be found in the form of high energy molecules. These molecules are required for energetically unfavourable reactions. Furthermore, the hydrolysis of these compounds drives a reaction forward. High energy molecules, such as ATP, have three phosphates. Often, the terminal phosphate is split off during hydrolysis and transferred to another molecule.
- **Catalysts:** These may be for example metal ions or coenzymes and they catalyze a reaction by increasing the rate of the reaction and lowering the activation energy.

5.1.3 Lipid

Many intricate macromolecules are synthesized in a pattern of simple, repeated structures. For example, the simplest structures of lipids are fatty acids. Fatty acids are hydrocarbon derivatives; they contain a carboxyl group "head" and a hydrocarbon chain "tail". These fatty acids create larger components, which in turn incorporate noncovalent interactions to form the lipid bilayer. Fatty acid chains are found in two major components of membrane lipids: phospholipids and sphingolipids. A third major membrane component, cholesterol, does not contain these fatty acid units.

Eukaryotic Phospholipids

The foundation of all biomembranes consists of a bilayer structure of phospholipids. The phospholipid molecule is amphipathic; it contains a hydrophilic polar head and a hydrophobic nonpolar tail. The phospholipid heads interact with each other and aqueous media, while the hydrocarbon tails orient themselves in the center, away from water. These latter interactions drive the bilayer structure that acts as a barrier for ions and molecules.

There are various types of phospholipids; consequently, their synthesis pathways differ. However, the first step in phospholipid synthesis involves the formation of phosphatidate or diacylglycerol 3-phosphate at the endoplasmic reticulum and outer mitochondrial membrane. The synthesis pathway is found below:

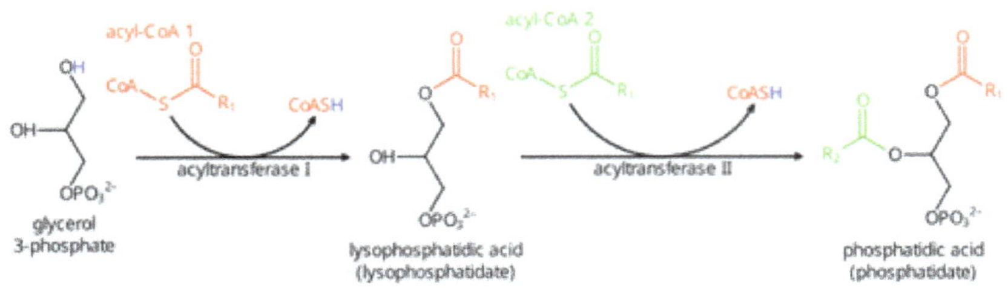

The pathway starts with glycerol 3-phosphate, which gets converted to lysophosphatidate via the addition of a fatty acid chain provided by acyl coenzyme A. Then, lysophosphatidate is converted to phosphatidate via the addition of another fatty acid chain contributed by a second acyl CoA; all of these steps are catalyzed by the glycerol phosphate acyltransferase enzyme. Phospholipid synthesis continues in the endoplasmic reticulum, and the biosynthesis pathway diverges depending on the components of the particular phospholipid.

Sphingolipids

Like phospholipids, these fatty acid derivatives have a polar head and nonpolar tails. Unlike phospholipids, sphingolipids have a sphingosine backbone. Sphingolipids exist in eukaryotic cells and are particularly abundant in the central nervous system. For example, sphingomyelin is part of the myelin sheath of nerve fibers.

Sphingolipids are formed from ceramides that consist of a fatty acid chain attached to the amino group of a sphingosine backbone. These ceramides are synthesized from the acylation of sphingosine. The biosynthetic pathway for sphingosine is found below:

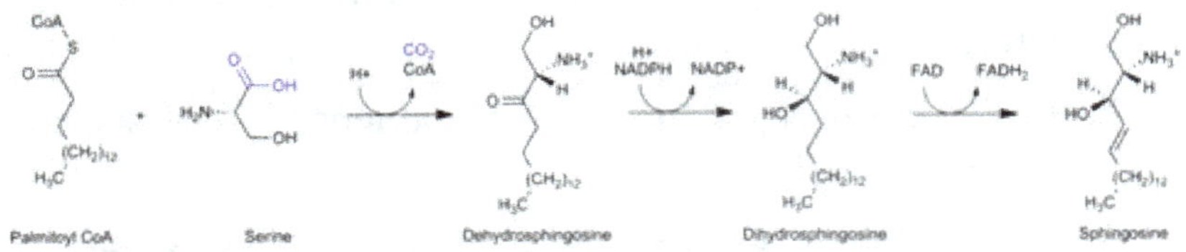

As the image denotes, during sphingosine synthesis, palmitoyl CoA and serine undergo a condensation reaction which results in the formation of 3-dehydrosphinganine. This product is then reduced to form dihydrospingosine, which is converted to sphingosine via the oxidation reaction by FAD.

Cholesterol

This lipid belongs to a class of molecules called sterols. Sterols have four fused rings and a hydroxyl group. Cholesterol is a particularly important molecule. Not only does it serve as a component of lipid membranes, it is also a precursor to several steroid hormones, including cortisol, testosterone, and estrogen. Cholesterol is synthesized from acetyl CoA. The pathway is shown below:

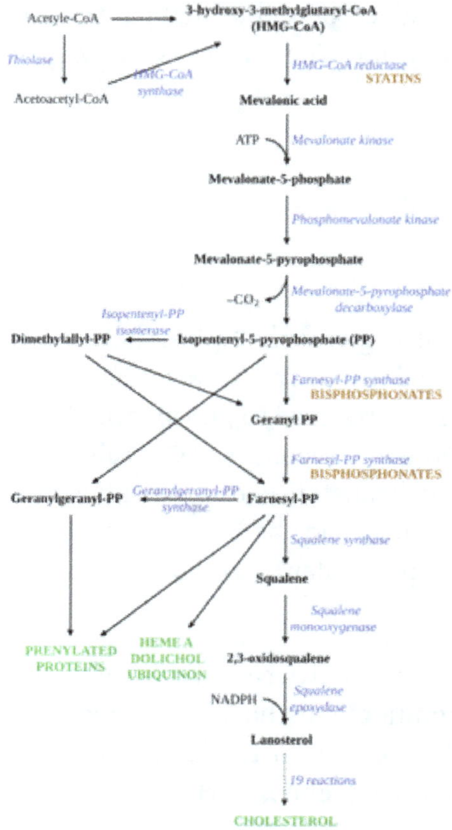

More generally, this synthesis occurs in three stages, with the first stage taking place in the cytoplasm and the second and third stages occurring in the endoplasmic reticulum. The stages are as follows:

- The synthesis of isopentenyl pyrophosphate, the "building block" of cholesterol.
- The formation of squalene via the condensation of six molecules of isopentenyl phosphate.
- The conversion of squalene into cholesterol via several enzymatic reactions.

5.1.4 Nucleotides

The biosynthesis of nucleotides involves enzyme-catalyzed reactions that convert substrates into more complex products. Nucleotides are the building blocks of DNA and RNA. Nucleotides are composed of a five-membered ring formed from ribose sugar in RNA, and deoxyribose sugar in DNA; these sugars are linked to a purine or pyrimidine base with a glycosidic bond and a phosphate group at the 5' location of the sugar.

Purine Nucleotides

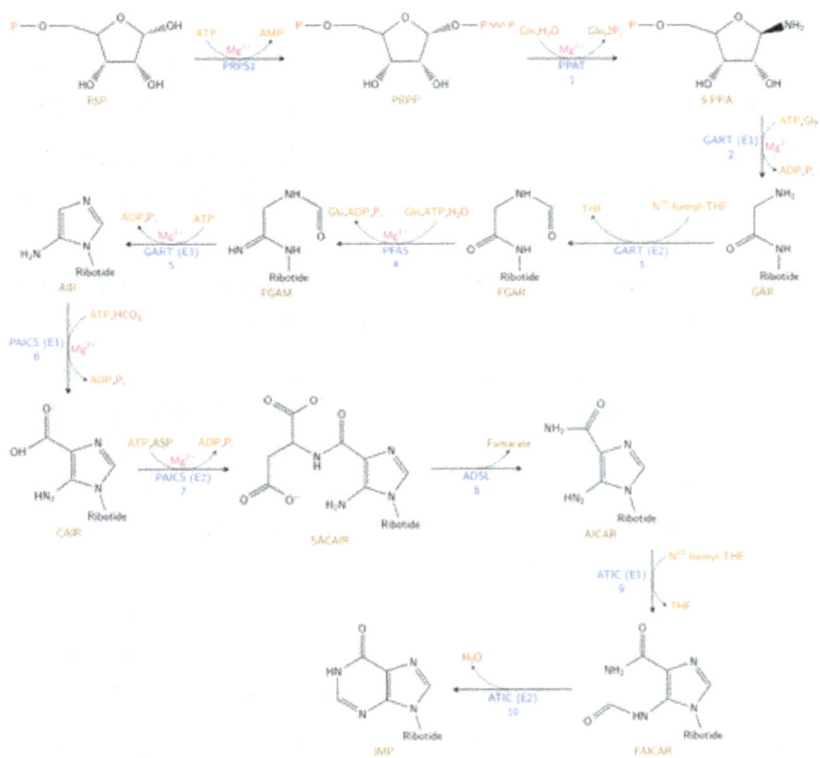

The Synthesis of IMP

The DNA nucleotides adenosine and guanosine consist of a purine base attached to a ribose sugar with a glycosidic bond. In the case of RNA nucleotides deoxyadenosine and deoxyguanosine, the purine bases are attached to a deoxyribose sugar with a glycosidic bond. The purine bases on DNA and RNA nucleotides are synthesized in a twelve-step reaction mechanism present in most single-celled organisms. Higher eukaryotes employ a similar reaction mechanism in ten reaction steps. Purine bases are synthesized by converting phosphoribosyl pyrophosphate (PRPP) to inosine monophosphate (IMP), which is the first key intermediate in purine base biosynthesis. Further enzymatic modification of IMP produces the adenosine and guanosine bases of nucleotides.

- The first step in purine biosynthesis is a condensation reaction, performed by glutamine-PRPP amidotransferase. This enzyme transfers the amino group from glutamine to PRPP, forming 5-phosphoribosylamine. The following step requires the activation of glycine by the addition of a phosphate group from ATP.
- GAR synthetase performs the condensation of activated glycine onto PRPP, forming glycineamide ribonucleotide (GAR).
- GAR transformylase adds a formyl group onto the amino group of GAR, forming formylglycinamide ribonucleotide (FGAR).
- FGAR amidotransferase catalyzes the addition of a nitrogen group to FGAR, forming formylglycinamidine ribonucleotide (FGAM).
- FGAM cyclase catalyzes ring closure, which involves removal of a water molecule, forming the 5-membered imidazole ring 5-aminoimidazole ribonucleotide (AIR).
- N5-CAIR synthetase transfers a carboxyl group, forming the intermediate N5-carboxyaminoimidazole ribonucleotide (N5-CAIR).
- N5-CAIR mutase rearranges the carboxyl functional group and transfers it onto the imidazole ring, forming carboxyamino- imidazole ribonucleotide (CAIR). The two step mechanism of CAIR formation from AIR is mostly found in single celled organisms. Higher eukaryotes contain the enzyme AIR carboxylase, which transfers a carboxyl group directly to AIR imidazole ring, forming CAIR.
- SAICAR synthetase forms a peptide bond between aspartate and the added carboxyl group of the imidazole ring, forming N-succinyl-5-aminoimidazole-4-carboxamide ribonucleotide (SAICAR).
- SAICAR lyase removes the carbon skeleton of the added aspartate, leaving the amino group and forming 5-aminoimidazole-4-carboxamide ribonucleotide (AICAR).
- AICAR transformylase transfers a carbonyl group to AICAR, forming N-formylaminoimidazole- 4-carboxamide ribonucleotide (FAICAR).
- The final step involves the enzyme IMP synthase, which performs the purine ring closure and forms the inosine monophosphate intermediate.

Pyrimidine Nucleotides

Other DNA and RNA nucleotide bases that are linked to the ribose sugar via a glycosidic bond are thymine, cytosine and uracil (which is only found in RNA). Uridine monophosphate biosynthesis involves an enzyme that is located in the mitochondrial inner membrane and multifunctional enzymes that are located in the cytosol.

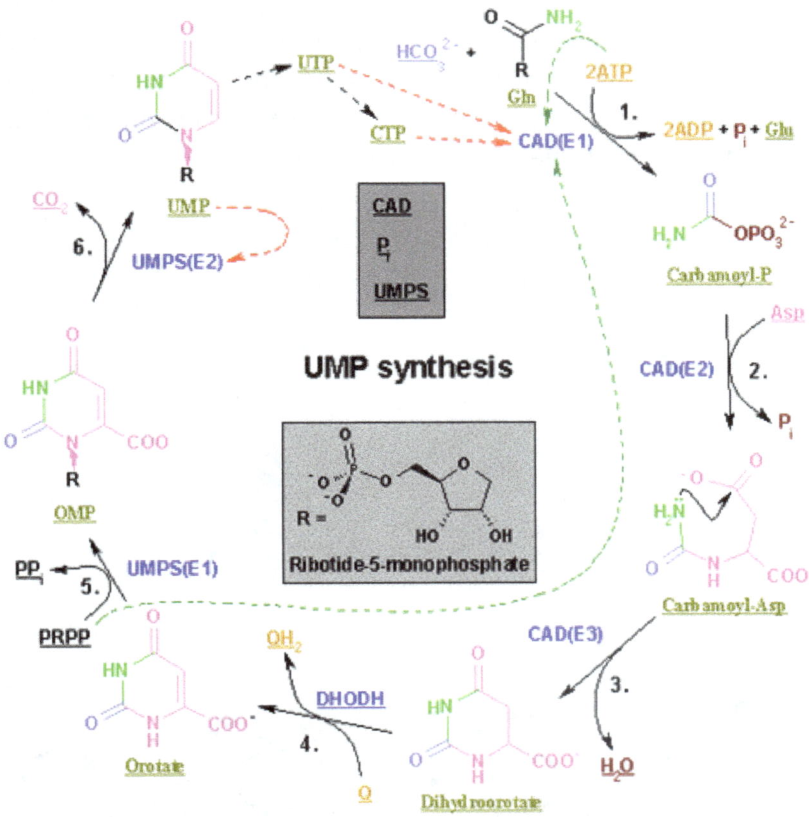

Fig 1: Uridine Monophosphate (UMP) Biosynthesis

- The first step involves the enzyme carbamoyl phosphate synthase combining glutamine with CO_2 in an ATP dependent reaction to form carbamoyl phosphate.
- Aspartate carbamoyltransferase condenses carbamoyl phosphate with aspartate to form uridosuccinate.
- Dihydroorotase performs ring closure, a reaction that loses water, to form dihydroorotate.
- Dihydroorotate dehydrogenase, located within the mitochondrial inner membrane, oxidizes dihydroorotate to orotate.
- Orotate phosphoribosyl hydrolase (OMP pyrophosphorylase) condenses orotate with PRPP to form orotidine-5'-phosphate.
- OMP decarboxylase catalyzes the conversion of orotidine-5'-phosphate to UMP.

After the uridine nucleotide base is synthesized, the other bases, cytosine and thymine are synthesized. Cytosine biosynthesis is a two-step reaction which involves the conversion of UMP to UTP. Phosphate addition to UMP is catalyzed by a kinase enzyme. The enzyme CTP synthase catalyzes the next reaction step: the conversion of UTP to CTP by transferring an amino group from glutamine to uridine; this forms the cytosine base of CTP. The mechanism, which depicts the reaction UTP + ATP + glutamine ⇔ CTP + ADP + glutamate, is below:

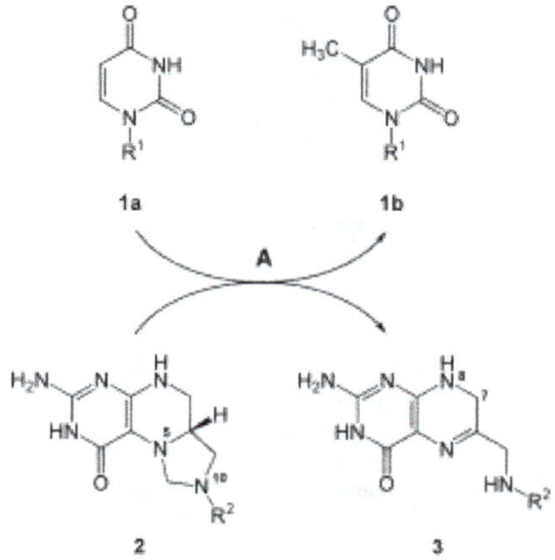

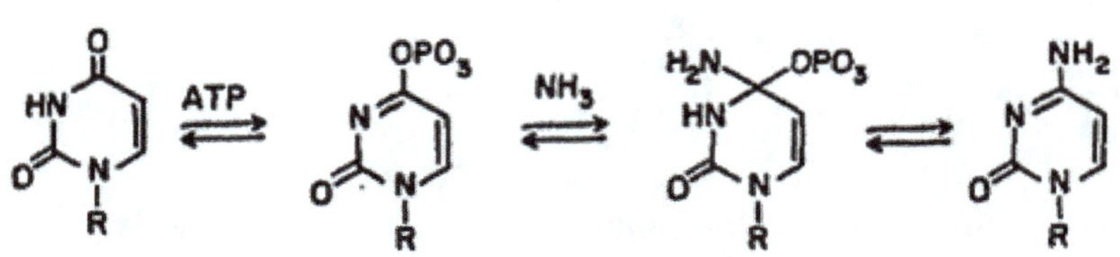

Cytosine is a nucleotide that is present in both DNA and RNA. However, uracil is only found in RNA. Therefore, after UTP is synthesized, it is must be converted into a deoxy form to be incorporated into DNA. This conversion involves the enzyme ribonucleoside triphosphate reductase. This reaction that removes the 2'-OH of the ribose sugar to generate deoxyribose is not affected by the bases attached to the sugar. This non-specificity allows ribonucleoside triphosphate reductase to convert all nucleotide triphosphates to deoxyribonucleotide by a similar mechanism.

In contrast to uracil, thymine bases are found mostly in DNA, not RNA. Cells do not normally contain thymine bases that are linked to ribose sugars in RNA, thus indicating that cells only synthesize deoxyribose-linked thymine. The enzyme thymidylate synthetase is responsible for synthesizing thymine residues from dUMP to dTMP. This reaction transfers a methyl group onto the uracil base of dUMP to generate dTMP. The thymidylate synthase reaction, dUMP + 5,10-methylenetetrahydrofolate ⇔ dTMP + dihydrofolate, is shown to the right.

5.1.5 DNA

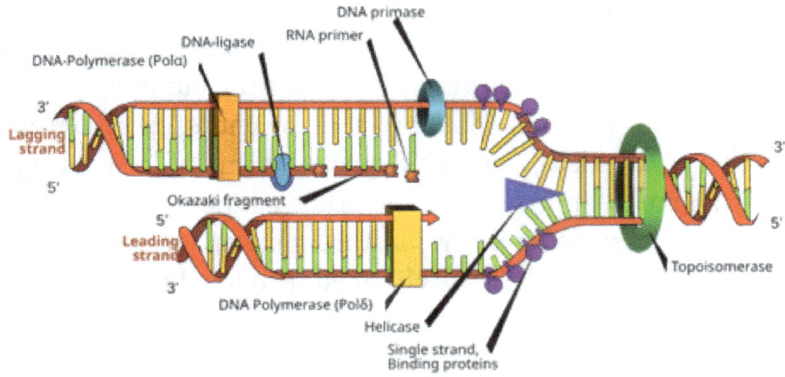

Fig 2: As DNA Polymerase Moves in a 3' to 5' Direction Along the Template Strand, it Synthesizes a New Strand in the 5' to 3' Direction

DNA is composed of nucleotides that are joined by phosphodiester bonds. DNA synthesis, which takes place in the nucleus, is a semiconservative process, which means that the resulting DNA molecule contains an original strand from the parent structure and a new strand. DNA synthesis is catalyzed by a family of DNA polymerases that require four deoxynucleoside triphosphates, a template strand, and a primer with a free 3'OH in which to incorporate nucleotides.

In order for DNA replication to occur, a replication fork is created by enzymes called helicases which unwind the DNA helix. Topoisomerases at the replication fork remove supercoils caused by DNA unwinding, and single-stranded DNA binding proteins maintain the two single-stranded DNA templates stabilized prior to replication. DNA synthesis is initiated by the RNA polymerase primase, which makes an RNA primer with a free 3'OH. This primer is attached to the single-stranded DNA template, and DNA polymerase elongates the chain by incorporating nucleotides; DNA polymerase also proofreads the newly synthesized DNA strand.

During the polymerization reaction catalyzed by DNA polymerase, a nucleophilic attack occurs by the 3'OH of the growing chain on the innermost phosphorus atom of a deoxynucleoside triphosphate; this yields the formation of a phosphodiester bridge that attaches a new nucleotide and releases pyrophosphate.

Two types of strands are created simultaneously during replication: the leading strand, which is synthesized continuously and grows towards the replication fork, and the lagging strand, which is made discontinuously in Okazaki fragments and grows away from the replication fork. Okazaki fragments are covalently joined by DNA ligase to form a continuous strand. Then, to complete DNA replication, RNA primers are removed, and the resulting gaps are replaced with DNA and joined via DNA ligase.

5.1.6 Amino Acids

A protein is a polymer that is composed from amino acids that are linked by peptide bonds. There are more than 300 amino acids found in nature of which only twenty two, known as the proteinogenic amino acids, are the building blocks for protein. Only green plants and most microbes are able to synthesize all of the 20 standard amino acids that are needed by all living species. Mammals can only synthesize ten of the twenty standard amino acids. The other amino acids, valine, methionine, leucine, isoleucine, phenylalanine, lysine, threonine and tryptophan for adults and histidine, and arginine for babies are obtained through diet.

Amino Acid Basic Structure

The general structure of the standard amino acids includes a primary amino group, a carboxyl group and the functional group attached to the α-carbon. The different amino acids are identified by the functional group. As a result of the three different groups attached to the α-carbon, amino acids are asymmetrical molecules. For all standard amino acids, except glycine, the α-carbon is a chiral center. In the case of glycine, the α-carbon has two hydrogen atoms, thus adding symmetry to this molecule. With the exception of proline, all of the amino acids found in life have the L-isoform conformation. Proline has a functional group on the α-carbon that forms a ring with the amino group.

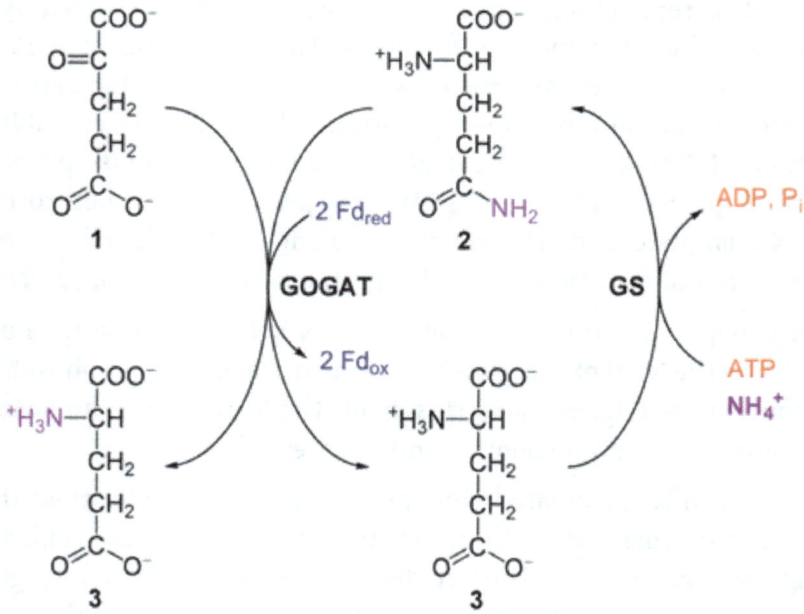

Nitrogen Source

One major step in amino acid biosynthesis involves incorporating a nitrogen group onto the α-carbon. In cells, there are two major pathways of incorporating nitrogen groups. One pathway involves the enzyme glutamine oxoglutarate aminotransferase (GOGAT) which removes the amide amino group of glutamine and transfers it onto 2-oxoglutarate, producing two glutamate molecules. In this catalysis reaction, glutamine serves as the nitrogen source. An image illustrating this reaction is found to the right. The other pathway for incorporating nitrogen onto the α-carbon of amino acids involves the enzyme glutamate dehydrogenase (GDH). GDH is able to transfer ammonia onto 2-oxoglutarate and form glutamate. Furthermore, the enzyme glutamine synthetase (GS) is able to transfer ammonia onto glutamate and synthesize glutamine, replenishing glutamine.

The Glutamate Family of Amino Acids

The glutamate family of amino acids includes the amino acids that derive from the amino acid glutamate. This family includes: glutamate, glutamine, proline, and arginine. This family also includes the amino acid lysine, which is derived from α-ketoglutarate.

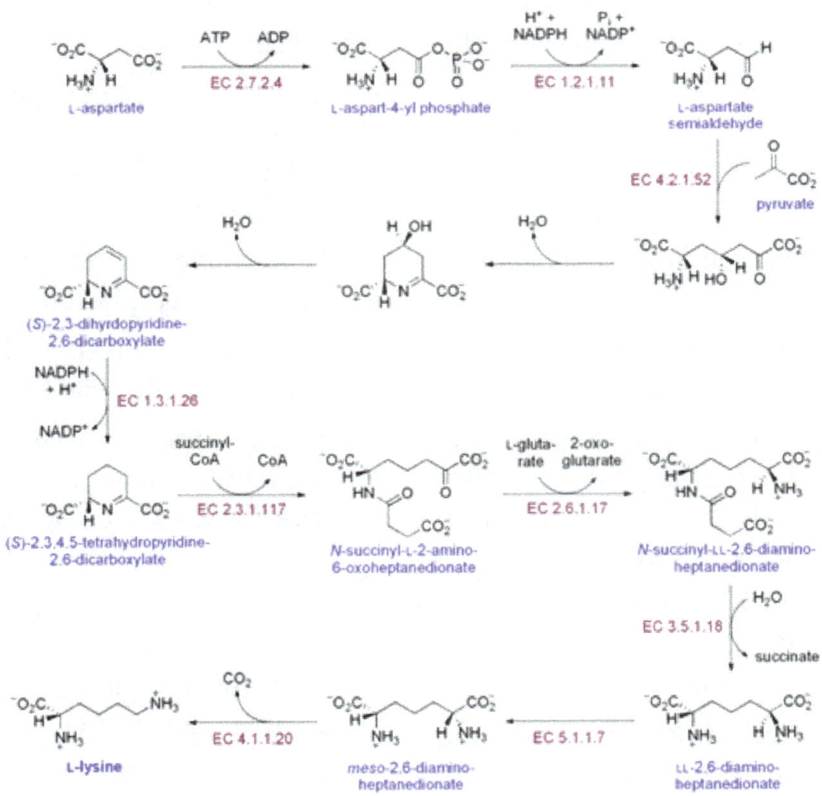

Fig 3: The Diaminopimelic Acid Pathway

The biosynthesis of glutamate and glutamine is a key step in the nitrogen assimilation discussed above. The enzymes GOGAT and GDH catalyze the nitrogen assimilation reactions. In bacteria, the enzyme glutamate 5-kinase initiates the biosynthesis of proline by transferring a phosphate group from ATP onto glutamate. The next reaction is catalyzed by the enzyme pyrroline-5-carboxylate synthase (P5CS), which catalyzes the reduction of the ϒ-carboxyl group of L-glutamate 5-phosphate.

This results in the formation of glutamate semialdehyde, which spontaneously cyclizes to pyrroline-5-carboxylate. Pyrroline-5-carboxylate is further reduced by the enzyme pyrroline-5-carboxylate reductase (P5CR) to yield a proline amino acid.

In the first step of arginine biosynthesis in bacteria, glutamate is acetylated by transferring the acetyl group from acetyl-CoA at the N-α position; this prevents spontaneous cyclization. The enzyme N-acetylglutamate synthase (glutamate N-acetyltransferase) is responsible for catalyzing the acetylation step. Subsequent steps are catalyzed by the enzymes N-acetylglutamate kinase, N-acetyl-gamma-glutamyl-phosphate reductase, and acetylornithine/succinyldiamino pimelate aminotransferase and yield the N-acetyl-L-ornithine. The acetyl group of acetylornithine is removed by the enzyme acetylornithinase (AO) or ornithine acetyltransferase (OAT), and this yields ornithine. Then, the enzymes citrulline and argininosuccinate convert ornithine to arginine.

There are two distinct lysine biosynthetic pathways: the diaminopimelic acid pathway and the α-aminoadipate pathway. The most common of the two synthetic pathways is the diaminopimelic acid pathway; it consists of several enzymatic reactions that add carbon groups to aspartate to yield lysine:

- Aspartate kinase initiates the diaminopimelic acid pathway by phosphorylating aspartate and producing aspartyl phosphate.
- Aspartate semialdehyde dehydrogenase catalyzes the NADPH-dependent reduction of aspartyl phosphate to yield aspartate semialdehyde.
- 4-hydroxy-tetrahydrodipicolinate synthase adds a pyruvate group to the β-aspartyl-4-semialdehyde, and a water molecule is removed. This causes cyclization and gives rise to (2S,4S)-4-hydroxy-2,3,4,5-tetrahydrodipicolinate.
- 4-hydroxy-tetrahydrodipicolinate reductase catalyzes the reduction of (2S,4S)-4-hydroxy-2,3,4,5-tetrahydrodipicolinate by NADPH to yield Δ'-piperideine-2,6-dicarboxylate (2,3,4,5-tetrahydrodipicolinate) and H_2O.
- Tetrahydrodipicolinate acyltransferase catalyzes the acetylation reaction that results in ring opening and yields N-acetyl α-amino-ε-ketopimelate.
- N-succinyl-α-amino-ε-ketopimelate-glutamate aminotransaminase catalyzes the transamination reaction that removes the keto group of N-acetyl α-amino-ε-ketopimelate and replaces it with an amino group to yield N-succinyl-L-diaminopimelate.

- N-acyldiaminopimelate deacylase catalyzes the deacylation of N-succinyl-L-diaminopimelate to yield L,L-diaminopimelate.
- DAP epimerase catalyzes the conversion of L,L-diaminopimelate to the meso form of L,L-diaminopimelate.
- DAP decarboxylase catalyzes the removal of the carboxyl group, yielding L-lysine.

The Serine Family of Amino Acids

The serine family of amino acid includes: serine, cysteine, and glycine. Most microorganisms and plants obtain the sulfur for synthesizing methionine from the amino acid cysteine. Furthermore, the conversion of serine to glycine provides the carbons needed for the biosynthesis of the methionine and histidine.

During serine biosynthesis, the enzyme phosphoglycerate dehydrogenase catalyzes the initial reaction that oxidizes 3-phospho-D-glycerate to yield 3-phosphonooxypyruvate. The following reaction is catalyzed by the enzyme phosphoserine aminotransferase, which transfers an amino group from glutamate onto 3-phosphonooxypyruvate to yield L-phosphoserine. The final step is catalyzed by the enzyme phosphoserine phosphatase, which dephosphorylates L-phosphoserine to yield L-serine.

There are two known pathways for the biosynthesis of glycine. Organisms that use ethanol and acetate as the major carbon source utilize the glyconeogenic pathway to synthesize glycine. The other pathway of glycine biosynthesis is known as the glycolytic pathway. This pathway converts serine synthesized from the intermediates of glycolysis to glycine. In the glycolytic pathway, the enzyme serine hydroxymethyltransferase catalyzes the cleavage of serine to yield glycine and transfers the cleaved carbon group of serine onto tetrahydrofolate, forming 5,10-methylene-tetrahydrofolate.

Cysteine biosynthesis is a two-step reaction that involves the incorporation of inorganic sulfur. In microorganisms and plants, the enzyme serine acetyltransferase catalyzes the transfer of acetyl group from acetyl-CoA onto L-serine to yield O-acetyl-L-serine. The following reaction step, catalyzed by the enzyme O-acetyl serine (thiol) lyase, replaces the acetyl group of O-acetyl-L-serine with sulfide to yield cysteine.

The Aspartate Family of Amino Acids

The aspartate family of amino acids includes: threonine, lysine, methionine, isoleucine, and aspartate. Lysine and isoleucine are considered part of the aspartate family even though part of their carbon skeleton is derived from pyruvate. In the case of methionine, the methyl carbon is derived from serine and the sulfur group, but in most organisms, it is derived from cysteine.

The biosynthesis of aspartate is a one step reaction that is catalyzed by a single enzyme. The enzyme aspartate aminotransferase catalyzes the transfer of an amino group from aspartate onto α-ketoglutarate to yield glutamate and oxaloacetate. Asparagine is synthesized by an ATP-dependent addition of an amino group onto aspartate; asparagine synthetase catalyzes the addition of nitrogen from glutamine or soluble ammonia to aspartate to yield asparagine.

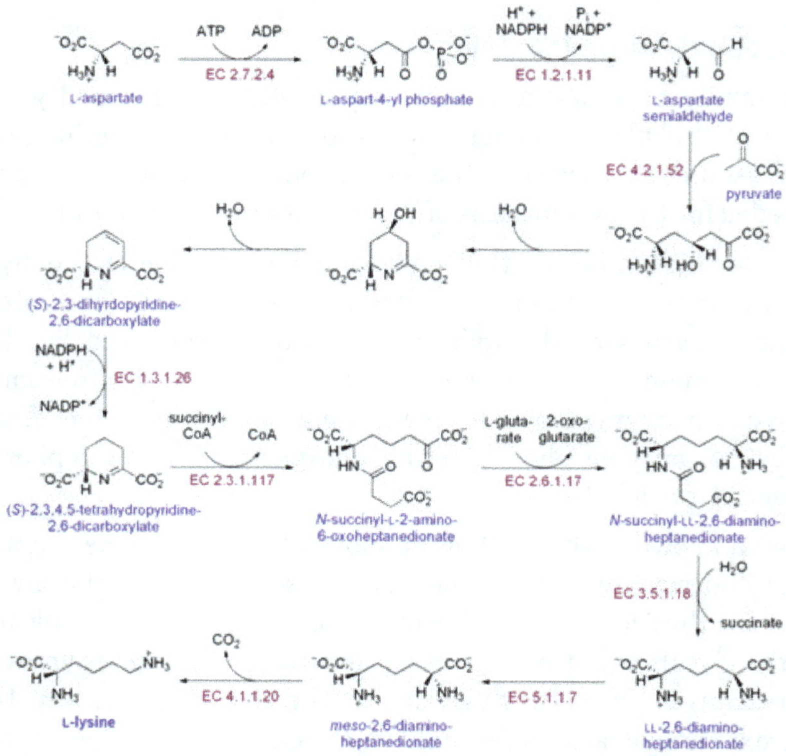

Fig 4: The Diaminopimelic Acid Lysine Biosynthetic Pathway

The diaminopimelic acid biosynthetic pathway of lysine belongs to the aspartate family of amino acids. This pathway involves nine enzyme-catalyzed reactions that convert aspartate to lysine.

- Aspartate kinase catalyzes the initial step in the diaminopimelic acid pathway by transferring a phosphoryl from ATP onto the carboxylate group of aspartate, which yields aspartyl-β-phosphate.
- Aspartate-semialdehyde dehydrogenase catalyzes the reduction reaction by dephosphorylation of aspartyl-β-phosphate to yield aspartate-β-semialdehyde.
- Dihydrodipicolinate synthase catalyzes the condensation reaction of aspartate-β-semialdehyde with pyruvate to yield dihydrodipicolinic acid.

- 4-hydroxy-tetrahydrodipicolinate reductase catalyzes the reduction of dihydrodipicolinic acid to yield tetrahydrodipicolinic acid.
- Tetrahydrodipicolinate N-succinyltransferase catalyzes the transfer of a succinyl group from succinyl-CoA on to tetrahydrodipicolinic acid to yield N-succinyl-L-2,6-diaminoheptanedioate.
- N-succinyldiaminopimelate aminotransferase catalyzes the transfer of an amino group from glutamate onto N-succinyl-L-2,6-diaminoheptanedioate to yield N-succinyl-L,L-diaminopimelic acid.
- Succinyl-diaminopimelate desuccinylase catalyzes the removal of acyl group from N-succinyl-L,L-diaminopimelic acid to yield L,L-diaminopimelic acid.
- Diaminopimelate epimerase catalyzes the inversion of the α-carbon of L,L-diaminopimelic acid to yield meso-diaminopimelic acid.
- Siaminopimelate decarboxylase catalyzes the final step in lysine biosynthesis that removes the carbon dioxide group from meso-diaminopimelic acid to yield L-lysine.

5.1.7 Proteins

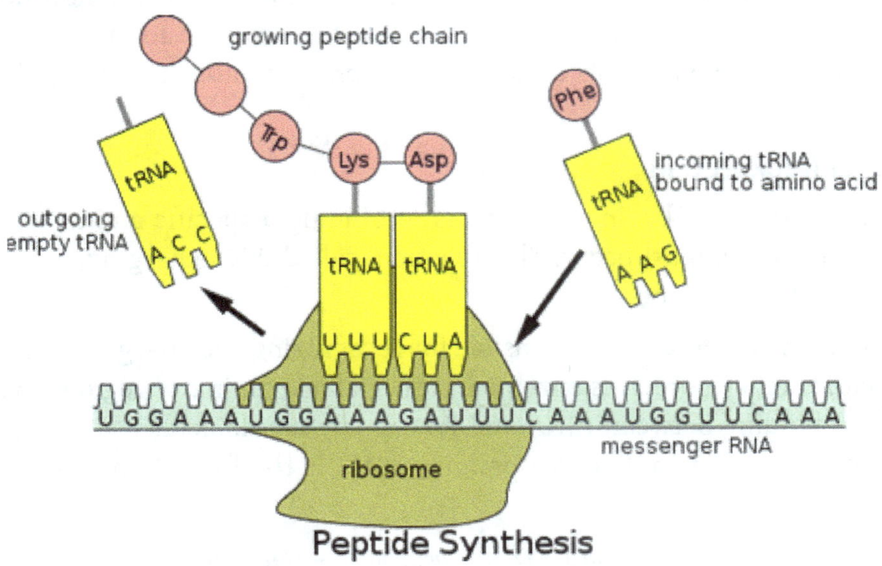

Fig 5: The tRNA Anticodon Interacts with the mRNA Codon in Order to Bind an Amino Acid to Growing Polypeptide Chain

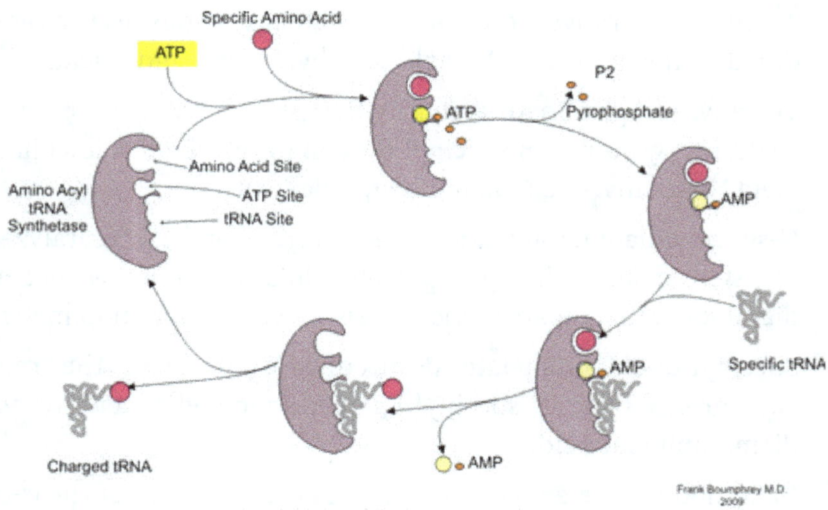

Loading a tRNA molecule

Fig 6: The Process of tRNA Charging

Protein synthesis occurs via a process called translation. During translation, genetic material called mRNA is read by ribosomes to generate a protein polypeptide chain. This process requires transfer RNA (tRNA) which serves as an adaptor by binding amino acids on one end and interacting with mRNA at the other end; the latter pairing between the tRNA and mRNA ensures that the correct amino acid is added to the chain. Protein synthesis occurs in three phases: initiation, elongation, and termination.

Additional Background

Before translation can begin, the process of binding a specific amino acid to its corresponding tRNA must occur. This reaction, called tRNA charging, is catalyzed by aminoacyl tRNA synthetase.

A specific tRNA synthetase is responsible for recognizing and charging a particular amino acid. Furthermore, this enzyme has special discriminator regions to ensure the correct binding between tRNA and its cognate amino acid. The first step for joining an amino acid to its corresponding tRNA is the formation of aminoacyl-AMP:

$$Amino\,acid + ATP <=> aminoacyl - AMP + PP_i$$

This is followed by the transfer of the aminoacyl group from aminoacyl-AMP to a tRNA molecule. The resulting molecule is aminoacyl-tRNA:

$$Aminoacyl - AMP + tRNA <=> aminoacyl - tRNA + AMP$$

The combination of these two steps, both of which are catalyzed by aminoacyl tRNA synthetase, produces a charged tRNA that is ready to add amino acids to the growing polypeptide chain.

In addition to binding an amino acid, tRNA has a three nucleotide unit called an anticodon that base pairs with specific nucleotide triplets on the mRNA called codons; codons encode a specific amino acid. This interaction is possible thanks to the ribosome, which serves as the site for protein synthesis. The ribosome possesses three tRNA binding sites: the aminoacyl site (A site), the peptidyl site (P site), and the exit site (E site).

There are numerous codons within an mRNA transcript, and it is very common for an amino acid to be specified by more than one codon; this phenomenon is called degeneracy. In all, there are 64 codons, 61 of each code for one of the 20 amino acids, while the remaining codons specify chain termination.

Translation in Steps

Translation occurs in three phases: initiation, elongation, and termination.

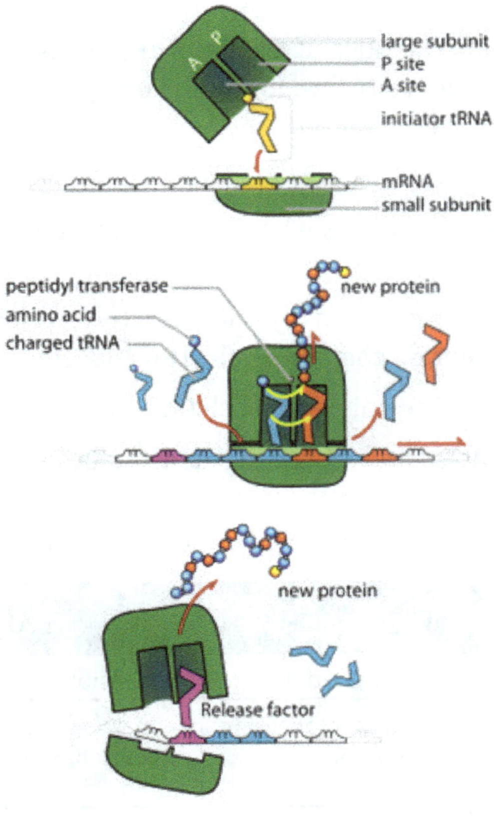

Fig 7: Translation

Step 1: Initiation

The completion of the initiation phase is dependent on the following three events:

- The recruitment of the ribosome to mRNA .

- The binding of a charged initiator tRNA into the P site of the ribosome.
- The proper alignment of the ribosome with mRNA's start codon.

Step 2: Elongation

Following initiation, the polypeptide chain is extended via anticodon:codon interactions, with the ribosome adding amino acids to the polypeptide chain one at a time.

The following steps must occur to ensure the correct addition of amino acids:

- The binding of the correct tRNA into the A site of the ribosome. .
- The formation of a peptide bond between the tRNA in the A site and the polypeptide chain attached to the tRNA in the P site. .
- Translocation or advancement of the tRNA-mRNA complex by three nucleotides.

Translocation "kicks off" the tRNA at the E site and shifts the tRNA from the A site into the P site, leaving the A site free for an incoming tRNA to add another amino acid.

Step 3: Termination

The last stage of translation occurs when a stop codon enters the A site. Then, the following steps occur:

- The recognition of codons by release factors, which causes the hydrolysis of the polypeptide chain from the tRNA located in the P site.
- The release of the polypeptide chain.
- The dissociation and "recycling" of the ribosome for future translation processes.

Table 1: Key Players in Translation

Key Players in Translation	Translation Stage	Purpose
tRNA synthetase	before initiation	Responsible for tRNA charging
mRNA	initiation, elongation, termination	Template for protein synthesis; contains regions named codons which encode amino acids
tRNA	initiation, elongation, termination	Binds ribosomes sites A, P, E; anticodon base pairs with mRNA codon to ensure that the correct amino acid is incorporated into the growing polypeptide chain
ribosome	initiation, elongation, termination	Directs protein synthesis and catalyzes the formation of the peptide bond

5.1.8 Diseases Associated with Macromolecule Deficiency

Errors in biosynthetic pathways can have deleterious consequences including the malformation of macromolecules or the underproduction of functional molecules. Below are examples that illustrate the disruptions that occur due to these inefficiencies.

- **Familial Hypercholesterolemia:** This disorder is characterized by the absence of functional receptors for LDL. Deficiencies in the formation of LDL receptors may cause faulty receptors which disrupt the endocytic pathway, inhibiting the entry of LDL into the liver and other cells. This causes a buildup of LDL in the blood plasma, which results in atherosclerotic plaques that narrow arteries and increase the risk of heart attacks.
- **Lesch–Nyhan Syndrome:** This genetic disease is characterized by self-mutilation, mental deficiency, and gout. It is caused by the absence of hypoxanthine-guanine phosphoribosyltransferase, which is a necessary enzyme for purine nucleotide formation. The lack of enzyme reduces the level of necessary nucleotides and causes the accumulation of biosynthesis intermediates, which results in the aforementioned unusual behavior.
- **Severe Combined Immunodeficiency (SCID):** SCID is characterized by a loss of T cells. Shortage of these immune system components increases the susceptibility to infectious agents because the affected individuals cannot develop immunological memory. This immunological disorder results from a deficiency in adenosine deaminase activity, which causes a buildup of dATP. These dATP molecules then inhibit ribonucleotide reductase, which prevents of DNA synthesis.
- **Huntington's Disease:** This neurological disease is caused from errors that occur during DNA synthesis. These errors or mutations lead to the expression of a mutant huntingtin protein, which contains repetitive glutamine residues that are encoded by expanding CAG trinucleotide repeats in the gene. Huntington's disease is characterized by neuronal loss and gliosis. Symptoms of the disease include: movement disorder, cognitive decline, and behavioral disorder.

5.2 PRIMARY METABOLITES AND THEIR BIOSYNTHETIC PATHWAYS

Primary metabolites are essential compounds that are directly involved in the growth, development, and reproduction of microorganisms. These metabolites are produced during the log phase of microbial growth and are often associated with vital processes such as energy production, cellular respiration, and synthesis of macromolecules.

5.2.1 Amino Acid Biosynthesis

Amino acids are key components of proteins and are synthesized through well-defined biosynthetic pathways. For example, the shikimate pathway is responsible for the biosynthesis of aromatic amino acids like phenylalanine, tyrosine, and tryptophan in microorganisms. This pathway involves the transformation of simple carbon sources into complex amino acids essential for protein synthesis and cellular functions. Microorganisms such as Corynebacterium glutamicum and Escherichia coli are utilized industrially to produce amino acids such as lysine and glutamate, which are used in the food industry as flavor enhancers and nutritional supplements.

5.2.2 Organic Acid Biosynthesis

Organic acids are another important group of primary metabolites. These acids, such as citric acid, lactic acid, and gluconic acid, are produced through pathways like glycolysis and the tricarboxylic acid (TCA) cycle. Citric acid production, primarily carried out by the mold Aspergillus niger, involves the breakdown of sugars via glycolysis into pyruvate, which is then further metabolized in the TCA cycle to generate citric acid. Citric acid is widely used in the food and beverage industry as a flavoring agent and preservative, while lactic acid finds applications in biodegradable plastics and the food industry as a preservative.

5.2.3 Nucleotide Biosynthesis

Nucleotides are the building blocks of nucleic acids (DNA and RNA) and are synthesized through the de novo pathway or salvage pathways. The de novo pathway synthesizes purines and pyrimidines from amino acids and carbon sources. For instance, the biosynthesis of adenine and guanine (purines) involves the formation of inosine monophosphate (IMP), while uridine and cytosine (pyrimidines) are formed through a pathway beginning with aspartate. Nucleotide production is essential for cell replication and repair, and some nucleotide analogs synthesized in microorganisms have found use as therapeutic agents in antiviral and cancer treatments.

5.2.4 Alcohol Fermentation Pathway

Alcohol fermentation is a well-known metabolic process used by yeasts and some bacteria to produce ethanol and other alcohols. The pathway begins with glucose undergoing glycolysis to produce pyruvate, which is then reduced to ethanol and carbon dioxide by the action of enzymes such as alcohol dehydrogenase. This fermentation process is utilized extensively in the production of alcoholic beverages, biofuels, and in various chemical industries. Saccharomyces cerevisiae is the primary organism employed in ethanol production, making it one of the most economically significant metabolic processes.

5.3 SECONDARY METABOLITES AND THEIR BIOSYNTHETIC PATHWAYS

Secondary metabolites are not directly involved in the growth or reproduction of microorganisms but play key roles in their ecological interactions, such as defense mechanisms, signaling, or competition for resources. These compounds, including antibiotics, alkaloids, pigments, and toxins, are produced during the stationary phase of microbial growth.

5.3.1 Antibiotic Biosynthesis

The production of antibiotics is one of the most well-known secondary metabolic processes. Many bacteria and fungi produce antibiotics as a defense mechanism against competing microorganisms. For example, penicillin is produced by the fungus Penicillium chrysogenum through a pathway involving the condensation of amino acid precursors such as cysteine and valine. This process yields a β-lactam structure, which is responsible for the antibiotic's mechanism of action. Industrial fermentation of microorganisms like Streptomyces and Penicillium is widely used to produce antibiotics such as streptomycin, tetracycline, and cephalosporins.

5.3.2 Alkaloid Biosynthesis

Alkaloids are nitrogen-containing secondary metabolites with significant bioactivity. The biosynthesis of alkaloids often involves the modification of amino acids like **tryptophan**.

For example, the production of **ergot alkaloids** by **Claviceps purpurea** is mediated by a complex pathway where tryptophan is transformed into pharmacologically active compounds such as **ergotamine**. Alkaloids are of great interest in pharmaceutical industries for their application as therapeutics in the treatment of various medical conditions, including migraines and hypertension.

5.3.3 Pigment Biosynthesis

Microorganisms produce a wide range of pigments that play roles in protecting cells from environmental stresses such as UV radiation, oxidative damage, or as virulence factors.

The biosynthetic pathway for pigments like **melanin** and **carotenoids** involves complex transformations of fatty acid or isoprenoid precursors. Microbial pigments, including **prodigiosin** from **Serratia marcescens**, have potential applications as natural dyes in food and textile industries and exhibit antimicrobial and anticancer properties.

5.4 REGULATION OF BIOSYNTHETIC PATHWAYS

Biosynthetic pathways in microorganisms are tightly regulated at multiple levels to ensure that the production of metabolites is aligned with the needs of the cell or environmental conditions. Key regulatory mechanisms include:

5.4.1 Feedback Inhibition

In many biosynthetic pathways, the end product of the pathway can inhibit the activity of enzymes earlier in the pathway, a mechanism known as **feedback inhibition**. For instance, in amino acid biosynthesis, the accumulation of an amino acid like **histidine** can inhibit the first enzyme in its biosynthetic pathway, preventing the overproduction of histidine and conserving resources.

5.4.2 Gene Regulation

Microorganisms regulate the expression of genes encoding enzymes involved in biosynthetic pathways through transcriptional control mechanisms. Environmental factors such as nutrient availability, stress, or population density can induce or repress the expression of genes that govern the production of secondary metabolites.

5.4.3 Quorum Sensing

Bacterial populations can regulate the production of secondary metabolites through **quorum sensing**, a cell-density dependent communication system. When a critical population size is reached, signaling molecules accumulate and activate the transcription of genes involved in the production of metabolites like antibiotics, allowing the bacteria to coordinate responses to external challenges.

5.5 INDUSTRIAL APPLICATIONS OF MICROBIAL BIOSYNTHETIC PATHWAYS

The manipulation of microbial biosynthetic pathways has opened up a wide range of industrial applications. Through techniques such as **metabolic engineering** and **synthetic biology**, microorganisms are engineered to produce higher yields of desired compounds, or even novel metabolites that are not naturally occurring.

5.5.1 Antibiotic Production

The production of antibiotics via microbial fermentation is one of the most successful applications of biosynthetic pathway manipulation. By altering regulatory networks or introducing new enzymes into antibiotic-producing strains, it is possible to enhance yields or broaden the spectrum of antibiotics produced.

5.5.2 Biofuel Production

Microbial biosynthetic pathways are increasingly used for the production of renewable energy sources like **bioethanol** and **biodiesel**. The pathways responsible for the conversion of sugars to ethanol or lipid precursors to biodiesel are modified in organisms such as **E. coli** and **yeast** to improve efficiency and scalability for industrial biofuel production.

5.5.3 Industrial Enzyme Production

Microorganisms are also engineered to produce industrial enzymes like **amylases**, **proteases**, and **lipases** that are widely used in food processing, detergent formulation, and pharmaceuticals. The optimization of biosynthetic pathways for these enzymes ensures high-yield production under fermentation conditions.

5.6 CONCLUSION: THE FUTURE OF BIOSYNTHETIC PATHWAYS IN BIOTECHNOLOGY

Biosynthetic pathways for metabolic products of microorganisms represent a key area of interest for industrial microbiology and biotechnology. By understanding and manipulating these pathways, scientists can optimize the production of essential compounds, develop new bio-based materials, and create sustainable solutions for industries. Advances in fields like synthetic biology, genomics, and systems biology will further enhance the capacity to engineer microbial metabolic networks for diverse applications, paving the way for new innovations in medicine, energy, and materials science.

CHAPTER 6

PROCESSES FOR OVERPRODUCTION OF MICROBIAL METABOLITES FOR INDUSTRIAL APPLICATIONS

Microbial metabolites are crucial to many industries, from pharmaceuticals to agriculture, food production, and biofuels. These bioactive compounds, produced by microorganisms like bacteria, yeast, and fungi, include antibiotics, enzymes, organic acids, amino acids, vitamins, and biofuels.

The demand for these products has prompted significant research into optimizing their production through microbial fermentation. However, microorganisms naturally produce metabolites in amounts insufficient for industrial-scale applications. To meet this demand, various processes are employed to enhance or overproduce microbial metabolites.

6.1 GENETIC ENGINEERING FOR OVERPRODUCTION OF METABOLITES

One of the most effective strategies for enhancing microbial metabolite production is genetic engineering. Genetic modifications of microorganisms enable the overexpression of genes involved in key biosynthetic pathways, suppression of competing pathways, and introduction of novel metabolic routes.

6.1.1 Gene Overexpression and Pathway Optimization

Gene overexpression is a critical technique to enhance microbial metabolite production. By increasing the expression of genes that encode enzymes involved in the biosynthesis of the desired metabolite, the metabolic pathway is "pushed" toward greater production levels. For example, overexpression of genes responsible for ethanol production in **Saccharomyces cerevisiae** has been instrumental in increasing bioethanol yield, widely used as a biofuel. Pathway optimization involves not just increasing enzyme levels but also balancing co-factor availability, substrate flow, and intermediate steps within the biosynthetic process. This comprehensive pathway tuning ensures that all steps in the metabolite's production are functioning optimally and in sync.

6.1.2 Knockout of Competing Pathways

In microorganisms, several metabolic pathways may compete for the same precursors or intermediates. To direct more resources to the production of the desired metabolite, gene knockouts are employed to inhibit pathways that compete with the target pathway. For instance, in **Corynebacterium glutamicum**, which is used for lysine production, knocking out the genes involved in byproduct formation ensures that most metabolic flux is directed towards lysine synthesis.

6.1.3 Heterologous Expression of Pathways

Sometimes, microorganisms do not naturally produce the desired metabolite. In such cases, researchers can introduce an entire biosynthetic pathway from another organism into the host microbe, enabling it to produce new compounds. A notable example is the production of the anti-malarial drug **artemisinin** in **yeast** and **E. coli**, where the metabolic pathway of the plant **Artemisia annua** was transferred to these microorganisms, leading to large-scale production of artemisinin.

6.2 METABOLIC ENGINEERING: REDIRECTING RESOURCES

Metabolic engineering is the practice of optimizing genetic and regulatory processes within cells to increase the cell's production of a certain substance. These processes are chemical networks that use a series of biochemical reactions and enzymes that allow cells to convert raw materials into molecules necessary for the cell's survival.

Metabolic engineering specifically seeks to mathematically model these networks, calculate a yield of useful products, and pin point parts of the network that constrain the production of these products. Genetic engineering techniques can then be used to modify the network in order to relieve these constraints. Once again this modified network can be modeled to calculate the new product yield.

The ultimate goal of metabolic engineering is to be able to use these organisms to produce valuable substances on an industrial scale in a cost-effective manner. Current examples include producing beer, wine, cheese, pharmaceuticals, and other biotechnology products. Another possible area of use is the development of oil crops whose composition has been modified to improve their nutritional value. Some of the common strategies used for metabolic engineering are :

- Overexpressing the gene encoding the rate-limiting enzyme of the biosynthetic pathway,
- Blocking the competing metabolic pathways,
- Heterologous gene expression, and
- Enzyme engineering.

Since cells use these metabolic networks for their survival, changes can have drastic effects on the cells' viability. Therefore, trade-offs in metabolic engineering arise between the cells ability to produce the desired substance and its natural survival needs. Therefore, instead of directly deleting and/or overexpressing the genes that encode for metabolic enzymes, the current focus is to target the regulatory networks in a cell to efficiently engineer the metabolism.

6.2.1 History and Applications

In the past, to increase the productivity of a desired metabolite, a microorganism was genetically modified by chemically induced mutation, and the mutant strain that overexpressed the desired metabolite was then chosen. However, one of the main problems with this technique was that the metabolic pathway for the production of that metabolite was not analyzed, and as a result, the constraints to production and relevant pathway enzymes to be modified were unknown.

In 1990s, a new technique called metabolic engineering emerged. This technique analyzes the metabolic pathway of a microorganism, and determines the constraints and their effects on the production of desired compounds. It then uses genetic engineering to relieve these constraints. Some examples of successful metabolic engineering are the following: (i) Identification of constraints to lysine production in *Corynebacterium glutamicum* and insertion of new genes to relieve these constraints to improve production (ii) Engineering of a new fatty acid biosynthesis pathway, called reversed beta oxidation pathway, that is more efficient than the native pathway in producing fatty acids and alcohols which can potentially be catalytically converted to chemicals and fuels (iii) Improved production of DAHP an aromatic metabolite produced by *E. coli* that is an intermediate in the production of aromatic amino acids. It was determined through metabolic flux analysis that the theoretical maximal yield of DAHP per glucose molecule utilized, was 3/7. This is because some of the carbon from glucose is lost as carbon dioxide, instead of being utilized to produce DAHP. Also, one of the metabolites (PEP, or phosphoenolpyruvate) that are used to produce DAHP, was being converted to pyruvate (PYR) to transport glucose into the cell, and therefore, was no longer available to produce DAHP. In order to relieve the shortage of PEP and increase yield, Patnaik et al. used genetic engineering on *E. coli* to introduce a reaction that converts PYR back to PEP. Thus, the PEP used to transport glucose into the cell is regenerated, and can be used to make DAHP. This resulted in a new theoretical maximal yield of 6/7 – double that of the native *E. coli* system.

At the industrial scale, metabolic engineering is becoming more convenient and cost-effective. According to the Biotechnology Industry Organization, "more than 50 biorefinery facilities are being built across North America to apply metabolic engineering to produce biofuels and chemicals from renewable biomass which can help reduce greenhouse gas emissions". Potential biofuels include short-chain alcohols and alkanes (to replace gasoline), fatty acid methyl esters and fatty alcohols (to replace diesel), and fatty acid-and isoprenoid-based biofuels (to replace diesel).

Metabolic engineering continues to evolve in efficiency and processes aided by breakthroughs in the field of synthetic biology and progress in understanding metabolite damage and its repair or preemption. Early metabolic engineering experiments showed that accumulation of reactive intermediates can limit flux in engineered pathways and be deleterious to host cells if matching damage control systems are missing or inadequate. Researchers in synthetic biology optimize genetic pathways, which in turn influence cellular metabolic outputs. Recent decreases in cost of synthesized DNA and developments in genetic circuits help to influence the ability of metabolic engineering to produce desired outputs.

6.2.2 Metabolic Flux Analysis

Setting up a Metabolic Pathway for Analysis

The first step in the process is to identify a desired goal to achieve through the improvement or modification of an organism's metabolism. Reference books and online databases are used to research reactions and metabolic pathways that are able to produce this product or result. These databases contain copious genomic and chemical information including pathways for metabolism and other cellular processes. Using this research, an organism is chosen that will be used to create the desired product or result. Considerations that are taken into account when making this decision are how close the organism's metabolic pathway is to the desired pathway, the maintenance costs associated with the organism, and how easy it is to modify the pathway of the organism. *Escherichia coli* (*E. coli*) is widely used in metabolic engineering to synthesize a wide variety of products such as amino acids because it is relatively easy to maintain and modify. If the organism does not contain the complete pathway for the desired product or result, then genes that produce the missing enzymes must be incorporated into the organism.

Analyzing a Metabolic Pathway

The completed metabolic pathway is modeled mathematically to find the theoretical yield of the product or the reaction fluxes in the cell. A flux is the rate at which a given reaction in the network occurs. Simple metabolic pathway analysis can be done by hand, but most require the use of software to perform the computations. These programs use complex linear algebra algorithms to solve these models. To solve a network using the equation for determined systems shown below, one must input the necessary information about the relevant reactions and their fluxes. Information about the reaction (such as the reactants and stoichiometry) are contained in the matrices G_x and G_m. Matrices V_m and V_x contain the fluxes of the relevant reactions. When solved, the equation yields the values of all the unknown fluxes (contained in V_x).

$$V_x = -(G_x)^{-1} * (G_m * V_m)$$

Determining the Optimal Genetic Manipulations

After solving for the fluxes of reactions in the network, it is necessary to determine which reactions may be altered in order to maximize the yield of the desired product. To determine what specific genetic manipulations to perform, it is necessary to use computational algorithms, such as OptGene or OptFlux. They provide recommendations for which genes should be overexpressed, knocked out, or introduced in a cell to allow increased production of the desired product. For example, if a given reaction has particularly low flux and is limiting the amount of product, the software may recommend that the enzyme catalyzing this reaction should be overexpressed in the cell to increase the reaction flux. The necessary genetic manipulations can be performed using standard molecular biology techniques. Genes may be overexpressed or knocked out from an organism, depending on their effect on the pathway and the ultimate goal.

Experimental Measurements

In order to create a solvable model, it is often necessary to have certain fluxes already known or experimentally measured. In addition, in order to verify the effect of genetic manipulations on the metabolic network (to ensure they align with the model), it is necessary to experimentally measure the fluxes in the network. To measure reaction fluxes, carbon flux measurements are made using carbon-13 isotopic labeling. The organism is fed a mixture that contains molecules where specific carbons are engineered to be carbon-13 atoms, instead of carbon-12. After these molecules are used in the network, downstream metabolites also become labeled with carbon-13, as they incorporate those atoms in their structures. The specific labeling pattern of the various metabolites is determined by the reaction fluxes in the network. Labeling patterns may be measured using techniques such as gas chromatography-mass spectrometry (GC-MS) along with computational algorithms to determine reaction fluxes.

6.3 FERMENTATION STRATEGIES FOR ENHANCED PRODUCTION

Fermentation is the most common method for large-scale microbial metabolite production. The efficiency of this process can be enhanced through optimization of growth conditions, choice of fermentation mode, and substrate utilization.

6.3.1 Batch vs. Continuous Fermentation

In industrial microbiology, **batch fermentation** is a closed system where the nutrients are provided at the beginning, and the microbial culture is allowed to grow until nutrients are depleted. While batch fermentation is suitable for short-term production, it may not be as efficient for metabolites that require long cultivation times. **Continuous fermentation**, on the other hand, involves constantly feeding nutrients into the culture and removing waste products and metabolites. This method is especially useful for products like biofuels, where sustained production is necessary. Continuous fermentation reduces downtime and increases overall yield by maintaining the microorganisms in their most productive phase for longer periods.

6.3.2 Optimization of Environmental Conditions

Environmental factors such as temperature, pH, and oxygen concentration greatly affect microbial metabolism. In large-scale fermentation, careful control of these factors can optimize production. For example, controlling the pH of a fermentation medium can significantly affect the production of organic acids. In **Aspergillus niger**, the production of citric acid is highly sensitive to pH, with optimal production occurring under slightly acidic conditions. Temperature control is also crucial. For example, **Thermophiles**, which thrive at high temperatures, are ideal for industrial fermentation processes that require high heat, such as biofuel production.

6.3.3 Substrate Selection and Fed-Batch Fermentation

The choice of substrates can impact the efficiency and cost of fermentation. Many industrial processes now use renewable or waste products, such as lignocellulosic biomass, as substrates for fermentation. This not only reduces production costs but also makes the process more sustainable. In **fed-batch fermentation**, nutrients are added gradually throughout the fermentation process. This strategy prevents the depletion of critical nutrients and avoids the accumulation of inhibitory byproducts. Fed-batch fermentation is often used for secondary metabolites like antibiotics, where the production phase occurs after the microbial growth phase.

6.4 BIOPROCESS SCALE-UP: FROM LAB TO INDUSTRIAL PRODUCTION

Scaling up from laboratory-level production to industrial-scale manufacturing presents numerous challenges, including maintaining consistency, optimizing yield, and controlling production costs.

6.4.1 Bioreactor Design

The design of the bioreactor used for fermentation is critical for scaling up metabolite production. Bioreactors must provide optimal conditions for microbial growth and metabolite production, including proper mixing, temperature regulation, and oxygen supply. For industrial-scale fermentation, **stirred-tank bioreactors** are commonly used due to their flexibility and ability to maintain homogenous conditions. However, for certain applications like biofuel production, alternative bioreactor designs such as **airlift** or **packed-bed bioreactors** may be more efficient.

6.4.2 Downstream Processing

After fermentation, the desired metabolites need to be separated and purified from the microbial culture. This step, known as **downstream processing**, often involves several steps such as filtration, centrifugation, and chromatography. Efficient downstream processing is crucial to maintaining the quality of the final product and minimizing costs. For instance, the production of antibiotics such as penicillin involves extracting the compound from the fermentation broth and purifying it to pharmaceutical standards. Innovations in downstream processing, such as **membrane filtration** and **affinity chromatography**, have improved the efficiency of metabolite recovery, leading to lower production costs.

6.5 SYNTHETIC BIOLOGY AND FUTURE PROSPECTS

The future of microbial metabolite overproduction lies in the integration of synthetic biology techniques, allowing scientists to design and construct entirely new metabolic pathways or improve existing ones.

6.5.1 Synthetic Gene Circuits

Synthetic biology enables the construction of gene circuits that can precisely control gene expression. These circuits allow for the timed or conditional expression of genes involved in metabolite production. For example, gene circuits can activate metabolite production only when the microbial population reaches a certain density or in response to specific environmental conditions. This ensures that microbial resources are optimally utilized for maximum production.

6.5.2 CRISPR and Genome Editing

The **CRISPR-Cas9** technology has revolutionized genetic engineering by providing a precise tool for editing microbial genomes. This technique allows for the targeted modification of genes, enabling researchers to fine-tune metabolic pathways with unprecedented precision. CRISPR has been used to improve the production of biofuels, antibiotics, and enzymes, making it a valuable tool for industrial microbiology.

6.6 CONCLUSION: OVERPRODUCTION OF MICROBIAL METABOLITES AND INDUSTRIAL APPLICATIONS

The overproduction of microbial metabolites plays an essential role in industries such as pharmaceuticals, biofuels, and agriculture. Advances in genetic engineering, metabolic pathway optimization, fermentation technologies, and synthetic biology have made it possible to enhance the production of a wide range of metabolites.

As biotechnology continues to evolve, the development of more efficient and sustainable processes for microbial metabolite production will be key to meeting the increasing global demand for bioactive compounds. By combining these strategies, industries can reduce production costs, improve yields, and create sustainable solutions for the future.

CHAPTER 7

SELECTION AND IMPROVEMENT OF INDUSTRIAL ORGANISMS FOR BIOTECHNOLOGICAL APPLICATIONS

Industrial organisms play a crucial role in the biotechnology industry, serving as biological factories for the production of a vast array of products, from pharmaceuticals and biofuels to enzymes and bioplastics. The selection and improvement of these organisms are fundamental to enhancing their performance in biotechnological processes, ensuring that they can operate under industrial conditions, produce high yields, and meet economic viability. Industrial microorganisms such as bacteria, yeast, and fungi are most commonly used due to their ability to grow rapidly and produce a wide range of biotechnologically important products. The process of selecting and improving these organisms involves a combination of traditional techniques and modern biotechnological approaches, such as genetic engineering, metabolic pathway optimization, and adaptive evolution.

7.1 CRITERIA FOR SELECTING INDUSTRIAL ORGANISMS

Before an organism can be used for industrial purposes, it must meet several essential criteria to ensure it is suitable for large-scale production. These criteria focus on the organism's ability to produce the desired product efficiently and cost-effectively, as well as its capacity to thrive in the industrial environment.

7.1.1 High Yield of Desired Products

One of the most critical criteria for selecting an industrial organism is its ability to produce high yields of the target product. Whether the goal is to produce an enzyme, a metabolite, or a biofuel, the organism must be capable of synthesizing the product in large quantities to make the process economically viable. For example, the yeast **Saccharomyces cerevisiae** is selected for bioethanol production due to its ability to convert sugars into ethanol at high yields.

7.1.2 Robustness and Tolerance to Industrial Conditions

Industrial processes often subject organisms to harsh conditions, such as extreme temperatures, high or low pH, and high concentrations of substrates and byproducts. The selected organism must be robust and able to tolerate these conditions while maintaining productivity. **Escherichia coli** and **Bacillus subtilis** are popular choices for industrial applications because they can thrive in diverse environments and withstand fluctuations in temperature and pH.

7.1.3 Fast Growth Rate

A fast growth rate is another important criterion, as it directly impacts the efficiency of the industrial process. Faster-growing organisms can quickly reach the desired cell density, reducing production time and costs. For instance, **Pichia pastoris**, a species of yeast, is commonly used in the production of recombinant proteins because of its rapid growth rate and ease of genetic manipulation.

7.1.4 Genetic Stability

Genetic stability is essential to ensure that the organism maintains its ability to produce the desired product over many generations. Instability in genetic traits can lead to a decrease in product yield and quality, undermining the efficiency of the industrial process. Organisms with stable genomes are preferred for industrial use, as they ensure consistent production across multiple batches.

7.2 TRADITIONAL METHODS OF IMPROVING INDUSTRIAL ORGANISMS

The improvement of industrial organisms has been a goal for many decades, and traditional methods such as strain selection, mutagenesis, and adaptive evolution have played a significant role in enhancing their performance.

7.2.1 Strain Selection

Strain selection involves screening naturally occurring strains of microorganisms to identify those with desirable traits, such as higher yields of the desired product or improved tolerance to industrial conditions. Strain selection has been used successfully in industries like brewing, where yeast strains have been selected for optimal fermentation and flavor production.

7.2.2 Mutagenesis

Mutagenesis is a traditional method used to induce random genetic mutations in microorganisms in the hope of discovering strains with improved characteristics. This can be achieved using chemical mutagens, such as ethyl methanesulfonate (EMS), or physical agents, such as ultraviolet (UV) radiation. Mutagenesis is followed by screening to identify mutant strains that exhibit enhanced performance. This method was used historically to increase the production of antibiotics like penicillin by **Penicillium** species.

7.2.3 Adaptive Evolution

Adaptive evolution, also known as directed evolution, is a method that involves growing microorganisms under selective pressure over extended periods. The goal is to naturally select for variants that are better adapted to the industrial environment. For instance, microorganisms can be evolved to tolerate high concentrations of toxic substrates or to improve their utilization of a particular nutrient. This method has been used to develop yeast strains that are more resistant to the toxic byproducts of bioethanol production.

7.3 GENETIC ENGINEERING FOR ORGANISM IMPROVEMENT

With advances in biotechnology, genetic engineering has revolutionized the improvement of industrial organisms. This approach involves the direct manipulation of an organism's genome to enhance desirable traits, such as increased productivity, stress tolerance, or the ability to produce novel compounds.

7.3.1 Recombinant DNA Technology

Recombinant DNA technology allows scientists to introduce foreign genes into industrial organisms, enabling them to produce new products or improve existing metabolic pathways.

For example, the introduction of genes from other organisms into **E. coli** has been used to produce human insulin on an industrial scale. Recombinant DNA technology has also been applied to the production of biofuels, where genes responsible for the production of ethanol, butanol, or biodiesel are introduced into microorganisms.

7.3.2 CRISPR-Cas9 and Genome Editing

CRISPR-Cas9 has become a game-changing tool in the field of genetic engineering. It allows for precise editing of an organism's genome, enabling targeted modifications of genes involved in metabolic pathways or stress responses. For instance, CRISPR-Cas9 has been used to modify yeast and bacterial strains to improve their performance in producing biofuels and pharmaceuticals. The precision and efficiency of CRISPR-Cas9 make it an invaluable tool for enhancing the capabilities of industrial organisms.

7.3.3 Metabolic Engineering

Metabolic engineering involves the redesign of an organism's metabolic pathways to optimize the production of a desired product. This can be done by overexpressing key enzymes, knocking out genes that lead to byproduct formation, or redirecting metabolic fluxes to the target pathway. For example, **Corynebacterium glutamicum** has been engineered to overproduce amino acids such as lysine, which is used as a feed additive in the agricultural industry.

7.4 SYNTHETIC BIOLOGY: ENGINEERING NOVEL ORGANISMS

Synthetic biology (SynBio) is a multidisciplinary field of science that focuses on living systems and organisms, and it applies engineering principles to develop new biological parts, devices, and systems or to redesign existing systems found in nature.

It is a branch of science that encompasses a broad range of methodologies from various disciplines, such as biochemistry, biotechnology, biomaterials, material science/engineering, genetic engineering, molecular biology, molecular engineering, systems biology, membrane science, biophysics, chemical and biological engineering, electrical and computer engineering, control engineering and evolutionary biology.

It includes designing and constructing biological modules, biological systems, and biological machines, or re-designing existing biological systems for useful purposes.

Additionally, it is the branch of science that focuses on the new abilities of engineering into existing organisms to redesign them for useful purposes.

In order to produce predictable and robust systems with novel functionalities that do not already exist in nature, it is also necessary to apply the engineering paradigm of systems design to biological systems. According to the European Commission, this possibly involves a molecular assembler based on biomolecular systems such as the ribosome.

7.4.1 History

- **1910:** First identifiable use of the term *synthetic biology* in Stéphane Leduc's publication *Théorie physico-chimique de la vie et générations spontanées*. He also noted this term in another publication, *La Biologie Synthétique* in 1912.
- **1944:** Canadian-American scientist Oswald Avery shows that DNA is the material of which genes and chromosomes are made. This becomes the bedrock on which all subsequent genetic research is built.
- **1953:** Francis Crick and James Watson publish the structure of the DNA in *Nature*.
- **1961:** Jacob and Monod postulate cellular regulation by molecular networks from their study of the *lac* operon in *E. coli* and envisioned the ability to assemble new systems from molecular components.
- **1973:** First molecular cloning and amplification of DNA in a plasmid is published in *P.N.A.S.* by Cohen, Boyer et al. constituting the dawn of synthetic biology.

- **1978:** Arber, Nathans and Smith win the Nobel Prize in Physiology or Medicine for the discovery of restriction enzymes, leading Szybalski to offer an editorial comment in the journal *Gene*:
  - The work on restriction nucleases not only permits us easily to construct recombinant DNA molecules and to analyze individual genes, but also has led us into the new era of synthetic biology where not only existing genes are described and analyzed but also new gene arrangements can be constructed and evaluated.
- **1988:** First DNA amplification by the polymerase chain reaction (PCR) using a thermostable DNA polymerase is published in *Science* by Mullis et al. This obviated adding new DNA polymerase after each PCR cycle, thus greatly simplifying DNA mutagenesis and assembly.
- **2000:** Two papers in *Nature* report synthetic biological circuits, a genetic toggle switch and a biological clock, by combining genes within *E. coli* cells.
- **2003:** The most widely used standardized DNA parts, BioBrick plasmids, are invented by Tom Knight. These parts will become central to the International Genetically Engineered Machine (iGEM) competition founded at MIT in the following year.
- **2003:** Researchers engineer an artemisinin precursor pathway in *E. coli.*
- **2004:** First international conference for synthetic biology, Synthetic Biology 1.0 (SB1.0) is held at MIT.
- **2005:** Researchers develop a light-sensing circuit in *E. coli.* Another group designs circuits capable of multicellular pattern formation.
- **2006:** Researchers engineer a synthetic circuit that promotes bacterial invasion of tumour cells.
- **2010:** Researchers publish in *Science* the first synthetic bacterial genome, called *M. mycoides* JCVI-syn1.0. The genome is made from chemically-synthesized DNA using yeast recombination.
- **2011:** Functional synthetic chromosome arms are engineered in yeast.
- **2012:** Charpentier and Doudna labs publish in *Science* the programming of CRISPR-Cas9 bacterial immunity for targeting DNA cleavage. This technology greatly simplified and expanded eukaryotic gene editing.
- **2019:** Scientists at ETH Zurich report the creation of the first bacterial genome, named *Caulobacter ethensis-2.0*, made entirely by a computer, although a related viable form of *C. ethensis-2.0* does not yet exist.

- **2019:** Researchers report the production of a new synthetic (possibly artificial) form of viable life, a variant of the bacteria *Escherichia coli*, by reducing the natural number of 64 codons in the bacterial genome to 59 codons instead, in order to encode 20 amino acids.
- **2020:** Scientists created the first xenobot, a programmable synthetic organism derived from frog cells and designed by AI.
- **2021:** Scientists reported that xenobots are able to self-replicate by gathering loose cells in the environment and then forming new xenobots.

7.4.2 Perspectives

It is a field whose scope is expanding in terms of systems integration, engineered organisms, and practical findings. Engineers view biology as *technology* (in other words, a given system includes *biotechnology* or its *biological engineering*). Synthetic biology includes the broad redefinition and expansion of biotechnology, with the ultimate goal of being able to design and build engineered live biological systems that process information, manipulate chemicals, fabricate materials and structures, produce energy, provide food, and maintain and enhance human health, as well as advance fundamental knowledge of biological systems and our environment.

Researchers and companies working in synthetic biology are using nature's power to solve issues in agriculture, manufacturing, and medicine. Due to more powerful genetic engineering capabilities and decreased DNA synthesis and sequencing costs, the field of synthetic biology is rapidly growing. In 2016, more than 350 companies across 40 countries were actively engaged in synthetic biology applications; all these companies had an estimated net worth of $3.9 billion in the global market. Synthetic biology currently has no generally accepted definition. Here are a few examples:

It is the science of emerging genetic and physical engineering to produce new (and, therefore, synthetic) life forms. To develop organisms with novel or enhanced characteristics, this emerging field of study combines biology, engineering, and related disciplines' knowledge and techniques to design chemically synthesised DNA.

Biomolecular engineering includes approaches that aim to create a toolkit of functional units that can be introduced to present new technological functions in living cells. Genetic engineering includes approaches to construct synthetic chromosomes or minimal organisms like *Mycoplasma laboratorium*.

Biomolecular design refers to the general idea of de novo design and additive combination of biomolecular components. Each of these approaches shares a similar task: to develop a more synthetic entity at a higher level of complexity by inventively manipulating a simpler part at the preceding level. Optimizing these exogenous pathways in unnatural systems takes iterative fine-tuning of the individual biomolecular components to select the highest concentrations of the desired product.

On the other hand, "re-writers" are synthetic biologists interested in testing the irreducibility of biological systems. Due to the complexity of natural biological systems, it would be simpler to rebuild the natural systems of interest from the ground up; to provide engineered surrogates that are easier to comprehend, control and manipulate. Re-writers draw inspiration from refactoring, a process sometimes used to improve computer software.

7.4.3 Categories

Bioengineering, synthetic genomics, protocell synthetic biology, unconventional molecular biology, and in silico techniques are the five categories of synthetic biology. It is necessary to review the distinctions and analogies between the categories of synthetic biology for its social and ethical assessment, to distinguish between issues affecting the whole field and particular to a specific one.

Bioengineering

The subfield of bioengineering concentrates on creating novel metabolic and regulatory pathways, and is currently the one that likely draws the attention of most researchers and funding. It is primarily motivated by the desire to establish biotechnology as a legitimate engineering discipline.

When referring to this area of synthetic biology, the word "bioengineering" should not be confused with "traditional genetic engineering", which involves introducing a single transgene into the intended organism. Bioengineers adapted synthetic biology to provide a substantially more integrated perspective on how to alter organisms or metabolic systems.

A typical example of single-gene genetic engineering is the insertion of the human insulin gene into bacteria to create transgenic proteins. The creation of whole new signalling pathways, containing numerous genes and regulatory components (such as an oscillator circuit to initiate the periodic production of green fluorescent protein (GFP) in mammalian cells), is known as bioengineering as part of synthetic biology.

By utilising simplified and abstracted metabolic and regulatory modules as well as other standardized parts that may be freely combined to create new pathways or creatures, bioengineering aims to create innovative biological systems. In addition to creating infinite opportunities for novel applications, this strategy is anticipated to make bioengineering more predictable and controllable than traditional biotechnology.

Synthetic Genomics

The formation of animals with a chemically manufactured (minimal) genome is another facet of synthetic biology that is highlighted by synthetic genomics. This area of synthetic biology has been made possible by ongoing advancements in DNA synthesis technology, which now makes it feasible to produce DNA molecules with thousands of base pairs at a reasonable cost. The goal is to combine these molecules into complete genomes and transplant them into living cells, replacing the host cell's genome and reprogramming its metabolism to perform different functions.

Scientists have previously demonstrated the potential of this approach by creating infectious viruses by synthesising the genomes of multiple viruses. These significant advances in science and technology triggered the initial public concerns concerning the risks associated with this technology.

A simple genome might also work as a "chassis genome" that could be enlarged quickly by gene inclusion created for particular tasks. Such "chassis creatures" would be more suited for the insertion of new functions than wild organisms since they would have fewer biological pathways that could potentially conflict with the new functionalities in addition to having specific insertion sites. Synthetic genomics strives to create creatures with novel "architectures," much like the bioengineering method. It adopts an integrative or holistic perspective of the organism. In this case, the objective is the creation of chassis genomes based on necessary genes and other required DNA sequences rather than the design of metabolic or regulatory pathways based on abstract criteria.

Protocell Synthetic Biology

The in vitro generation of synthetic cells is the protocell branch of synthetic biology. Lipid vesicles, which have all the necessary components to function as a complete system, can be used to create these artificial cells. In the end, these synthetic cells should meet the requirements for being deemed alive, namely the capacity for self-replication, self-maintenance, and evolution. The protocell technique has this as its end aim, however there are other intermediary steps that fall short of meeting all the criteria for a living cell. In order to carry out a specific function, these lipid vesicles contain cell extracts or more specific sets of biological macromolecules and complex structures, such as enzymes, nucleic acids, or ribosomes. For instance, liposomes may carry out particular polymerase chain reactions or synthesise a particular protein.

Protocell synthetic biology takes artificial life one step closer to reality by eventually synthesizing not only the genome but also every component of the cell in vitro, as opposed to the synthetic genomics approach, which relies on coercing a natural cell to carry out the instructions encoded by the introduced synthetic genome. Synthetic biologists in this field view their work as basic study into the conditions necessary for life to exist and its origin more than in any of the other techniques. The protocell technique, however, also lends itself well to applications; similar to other synthetic biology byproducts, protocells could be employed for the manufacture of biopolymers and medicines.

Unconventional Molecular Biology

The objective of the "unnatural molecular biology" strategy is to create new varieties of life that are based on a different kind of molecular biology, such as new types of nucleic acids or a new genetic code. The creation of new types of nucleotides that can be built into unique nucleic acids could be accomplished by changing certain DNA or RNA constituents, such as the bases or the backbone sugars.

The normal genetic code is being altered by inserting quadruplet codons or changing some codons to encode new amino acids, which would subsequently permit the use of non-natural amino acids with unique features in protein production. It is a scientific and technological problem to adjust the enzymatic machinery of the cell for both approaches.

A new sort of life would be formed by organisms with a genome built on synthetic nucleic acids or on a totally new coding system for synthetic amino acids. This new style of life would have some benefits but also some new dangers. On release into the environment, there would be no horizontal gene transfer or outcrossing of genes with natural species. Furthermore, these kinds of synthetic organisms might be created to require non-natural materials for protein or nucleic acid synthesis, rendering them unable to thrive in the wild if they accidentally escaped.

On the other hand, if these organisms ultimately were able to survive outside of controlled space, they might have a particular benefit over natural organisms because they would be resistant to predatory living organisms or natural viruses, that could lead to an unmanaged spread of the synthetic organisms.

In Silico Technique

Synthetic biology in silico and the various strategies are interconnected. The development of complex designs, whether they are metabolic pathways, fundamental cellular processes, or chassis genomes, is one of the major difficulties faced by the four synthetic-biology methods outlined above. Because of this, synthetic biology has a robust in silico branch, similar to systems biology, that aims to create computational models for the design of common biological components or synthetic circuits, which are essentially simulations of synthetic organisms.

The practical application of simulations and models through bioengineering or other fields of synthetic biology is the long-term goal of in silico synthetic biology.

Many of the computational simulations of synthetic organisms up to this point possess little to no direct analogy to living things. Due to this, in silico synthetic biology is regarded as a separate group in this article.

It is sensible to integrate the five areas under the umbrella of synthetic biology as an unified area of study. Even though they focus on various facets of life, such as metabolic regulation, essential elements, or biochemical makeup, these five strategies all work toward the same end: creating new types of living organisms. Additionally, the varied methodologies begin with numerous methodological approaches, which leads to the diversity of synthetic biology approaches.

Synthetic biology is an interdisciplinary field that draws from and is inspired by many different scientific disciplines, not one single field or technique. Synthetic biologists all have the same underlying objective of designing and producing new forms of life, despite the fact that they may employ various methodologies, techniques, and research instruments. Any evaluation of synthetic biology, whether it examines ethical, legal, or safety considerations, must take into account the fact that while some questions, risks, and issues are unique to each technique, in other circumstances, synthetic biology as a whole must be taken into consideration.

7.4.4 Four Engineering Approaches

Synthetic biology has traditionally been divided into four different engineering approaches: top down, parallel, orthogonal and bottom up. To replicate emergent behaviours from natural biology and build artificial life, unnatural chemicals are used. The other looks for interchangeable components from biological systems to put together and create systems that do not work naturally. In either case, a synthetic objective compels researchers to venture into new area in order to engage and resolve issues that cannot be readily resolved by analysis. Due to this, new paradigms are driven to arise in ways that analysis cannot easily do. In addition to equipments that oscillate, creep, and play tic-tac-toe, synthetic biology has produced diagnostic instruments that enhance the treatment of patients with infectious diseases.

Top-Down Approach

It involves using metabolic and genetic engineering techniques to impart new functions to living cells. By comparing universal genes and eliminating non-essential ones to create a basic genome, this method seeks to lessen the complexity of existing cells. These initiatives are founded on the hypothesis of a single genesis for cellular life, the so-called Last Universal Common Ancestor, which supports the presence of a universal minimal genome that gave rise to all living things. Recent studies, however, raise the possibility that the eukaryotic and prokaryotic cells that make up the tree of life may have evolved from a group of primordial cells rather than from a single cell. As a result, even while the Holy Grail-like pursuit of the "minimum genome" has grown elusive, cutting out a number of non-essential functions impairs an organism's fitness and leads to "fragile" genomes.

Bottom-Up Approach

This approach involves creating new biological systems in vitro by bringing together 'non-living' biomolecular components, often with the aim of constructing an artificial cell. Reproduction, replication, and assembly are three crucial self-organizational principles that are taken into account in order to accomplish this. Cells, which are made up of a container and a metabolism, are considered "hardware" in the definition of reproduction, whereas replication occurs when a system duplicates a perfect copy of itself, as in the case of DNA, which is considered "software." When vesicles or containers (such as Oparin's coacervates) formed of tiny droplets of molecules that are organic like lipids or liposomes, membrane-like structures comprising phospholipids, aggregate, assembly occur.

The study of protocells exists along with other in vitro synthetic biology initiatives that seek to produce minimum cells, metabolic pathways, or "never-born proteins" as well as to mimic physiological functions including cell division and growth. The in vitro enhancement of synthetic pathways does have the potential to have an effect on some other synthetic biology sectors, including metabolic engineering, despite the fact that it no longer classified as synthetic biology research. This research, which is primarily essential, deserves proper recognition as synthetic biology research.

Parallel Approach

Parallel engineering is also known as bioengineering. The basic genetic code is the foundation for parallel engineering research, which uses conventional biomolecules like nucleic acids and the 20 amino acids to construct biological systems. For a variety of applications in biocomputing, bioenergy, biofuels, bioremediation, optogenetics, and medicine, it involves the standardisation of DNA components, engineering of switches, biosensors, genetic circuits, logic gates, and cellular communication operators. For directing the expression of two or more genes and/or proteins, the majority of these applications often rely on the use of one or more vectors (or plasmids). Small, circular, double-strand DNA units known as plasmids, which are primarily found in prokaryotic but can also occasionally be detected in eukaryotic cells, may replicate autonomously of chromosomal DNA.

Orthogonal Approach

It is also known as perpendicular engineering. This strategy, also referred to as "chemical synthetic biology," principally seeks to alter or enlarge the genetic codes of living systems utilising artificial DNA bases and/or amino acids. This subfield is also connected to xenobiology, a newly developed field that combines systems chemistry, synthetic biology, exobiology, and research into the origins of life. In recent decades, researchers have created compounds that are structurally similar to the DNA canonical bases to see if those "alien" or xeno (XNA) molecules may be employed as genetic information carriers. Similar to this, noncanonical moieties have taken the place of the DNA sugar (deoxyribose). In order to express information other than the 20 conventional amino acids of proteins, the genetic code can be altered or enlarged. One method involves incorporating a specified unnatural, noncanonical, or xeno amino acid (XAA) into one or more proteins at one or more precise places using orthogonal enzymes and a transfer RNA adaptor from an other organism. By using "directed evolution," which entails repeated cycles of gene mutagenesis (genotypic diversity production), screening or selection (of a specific phenotypic trait), and amplification of a better variant for the following iterative round, orthogonal enzymes are produced Numerous XAAs have been effectively incorporated into proteins in more complex creatures like worms and flies as well as in bacteria, yeast, and human cell lines. As a result of canonical DNA sequence changes, directed evolution also enables the development of orthogonal ribosomes, which make it easier to incorporate XAAs into proteins or create "mirror life," or biological systems that contain biomolecules made up of enantiomers with different chiral orientations.

7.4.5 Enabling Technologies

Several novel enabling technologies were critical to the success of synthetic biology. Concepts include standardization of biological parts and hierarchical abstraction to permit using those parts in synthetic systems. DNA serves as the guide for how biological processes should function, like the score to a complex symphony of life. Our ability to comprehend and design biological systems has undergone significant modifications as a result of developments in the previous few decades in both reading (sequencing) and writing (synthesis) DNA sequences. These developments have produced ground-breaking techniques for designing, assembling, and modifying DNA-encoded genes, materials, circuits, and metabolic pathways, enabling an ever-increasing amount of control over biological systems and even entire organisms. Basic technologies include reading and writing DNA (sequencing and fabrication). Measurements under multiple conditions are needed for accurate modeling and computer-aided design (CAD).

DNA and Gene Synthesis

Driven by dramatic decreases in costs of oligonucleotide ("oligos") synthesis and the advent of PCR, the sizes of DNA constructions from oligos have increased to the genomic level. In 2000, researchers reported synthesis of the 9.6 kbp (kilo bp) Hepatitis C virus genome from chemically synthesized 60 to 80-mers. In 2002, researchers at Stony Brook University succeeded in synthesizing the 7741 bp poliovirus genome from its published sequence, producing the second synthetic genome, spanning two years. In 2003, the 5386 bp genome of the bacteriophage Phi X 174 was assembled in about two weeks. In 2006, the same team, at the J. Craig Venter Institute, constructed and patented a synthetic genome of a novel minimal bacterium, *Mycoplasma laboratorium* and were working on getting it functioning in a living cell.

In 2007, it was reported that several companies were offering synthesis of genetic sequences up to 2000 base pairs (bp) long, for a price of about $1 per bp and a turnaround time of less than two weeks. Oligonucleotides harvested from a photolithographic- or inkjet-manufactured DNA chip combined with PCR and DNA mismatch error-correction allows inexpensive large-scale changes of codons in genetic systems to improve gene expression or incorporate novel amino-acids. This favors a synthesis-from-scratch approach.

Additionally, the CRISPR/Cas system has emerged as a promising technique for gene editing. It was described as "the most important innovation in the synthetic biology space in nearly 30 years". While other methods take months or years to edit gene sequences, CRISPR speeds that time up to weeks. Due to its ease of use and accessibility, however, it has raised ethical concerns, especially surrounding its use in biohacking.

Sequencing

DNA sequencing determines the order of nucleotide bases in a DNA molecule. Synthetic biologists use DNA sequencing in their work in several ways. First, large-scale genome sequencing efforts continue to provide information on naturally occurring organisms. This information provides a rich substrate from which synthetic biologists can construct parts and devices. Second, sequencing can verify that the fabricated system is as intended. Third, fast, cheap, and reliable sequencing can facilitate rapid detection and identification of synthetic systems and organisms.

Modularity

This is the ability of a system or component to operate without reference to its context. The most used standardized DNA parts are BioBrick plasmids, invented by Tom Knight in 2003. Biobricks are stored at the Registry of Standard Biological Parts in Cambridge, Massachusetts. The BioBrick standard has been used by tens of thousands of students worldwide in the international Genetically Engineered Machine (iGEM) competition. BioBrick Assembly Standard 10 promotes modularity by allowing BioBrick coding sequences to be spliced out and exchanged using restriction enzymes EcoRI or XbaI (BioBrick prefix) and SpeI and PstI (BioBrick suffix).

Sequence overlap between two genetic elements (genes or coding sequences), called overlapping genes, can prevent their individual manipulation. To increase genome modularity, the practice of genome refactoring or improving "the internal structure of an existing system for future use, while simultaneously maintaining external system function" has been adopted across synthetic biology disciplines. Some notable examples of refactoring including the nitrogen fixation cluster and type III secretion system along with bacteriophages T7 and ΦX174.

While DNA is most important for information storage, a large fraction of the cell's activities are carried out by proteins. Tools can send proteins to specific regions of the cell and to link different proteins together. The interaction strength between protein partners should be tunable between a lifetime of seconds (desirable for dynamic signaling events) up to an irreversible interaction (desirable for device stability or resilient to harsh conditions). Interactions such as coiled coils, SH3 domain-peptide binding or SpyTag/SpyCatcher offer such control. In addition, it is necessary to regulate protein-protein interactions in cells, such as with light (using light-oxygen-voltage-sensing domains) or cell-permeable small molecules by chemically induced dimerization.

In a living cell, molecular motifs are embedded in a bigger network with upstream and downstream components. These components may alter the signaling capability of the modeling module. In the case of ultrasensitive modules, the sensitivity contribution of a module can differ from the sensitivity that the module sustains in isolation.

Modeling

Models inform the design of engineered biological systems by better predicting system behavior prior to fabrication. Synthetic biology benefits from better models of how biological molecules bind substrates and catalyze reactions, how DNA encodes the information needed to specify the cell and how multi-component integrated systems behave.

Multiscale models of gene regulatory networks focus on synthetic biology applications. Simulations can model all biomolecular interactions in transcription, translation, regulation and induction of gene regulatory networks. Only extensive modelling can enable the exploration of dynamic gene expression in a form suitable for research and design due to the numerous involved species and the intricacy of their relationships. Dynamic simulations of the entire biomolecular interconnection involved in regulation, transport, transcription, induction, and translation enable the molecular level detailing of designs. As opposed to modelling artificial networks a posteriori, this is contrasted.

Microfluidics

Microfluidics, in particular droplet microfluidics, is an emerging tool used to construct new components, and to analyze and characterize them. It is widely employed in screening assays.

Synthetic Transcription Factors

Studies have considered the components of the DNA transcription mechanism. One desire of scientists creating synthetic biological circuits is to be able to control the transcription of synthetic DNA in unicellular organisms (prokaryotes) and in multicellular organisms (eukaryotes). One study tested the adjustability of synthetic transcription factors (sTFs) in areas of transcription output and cooperative ability among multiple transcription factor complexes. Researchers were able to mutate functional regions called zinc fingers, the DNA specific component of sTFs, to decrease their affinity for specific operator DNA sequence sites, and thus decrease the associated site-specific activity of the sTF (usually transcriptional regulation). They further used the zinc fingers as components of complex-forming sTFs, which are the eukaryotic translation mechanisms.

7.4.6 Applications

Synthetic biology initiatives frequently aim to redesign organisms so that they can create a material, such as a drug or fuel, or acquire a new function, such as the ability to sense something in the environment. Examples of what researchers are creating using synthetic biology include:

- Utilizing microorganisms for bioremediation to remove contaminants from our water, soil, and air.
- Production of complex natural products that are usually extracted from plants but cannot be obtained in sufficient amounts, e.g. drugs of natural origin, such as artemisinin and paclitaxel.
- Beta-carotene, a substance typically associated with carrots that prevents vitamin A deficiency, is produced by rice that has been modified. Every year, between 250,000 and 500,000 children lose their vision due to vitamin A deficiency, which also significantly raises their chance of dying from infectious infections.
- As a sustainable and environmentally benign alternative to the fresh roses that perfumers use to create expensive smells, yeast has been created to produce rose oil.

Biosensors

A biosensor refers to an engineered organism, usually a bacterium, that is capable of reporting some ambient phenomenon such as the presence of heavy metals or toxins. One such system is the Lux operon of *Aliivibrio fischeri,* which codes for the enzyme that is the source of bacterial bioluminescence, and can be placed after a respondent promoter to express the luminescence genes in response to a specific environmental stimulus.

One such sensor created, consisted of a bioluminescent bacterial coating on a photosensitive computer chip to detect certain petroleum pollutants. When the bacteria sense the pollutant, they luminesce. Another example of a similar mechanism is the detection of landmines by an engineered *E.coli* reporter strain capable of detecting TNT and its main degradation product DNT, and consequently producing a green fluorescent protein (GFP).

Modified organisms can sense environmental signals and send output signals that can be detected and serve diagnostic purposes. Microbe cohorts have been used. Biosensors could also be used to detect pathogenic signatures—such as of SARS-CoV-2—and can be wearable.

For the purpose of detecting and reacting to various and temporary environmental factors, cells have developed a wide range of regulatory circuits, ranging from transcriptional to post-translational.

These circuits are made up of transducer modules that filter the signals and activate a biological response, as well as carefully designed sensitive sections that attach analytes and regulate signal-detection thresholds. Modularity and selectivity are programmed to biosensor circuits at the transcriptional, translational, and post-translational levels, to achieve the delicate balancing of the two basic sensing modules.

Food and Drink

Cellular agriculture focuses on the production of agricultural products from cell cultures using a combination of biotechnology, tissue engineering, molecular biology, and synthetic biology to create and design new methods of producing proteins, fats, and tissues that would otherwise come from traditional agriculture. Most of the industry is focused on animal products such as meat, milk, and eggs, produced in cell culture rather than raising and slaughtering farmed livestock which is associated with substantial global problems of detrimental environmental impacts (e.g., of meat production), animal welfare, food security and human health. Cellular agriculture is a field of the biobased economy. The most well known cellular agriculture concept is cultured meat.

However, not all synthetic nutrition products are animal food products - for instance, as of 2021, there are also products of synthetic coffee that are reported to be close to commercialization. Similar fields of research and production based on synthetic biology that can be used for the production of food and drink are:

- Genetically engineered microbial food cultures (e.g., for solar-energy-based protein powder).
- Cell-free artificial synthesis (e.g., synthetic starch).

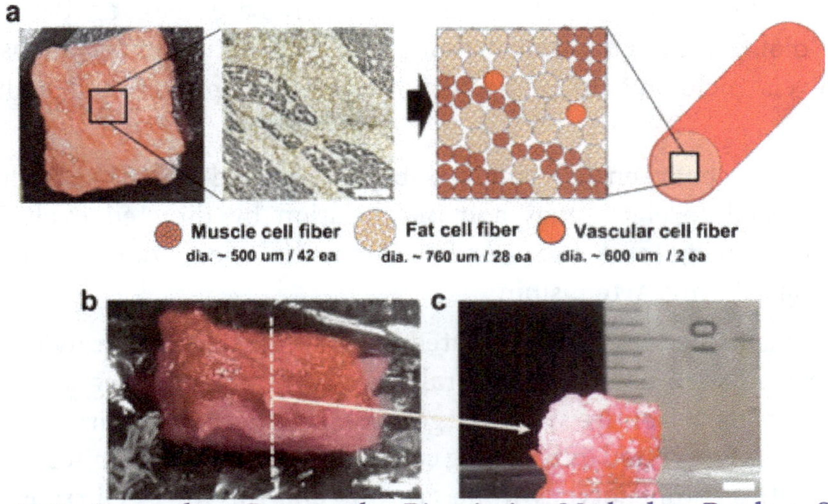

Fig 1: In 2021, Researchers Presented a Bioprinting Method to Produce Steak-Like Cultured Meat

Materials

Photosynthetic microbial cells have been used as a step to synthetic production of spider silk.

Biological Computers

A biological computer refers to an engineered biological system that can perform computer-like operations, which is a dominant paradigm in synthetic biology. Researchers built and characterized a variety of logic gates in a number of organisms, and demonstrated both analog and digital computation in living cells.

They demonstrated that bacteria can be engineered to perform both analog and/or digital computation. In 2007, in human cells, research demonstrated a universal logic evaluator that operates in mammalian cells.

Subsequently, researchers utilized this paradigm to demonstrate a proof-of-concept therapy that uses biological digital computation to detect and kill human cancer cells in 2011. In 2016, another group of researchers demonstrated that principles of computer engineering can be used to automate digital circuit design in bacterial cells.

In 2017, researchers demonstrated the 'Boolean logic and arithmetic through DNA excision' (BLADE) system to engineer digital computation in human cells. In 2019, researchers implemented a perceptron in biological systems opening the way for machine learning in these systems.

Cell Transformation

Cells use interacting genes and proteins, which are called gene circuits, to implement diverse function, such as responding to environmental signals, decision making and communication. Three key components are involved: DNA, RNA and Synthetic biologist designed gene circuits that can control gene expression from several levels including transcriptional, post-transcriptional and translational levels.

Traditional metabolic engineering has been bolstered by the introduction of combinations of foreign genes and optimization by directed evolution. This includes engineering *E. coli* and yeast for commercial production of a precursor of the antimalarial drug, Artemisinin.

Entire organisms have yet to be created from scratch, although living cells can be transformed with new DNA. Several ways allow constructing synthetic DNA components and even entire synthetic genomes, but once the desired genetic code is obtained, it is integrated into a living cell that is expected to manifest the desired new capabilities or phenotypes while growing and thriving. Cell transformation is used to create biological circuits, which can be manipulated to yield desired outputs.

By integrating synthetic biology with materials science, it would be possible to use cells as microscopic molecular foundries to produce materials whose properties were genetically encoded. Re-engineering has produced Curli fibers, the amyloid component of extracellular material of biofilms, as a platform for programmable nanomaterial. These nanofibers were genetically constructed for specific functions, including adhesion to substrates, nanoparticle templating and protein immobilization.

Designed Proteins

Natural proteins can be engineered, for example, by directed evolution, novel protein structures that match or improve on the functionality of existing proteins can be produced. One group generated a helix bundle that was capable of binding oxygen with similar properties as hemoglobin, yet did not bind carbon monoxide. A similar protein structure was generated to support a variety of oxidoreductase activities while another formed a structurally and sequentially novel ATPase. Another group generated a family of G-protein coupled receptors that could be activated by the inert small molecule clozapine N-oxide but insensitive to the native ligand, acetylcholine; these receptors are known as DREADDs. Novel functionalities or protein specificity can also be engineered using computational approaches. One study was able to use two different computational methods: a bioinformatics and molecular modeling method to mine sequence databases, and a computational enzyme design method to reprogram enzyme specificity. Both methods resulted in designed enzymes with greater than 100 fold specificity for production of longer chain alcohols from sugar.

Another common investigation is expansion of the natural set of 20 amino acids. Excluding stop codons, 61 codons have been identified, but only 20 amino acids are coded generally in all organisms. Certain codons are engineered to code for alternative amino acids including: nonstandard amino acids such as O-methyl tyrosine; or exogenous amino acids such as 4-fluorophenylalanine. Typically, these projects make use of re-coded nonsense suppressor tRNA-Aminoacyl tRNA synthetase pairs from other organisms, though in most cases substantial engineering is required.

Other researchers investigated protein structure and function by reducing the normal set of 20 amino acids. Limited protein sequence libraries are made by generating proteins where groups of amino acids may be replaced by a single amino acid. For instance, several non-polar amino acids within a protein can all be replaced with a single non-polar amino acid. One project demonstrated that an engineered version of Chorismate mutase still had catalytic activity when only nine amino acids were used.

Researchers and companies practice synthetic biology to synthesize industrial enzymes with high activity, optimal yields and effectiveness. These synthesized enzymes aim to improve products such as detergents and lactose-free dairy products, as well as make them more cost effective. The improvements of metabolic engineering by synthetic biology is an example of a biotechnological technique utilized in industry to discover pharmaceuticals and fermentive chemicals. Synthetic biology may investigate modular pathway systems in biochemical production and increase yields of metabolic production. Artificial enzymatic activity and subsequent effects on metabolic reaction rates and yields may develop "efficient new strategies for improving cellular properties ... for industrially important biochemical production".

Designed Nucleic Acid Systems

Scientists can encode digital information onto a single strand of synthetic DNA. In 2012, George M. Church encoded one of his books about synthetic biology in DNA. The 5.3 Mb of data was more than 1000 times greater than the previous largest amount of information to be stored in synthesized DNA. A similar project encoded the complete sonnets of William Shakespeare in DNA. More generally, algorithms such as NUPACK, ViennaRNA, Ribosome Binding Site Calculator, Cello, and Non-Repetitive Parts Calculator enables the design of new genetic systems. Many technologies have been developed for incorporating unnatural nucleotides and amino acids into nucleic acids and proteins, both in vitro and in vivo. For example, in May 2014, researchers announced that they had successfully introduced two new artificial nucleotides into bacterial DNA. By including individual artificial nucleotides in the culture media, they were able to exchange the bacteria 24 times; they did not generate mRNA or proteins able to use the artificial nucleotides.

Space Exploration

Synthetic biology raised NASA's interest as it could help to produce resources for astronauts from a restricted portfolio of compounds sent from Earth. On Mars, in particular, synthetic biology could lead to production processes based on local resources, making it a powerful tool in the development of occupied outposts with less dependence on Earth. Work has gone into developing plant strains that are able to cope with the harsh Martian environment, using similar techniques to those employed to increase resilience to certain environmental factors in agricultural crops.

Synthetic Life

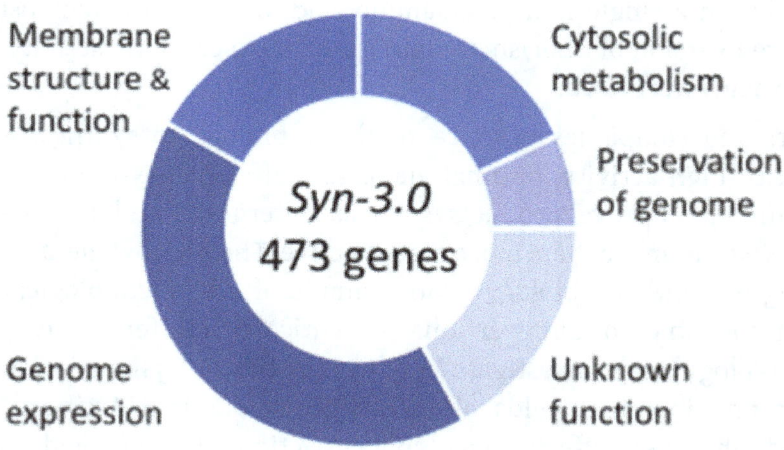

Fig 2: Gene Functions in the Minimal Genome of the Synthetic Organism

One important topic in synthetic biology is *synthetic life*, that is concerned with hypothetical organisms created *in vitro* from biomolecules and/or chemical analogues thereof.

Synthetic life experiments attempt to either probe the origins of life, study some of the properties of life, or more ambitiously to recreate life from non-living (abiotic) components. Synthetic life biology attempts to create living organisms capable of carrying out important functions, from manufacturing pharmaceuticals to detoxifying polluted land and water. In medicine, it offers prospects of using designer biological parts as a starting point for new classes of therapies and diagnostic tools.

A living "artificial cell" has been defined as a completely synthetic cell that can capture energy, maintain ion gradients, contain macromolecules as well as store information and have the ability to mutate. Nobody has been able to create such a cell. A completely synthetic bacterial chromosome was produced in 2010 by Craig Venter, and his team introduced it to genomically emptied bacterial host cells. The host cells were able to grow and replicate. The *Mycoplasma laboratorium* is the only living organism with completely engineered genome.

The first living organism with 'artificial' expanded DNA code was presented in 2014; the team used *E. coli* that had its genome extracted and replaced with a chromosome with an expanded genetic code. The nucleosides added are d5SICS and dNaM. In May 2019, in a milestone effort, researchers reported the creation of a new synthetic (possibly artificial) form of viable life, a variant of the bacteria *Escherichia coli*, by reducing the natural number of 64 codons in the bacterial genome to 59 codons instead, in order to encode 20 amino acids. In 2017, the international Build-a-Cell large-scale open-source research collaboration for the construction of synthetic living cells was started, followed by national synthetic cell organizations in several countries, including FabriCell, MaxSynBio and BaSyC. The European synthetic cell efforts were unified in 2019 as SynCellEU initiative. In 2023, researchers were able to create the first synthetically made human embryos derived from stem cells.

Drug Delivery Platforms

In therapeutics, synthetic biology has achieved significant advancements in altering and simplifying the therapeutics scope in a relatively short period of time. In fact, new therapeutic platforms, from the discovery of disease mechanisms and drug targets to the manufacture and transport of small molecules, are made possible by the logical and model-guided design construction of biological components.

Synthetic biology devices have been designed to act as therapies in therapeutic treatment. It is possible to control complete created viruses and organisms to target particular pathogens and diseased pathways. Thus, in two independent studies 91,92, researchers utilised genetically modified bacteriophages to fight antibiotic-resistant bacteria by giving them genetic features that specifically target and hinder bacterial defences against antibiotic activity.

In the therapy of cancer, since conventional medicines frequently indiscriminately target tumours and normal tissues, artificially created viruses and organisms that can identify and connect their therapeutic action to pathological signals may be helpful. For example, p53 pathway activity in human cells was put into adenoviruses to control how they replicated.

Engineered Bacteria-Based Platform

Bacteria have long been used in cancer treatment. *Bifidobacterium* and *Clostridium* selectively colonize tumors and reduce their size. Recently synthetic biologists reprogrammed bacteria to sense and respond to a particular cancer state. Most often bacteria are used to deliver a therapeutic molecule directly to the tumor to minimize off-target effects.

To target the tumor cells, peptides that can specifically recognize a tumor were expressed on the surfaces of bacteria. Peptides used include an affibody molecule that specifically targets human epidermal growth factor receptor 2 and a synthetic adhesin. The other way is to allow bacteria to sense the tumor microenvironment, for example hypoxia, by building an AND logic gate into bacteria. Then the bacteria only release target therapeutic molecules to the tumor through either lysis or the bacterial secretion system. Lysis has the advantage that it can stimulate the immune system and control growth. Multiple types of secretion systems can be used and other strategies as well. The system is inducible by external signals. Inducers include chemicals, electromagnetic or light waves.

Multiple species and strains are applied in these therapeutics. Most commonly used bacteria are *Salmonella typhimurium*, *Escherichia coli*, *Bifidobacteria*, *Streptococcus*, *Lactobacillus*, *Listeria* and *Bacillus subtilis*. Each of these species have their own property and are unique to cancer therapy in terms of tissue colonization, interaction with immune system and ease of application.

Engineered Yeast-Based Platform

Synthetic biologists are developing genetically modified live yeast that can deliver therapeutic biologic medicines. When orally delivered, these live yeast act like micro-factories and will make therapeutic molecules directly in the gastrointestinal tract. Because yeast are eukaryotic, a key benefit is that they can be administered together with antibiotics. Probiotic yeast expressing human P2Y2 purinergic receptor suppressed intestinal inflammation in mouse models of inflammatory bowel disease. A live S. boulardii yeast delivering a tetra-specific anti-toxin that potently neutralizes Toxin A and Toxin B of Clostridioides difficile has been developed. This therapeutic anti-toxin is a fusion of four single-domain antibodies (nanobodies) that potently and broadly neutralize the two major virulence factors of C. difficile at the site of infection in preclinical models. The first in human clinical trial of engineered live yeast for the treatment of Clostridium difficile infection is anticipated in 2024 and will be sponsored by the developer Fzata, Inc.

Cell-Based Platform

The immune system plays an important role in cancer and can be harnessed to attack cancer cells. Cell-based therapies focus on immunotherapies, mostly by engineering T cells. T cell receptors were engineered and 'trained' to detect cancer epitopes. Chimeric antigen receptors (CARs) are composed of a fragment of an antibody fused to intracellular T cell signaling domains that can activate and trigger proliferation of the cell. Multiple second generation CAR-based therapies have been approved by FDA. Gene switches were designed to enhance safety of the treatment. Kill switches were developed to terminate the therapy should the patient show severe side effects. Mechanisms can more finely control the system and stop and reactivate it. Since the number of T-cells are important for therapy persistence and severity, growth of T-cells is also controlled to dial the effectiveness and safety of therapeutics. Although several mechanisms can improve safety and control, limitations include the difficulty of inducing large DNA circuits into the cells and risks associated with introducing foreign components, especially proteins, into cells.

Biofuels, Pharmaceuticals and Biomaterials

The most popular biofuel is ethanol produced from corn or sugar cane, but this method of producing biofuels is troublesome and constrained due to the high agricultural cost and inadequate fuel characteristics of ethanol. An substitute and potential source of renewable energy is microbes that have had their metabolic pathways altered to be more efficient at converting biomass into biofuels. Only if their production costs could be made to match or even exceed those of present fuel production can these techniques be expected to be successful. Related to this, there are several medicines whose pricey manufacturing procedures prevent them from having a larger therapeutic range. The creation of new materials and the microbiological manufacturing of biomaterials would both benefit substantially from novel artificial biology tools.

CRISPR/Cas9

The clustered frequently interspaced short palindromic repetitions (CRISPR)/CRISPR associated (Cas) system is a powerful method of genome engineering in a range of organisms because of its simplicity, modularity, and scalability. In this technique, a guide RNA (gRNA) attracts the CRISPR nuclease Cas9 to a particular spot in the genome, causing a double strand break. Several DNA repair processes, including homology-directed recombination and non-homology end joining, can be used to accomplish the desired genome change (i.e., gene deletion or insertion). Additionally, dCas9 (dead Cas9 or nuclease-deficient Cas9), a Cas9 double mutant (H840A, D10A), has been utilised to control gene expression in bacteria or when linked to a stimulation of suppression site in yeast.

Regulatory Elements

To build and develop biological systems, regulating components including regulators, ribosome-binding sites (RBSs), and terminators are crucial. Despite years of study, there are many various varieties and numbers of promoters and terminators for Escherichia coli, but also for the well-researched model organism

Saccharomyces cerevisiae, as well as for other organisms of interest, these tools are quite scarce. Numerous techniques have been invented for the finding and identification of promoters and terminators in order to overcome this constraint, including genome mining, random mutagenesis, hybrid engineering, biophysical modelling, combinatorial design, and rational design.

Organoids

Synthetic biology has been used for organoids, which are lab-grown organs with application to medical research and transplantation.

Bioprinted Organs

3D bioprinting can be used to reconstruct tissue from various regions of the body. The precursor to the adoption of 3D printing in healthcare was a series of trials conducted by researchers at Boston Children's Hospital. The team built replacement urinary bladders by hand for seven patients by constructing scaffolds, then layering the scaffolds with cells from the patients and allowing them to grow. The trials were a success as the patients remained in good health 7 years after implantation, which led a research fellow named Anthony Atala, MD, to search or ways to automate the process.

Patients with end-stage bladder disease can now be treated by using bio-engineered bladder tissues to rebuild the damaged organ. This technology can also potentially be applied to bone, skin, cartilage and muscle tissue. Though one long-term goal of 3D bioprinting technology is to reconstruct an entire organ as well as minimize the problem of the lack of organs for transplantation. There has been little success in bioprinting of fully functional organs e.g., liver, skin, meniscus or pancreas. Unlike implantable stents, organs have complex shapes and are significantly harder to bioprint. A bioprinted heart, for example, must not only meet structural requirements, but also vascularization, mechanical load, and electrical signal propagation requirements. In 2022, the first success of a clinical trial for a 3D bioprinted transplant that is made from the patient's own cells, an external ear to treat microtia, was reported.

3D bioprinting contributes to significant advances in the medical field of tissue engineering by allowing for research to be done on innovative materials called biomaterials. Some of the most notable bioengineered substances are usually stronger than the average bodily materials, including soft tissue and bone. These constituents can act as future substitutes, even improvements, for the original body materials. In addition, the Defense Threat Reduction Agency aims to print mini organs such as hearts, livers, and lungs as the potential to test new drugs more accurately and perhaps eliminate the need for testing in animals.

Other Transplants and Induced Regeneration

There is ongoing research and development into synthetic biology based methods for inducing regeneration in humans as well the creation of transplantable artificial organs.

Nanoparticles, Artificial Cells and Micro-Droplets

Synthetic biology can be used for creating nanoparticles which can be used for drug-delivery as well as for other purposes. Complementing research and development seeks to and has created synthetic cells that mimics functions of biological cells. Applications include medicine such as designer-nanoparticles that make blood cells eat away—from the inside out—portions of atherosclerotic plaque that cause heart attacks. Synthetic micro-droplets for algal cells or synergistic algal-bacterial multicellular spheroid microbial reactors, for example, could be used to produce hydrogen as hydrogen economy biotechnology.

Electrogenetics

Mammalian designer cells are engineered by humans to behave a specific way, such as an immune cell that expresses a synthetic receptor designed to combat a specific disease. Electrogenetics is an application of synthetic biology that involves utilizing electrical fields to stimulate a response in engineered cells. Controlling the designer cells can be done with relative ease through the use of common electronic devices, such as smartphones. Additionally, electrogenetics allows for the possibility of creating devices that are much smaller and compact than devices that use other stimulus through the use of microscopic electrodes. One example of how electrogenetics is used to benefit public health is through stimulating designer cells that are able to produce/deliver therapeutics. This was implemented in $_{\text{Electro}}$HEK cells, cells that contain voltage-gated calcium channels that are electrosensitive, meaning that the ion channel can be controlled by electrical conduction between electrodes and the $_{\text{Electro}}$HEK cells. The expression levels of the artificial gene that these $_{\text{Electro}}$HEK cells contained was shown to be able to be controlled by changing the voltage or electrical pulse length. Further studies have expanded on this robust system, one of which is a beta cell line system designed to control the release of insulin based on electric signals.

7.4.7 Ethics

The creation of new life and the tampering of existing life has raised ethical concerns in the field of synthetic biology and are actively being discussed.

Common ethical questions include:

- Is it morally right to tamper with nature?
- Is one playing God when creating new life?

- What happens if a synthetic organism accidentally escapes?
- What if an individual misuses synthetic biology and creates a harmful entity (e.g., a biological weapon)?
- Who will have control of and access to the products of synthetic biology?
- Who will gain from these innovations? Investors? Medical patients? Industrial farmers?
- Does the patent system allow patents on living organisms? What about parts of organisms, like HIV resistance genes in humans?
- What if a new creation is deserving of moral or legal status?

The ethical aspects of synthetic biology has three main features: biosafety, biosecurity, and the creation of new life forms. Other ethical issues mentioned include the regulation of new creations, patent management of new creations, benefit distribution, and research integrity.

Ethical issues have surfaced for recombinant DNA and genetically modified organism (GMO) technologies and extensive regulations of genetic engineering and pathogen research were in place in many jurisdictions. Amy Gutmann, former head of the Presidential Bioethics Commission, argued that we should avoid the temptation to over-regulate synthetic biology in general, and genetic engineering in particular. According to Gutmann, "Regulatory parsimony is especially important in emerging technologies...where the temptation to stifle innovation on the basis of uncertainty and fear of the unknown is particularly great. The blunt instruments of statutory and regulatory restraint may not only inhibit the distribution of new benefits, but can be counterproductive to security and safety by preventing researchers from developing effective safeguards.".

The "Creation" of Life

One ethical question is whether or not it is acceptable to create new life forms, sometimes known as "playing God". Currently, the creation of new life forms not present in nature is at small-scale, the potential benefits and dangers remain unknown, and careful consideration and oversight are ensured for most studies.

Many advocates express the great potential value—to agriculture, medicine, and academic knowledge, among other fields—of creating artificial life forms. Creation of new entities could expand scientific knowledge well beyond what is currently known from studying natural phenomena. Yet there is concern that artificial life forms may reduce nature's "purity" (i.e., nature could be somehow corrupted by human intervention and manipulation) and potentially influence the adoption of more engineering-like principles instead of biodiversity- and nature-focused ideals. Some are also concerned that if an artificial life form were to be released into nature, it could hamper biodiversity by beating out natural species for resources (similar to how algal blooms kill marine species). Another concern involves the ethical treatment of newly created entities if they happen to sense pain, sentience, and self-perception. There is an ongoing debate as to whether such life forms should be granted moral or legal rights, though no consensus exists as to how these rights would be administered or enforced.

Ethical Support for Synthetic Biology

Ethics and moral rationales that support certain applications of synthetic biology include their potential mitigation of substantial global problems of detrimental environmental impacts of conventional agriculture (including meat production), animal welfare, food security, and human health, as well as potential reduction of human labor needs and, via therapies of diseases, reduction of human suffering and prolonged life.

Biosafety and Biocontainment

What is most ethically appropriate when considering biosafety measures? How can accidental introduction of synthetic life in the natural environment be avoided? Much ethical consideration and critical thought has been given to these questions. Biosafety not only refers to biological containment; it also refers to strides taken to protect the public from potentially hazardous biological agents. Even though such concerns are important and remain unanswered, not all products of synthetic biology present concern for biological safety or negative consequences for the environment. It is argued that most synthetic technologies are benign and are incapable of flourishing in the outside world due to their "unnatural" characteristics as there is yet to be an example of a transgenic microbe conferred with a fitness advantage in the wild.

In general, existing hazard controls, risk assessment methodologies, and regulations developed for traditional genetically modified organisms (GMOs) are considered to be sufficient for synthetic organisms. "Extrinsic" biocontainment methods in a laboratory context include physical containment through biosafety cabinets and gloveboxes, as well as personal protective equipment.

In an agricultural context, they include isolation distances and pollen barriers, similar to methods for biocontainment of GMOs. Synthetic organisms may offer increased hazard control because they can be engineered with "intrinsic" biocontainment methods that limit their growth in an uncontained environment, or prevent horizontal gene transfer to natural organisms. Examples of intrinsic biocontainment include auxotrophy, biological kill switches, inability of the organism to replicate or to pass modified or synthetic genes to offspring, and the use of xenobiological organisms using alternative biochemistry, for example using artificial xeno nucleic acids (XNA) instead of DNA. Regarding auxotrophy, bacteria and yeast can be engineered to be unable to produce histidine, an important amino acid for all life. Such organisms can thus only be grown on histidine-rich media in laboratory conditions, nullifying fears that they could spread into undesirable areas.

Biosecurity and Bioterrorism

Some ethical issues relate to biosecurity, where biosynthetic technologies could be deliberately used to cause harm to society and/or the environment. Since synthetic biology raises ethical issues and biosecurity issues, humanity must consider and plan on how to deal with potentially harmful creations, and what kinds of ethical measures could possibly be employed to deter nefarious biosynthetic technologies. With the exception of regulating synthetic biology and biotechnology companies, however, the issues are not seen as new because they were raised during the earlier recombinant DNA and genetically modified organism (GMO) debates, and extensive regulations of genetic engineering and pathogen research are already in place in many jurisdictions.

Additionally, the development of synthetic biology tools has made it easier for individuals with less education, training, and access to equipment to modify and use pathogenic organisms as bioweapons. This increases the threat of bioterrorism, especially as terrorist groups become aware of the significant social, economic, and political disruption caused by pandemics like COVID-19. As new techniques are developed in the field of synthetic biology, the risk of bioterrorism is likely to continue to grow. Juan Zarate, who served as Deputy National Security Advisor for Combating Terrorism from 2005 to 2009, noted that "the severity and extreme disruption of a novel coronavirus will likely spur the imagination of the most creative and dangerous groups and individuals to reconsider bioterrorist attacks."

European Union

The European Union-funded project SYNBIOSAFE has issued reports on how to manage synthetic biology. A 2007 paper identified key issues in safety, security, ethics, and the science-society interface, which the project defined as public education and ongoing dialogue among scientists, businesses, government and ethicists. The key security issues that SYNBIOSAFE identified involved engaging companies that sell synthetic DNA and the biohacking community of amateur biologists. Key ethical issues concerned the creation of new life forms.

A subsequent report focused on biosecurity, especially the so-called dual-use challenge. For example, while synthetic biology may lead to more efficient production of medical treatments, it may also lead to synthesis or modification of harmful pathogens (e.g., smallpox). The biohacking community remains a source of special concern, as the distributed and diffuse nature of open-source biotechnology makes it difficult to track, regulate or mitigate potential concerns over biosafety and biosecurity.

COSY, another European initiative, focuses on public perception and communication. To better communicate synthetic biology and its societal ramifications to a broader public, COSY and SYNBIOSAFE published *SYNBIOSAFE*, a 38-minute documentary film, in October 2009.

The International Association Synthetic Biology has proposed self-regulation. This proposes specific measures that the synthetic biology industry, especially DNA synthesis companies, should implement. In 2007, a group led by scientists from leading DNA-synthesis companies published a "practical plan for developing an effective oversight framework for the DNA-synthesis industry".

United States

In January 2009, the Alfred P. Sloan Foundation funded the Woodrow Wilson Center, the Hastings Center, and the J. Craig Venter Institute to examine the public perception, ethics and policy implications of synthetic biology.

On July 9–10, 2009, the National Academies' Committee of Science, Technology & Law convened a symposium on "Opportunities and Challenges in the Emerging Field of Synthetic Biology".

After the publication of the first synthetic genome and the accompanying media coverage about "life" being created, President Barack Obama established the Presidential Commission for the Study of Bioethical Issues to study synthetic biology. The commission convened a series of meetings, and issued a report in December 2010 titled "New Directions: The Ethics of Synthetic Biology and Emerging Technologies." The commission stated that "while Venter's achievement marked a significant technical advance in demonstrating that a relatively large genome could be accurately synthesized and substituted for another, it did not amount to the "creation of life". It noted that synthetic biology is an emerging field, which creates potential risks and rewards. The commission did not recommend policy or oversight changes and called for continued funding of the research and new funding for monitoring, study of emerging ethical issues and public education.

Synthetic biology, as a major tool for biological advances, results in the "potential for developing biological weapons, possible unforeseen negative impacts on human health ... and any potential environmental impact". The proliferation of such technology could also make the production of biological and chemical weapons available to a wider array of state and non-state actors. These security issues may be avoided by regulating industry uses of biotechnology through policy legislation. Federal guidelines on genetic manipulation are being proposed by "the President's

Bioethics Commission ... in response to the announced creation of a self-replicating cell from a chemically synthesized genome, put forward 18 recommendations not only for regulating the science ... for educating the public".

Opposition

On March 13, 2012, over 100 environmental and civil society groups, including Friends of the Earth, the International Center for Technology Assessment, and the ETC Group, issued the manifesto *The Principles for the Oversight of Synthetic Biology*. This manifesto calls for a worldwide moratorium on the release and commercial use of synthetic organisms until more robust regulations and rigorous biosafety measures are established. The groups specifically call for an outright ban on the use of synthetic biology on the human genome or human microbiome. Richard Lewontin wrote that some of the safety tenets for oversight discussed in *The Principles for the Oversight of Synthetic Biology* are reasonable, but that the main problem with the recommendations in the manifesto is that "the public at large lacks the ability to enforce any meaningful realization of those recommendations".

7.4.8 Health and Safety

The hazards of synthetic biology include biosafety hazards to workers and the public, biosecurity hazards stemming from deliberate engineering of organisms to cause harm, and environmental hazards. The biosafety hazards are similar to those for existing fields of biotechnology, mainly exposure to pathogens and toxic chemicals, although novel synthetic organisms may have novel risks. For biosecurity, there is concern that synthetic or redesigned organisms could theoretically be used for bioterrorism. Potential risks include recreating known pathogens from scratch, engineering existing pathogens to be more dangerous, and engineering microbes to produce harmful biochemicals. Lastly, environmental hazards include adverse effects on biodiversity and ecosystem services, including potential changes to land use resulting from agricultural use of synthetic organisms. Synthetic biology is an example of a dual-use technology with the potential to be used in ways that could intentionally or unintentionally harm humans and/or damage the environment. Often "scientists, their host institutions and funding bodies" consider whether the planned research could be misused and sometimes implement measures to reduce the likelihood of misuse.

Existing risk analysis systems for GMOs are generally considered sufficient for synthetic organisms, although there may be difficulties for an organism built "bottom-up" from individual genetic sequences. Synthetic biology generally falls under existing regulations for GMOs and biotechnology in general, and any regulations that exist for downstream commercial products, although there are generally no regulations in any jurisdiction that are specific to synthetic biology.

7.5 CONCLUSION: THE FUTURE OF INDUSTRIAL ORGANISM IMPROVEMENT

The selection and improvement of industrial organisms are critical for the success of biotechnological applications. While traditional methods such as strain selection, mutagenesis, and adaptive evolution have paved the way, modern techniques like genetic engineering, CRISPR-Cas9, and synthetic biology are revolutionizing the field. These advanced techniques allow for precise control over an organism's metabolic pathways, enabling the production of novel compounds and improving the efficiency of existing processes.

As biotechnology continues to evolve, the development of new methods for selecting and improving industrial organisms will be key to meeting the growing global demand for bio-based products. From pharmaceuticals and biofuels to enzymes and bioplastics, the future of industrial biotechnology depends on the continued innovation and optimization of microbial cell factories.

CHAPTER 8

METHODS OF PRESERVATION OF THE GENE POOL IN INDUSTRIAL ORGANISMS AND CULTURE COLLECTIONS

The gene pool of industrial organisms, including bacteria, yeast, fungi, and other microorganisms, represents a valuable resource for biotechnological applications. These organisms are employed in various industries, including pharmaceuticals, agriculture, food production, and biofuels, and their genetic diversity is essential for enhancing productivity, developing new strains, and ensuring the sustainability of biotechnological processes. Preserving the gene pool is critical to maintaining this diversity and ensuring the availability of organisms with desirable traits for future research and industrial applications.

8.1 THE SIGNIFICANCE OF GENE POOL PRESERVATION

Preserving the gene pool of industrial organisms is vital for several reasons. First, the genetic diversity within microbial populations contributes to their adaptability and resilience, allowing them to thrive in changing environmental conditions. This adaptability is particularly important in industrial settings, where organisms may be subjected to fluctuating substrates, temperatures, and other stresses.

Second, the gene pool serves as a reservoir of genetic traits that can be harnessed through breeding, genetic engineering, and synthetic biology to improve production efficiency, yield, and product quality. For example, certain strains of *Saccharomyces cerevisiae* have been identified as having superior fermentation characteristics, making them invaluable for the production of biofuels and alcoholic beverages.

Lastly, the preservation of the gene pool is critical for maintaining biodiversity and protecting against the extinction of valuable strains that may have unique metabolic capabilities. As industries evolve and new challenges emerge, the need for innovative solutions necessitates the availability of diverse genetic resources.

8.2 METHODS OF PRESERVATION OF THE GENE POOL

Preserving the gene pool of industrial organisms can be achieved through several methods, each with its own advantages and limitations. These methods can be broadly categorized into two main types: short-term and long-term preservation techniques.

8.2.1 Short-Term Preservation Methods

Short-term preservation methods are used to maintain the viability and functionality of microbial cultures for relatively brief periods. These techniques typically involve storing cultures under conditions that minimize metabolic activity and prevent contamination.

Refrigeration

Refrigeration is one of the most common methods for short-term preservation of microbial cultures. By storing cultures at low temperatures (typically 4°C), metabolic activity is significantly reduced, allowing for the preservation of microbial strains for weeks to months. However, this method is not suitable for long-term storage, as prolonged refrigeration can lead to loss of viability and genetic stability over time.

Cryopreservation in Liquid Nitrogen

Cryopreservation involves storing microbial cultures in liquid nitrogen at temperatures around -196°C. This method is effective in preserving the viability of microorganisms for extended periods. By rapidly freezing the culture, ice crystals that can damage cellular structures are minimized, and metabolic processes are halted. Cryopreservation is widely used in laboratories and culture collections to ensure the long-term preservation of microbial strains.

Lyophilization (Freeze-Drying)

Lyophilization is a technique that involves freezing microbial cultures and then reducing the pressure to allow the ice to sublimate directly into vapor. This process removes moisture from the cells, making them stable and preserving their viability for extended periods. Lyophilized cultures can be stored at ambient temperatures, making them convenient for transport and long-term storage. However, the initial preparation of lyophilized cultures requires careful optimization to ensure high recovery rates upon rehydration.

8.2.2 Long-Term Preservation Methods

Long-term preservation methods are essential for ensuring the continued availability of microbial strains over decades or even centuries. These methods often involve more sophisticated techniques that minimize the risk of genetic drift and loss of viability.

Cryogenic Storage

In addition to short-term cryopreservation, cryogenic storage at ultra-low temperatures (-80°C or -196°C) is commonly used for long-term preservation of microbial cultures. This technique is particularly effective for preserving the gene pool, as it prevents cellular metabolism and preserves genetic stability. Culture collections often employ cryogenic storage to safeguard valuable strains for future research and industrial applications.

Deep-Freezing

Deep-freezing involves storing microbial cultures at temperatures between -20°C and -80°C. While not as extreme as cryogenic storage, deep-freezing can effectively maintain the viability of microorganisms for several years. This method is commonly used in research laboratories to preserve strains that are not frequently used but still require long-term storage.

Culture Collections

Culture collections play a vital role in the preservation of the gene pool of industrial organisms. These collections serve as repositories for microbial strains, providing a systematic approach to their preservation and characterization.

International Culture Collections

International culture collections, such as the American Type Culture Collection (ATCC) and the Deutsche Sammlung von Mikroorganismen und Zellkulturen (DSMZ), maintain extensive repositories of microbial strains. These collections employ standardized protocols for the preservation, characterization, and distribution of microbial cultures, ensuring the availability of well-documented strains for researchers and industry professionals.

Local and Regional Culture Collections

Local and regional culture collections also contribute to the preservation of microbial diversity by maintaining collections of indigenous and economically important strains. These collections often focus on preserving organisms that are particularly relevant to local industries or ecosystems.

8.3 CHALLENGES IN PRESERVING THE GENE POOL

While various methods exist for preserving the gene pool of industrial organisms, several challenges must be addressed to ensure the effectiveness of these preservation techniques.

8.3.1 Genetic Drift and Loss of Viability

Over time, preserved microbial cultures may experience genetic drift, leading to changes in traits and loss of viability. This can be particularly problematic for strains that are preserved through repeated sub-culturing or passaging. To mitigate this risk, culture collections often implement strategies such as periodic re-isolation of strains from environmental sources or maintaining multiple stock cultures of the same strain.

8.3.2 Contamination

Contamination poses a significant threat to the preservation of microbial cultures. Contaminating organisms can outcompete the desired strain for nutrients and resources, leading to its decline or extinction. To prevent contamination, strict aseptic techniques are employed during culture handling, and cultures are regularly monitored for signs of contamination.

8.3.3 Technological Limitations

Despite advancements in preservation techniques, technological limitations may still hinder the effective preservation of certain organisms. For example, some strains may be sensitive to freezing and thawing processes, leading to reduced viability upon rehydration. Research into optimizing preservation methods for specific organisms is essential to overcoming these challenges and ensuring successful long-term storage.

8.4 FUTURE DIRECTIONS IN GENE POOL PRESERVATION

As the field of biotechnology continues to evolve, innovative approaches to preserving the gene pool of industrial organisms are emerging. These approaches aim to enhance the efficiency and effectiveness of preservation methods while ensuring the long-term availability of microbial strains.

8.4.1 Genomic and Metagenomic Techniques

Advancements in genomic and metagenomic technologies offer new opportunities for preserving microbial diversity. By sequencing the genomes of preserved strains, researchers can create comprehensive databases that facilitate the identification and characterization of valuable genetic traits. Additionally, metagenomic approaches enable the exploration of microbial communities, providing insights into novel strains with potential biotechnological applications.

8.4.2 Bioinformatics and Data Management

The integration of bioinformatics tools and data management systems is essential for effectively managing culture collections and preserving microbial diversity. These systems can streamline the documentation and characterization of microbial strains, ensuring that valuable information is readily accessible for researchers and industry professionals.

8.4.3 Public Awareness and Collaboration

Raising public awareness about the importance of preserving microbial diversity and the gene pool is crucial for fostering collaboration among researchers, industry professionals, and policymakers. Collaborative efforts between culture collections, academic institutions, and industries can facilitate the sharing of resources, knowledge, and best practices for preserving microbial strains.

8.5 CONCLUSION: THE CRITICAL ROLE OF GENE POOL PRESERVATION

The preservation of the gene pool in industrial organisms is essential for the sustainability and advancement of biotechnological applications. Various methods, including short-term and long-term preservation techniques, play a vital role in ensuring the viability and genetic stability of microbial strains. Culture collections serve as invaluable repositories for microbial biodiversity, safeguarding valuable genetic resources for future research and industrial use.

As the biotechnology landscape continues to evolve, addressing the challenges of genetic drift, contamination, and technological limitations will be crucial in preserving the gene pool. By embracing innovative approaches and fostering collaboration, the biotechnology community can ensure that the genetic diversity of industrial organisms remains a valuable asset for generations to come.

SECTION 3:

BASIC OPERATIONS IN INDUSTRIAL FERMENTATION

CHAPTER 9

FERMENTORS AND OPERATION OF FERMENTATION EQUIPMENT

Fermentation is a biological process that converts sugars into acids, gases, or alcohol using microorganisms like bacteria, yeast, or fungi. It is a crucial aspect of various industries, including food and beverage, pharmaceuticals, and biofuels. The efficiency of fermentation processes largely depends on the design and operation of fermentation equipment, commonly referred to as fermentors. These specialized vessels create controlled environments for microbial growth and product formation, significantly influencing yield and quality. This article delves into the types of fermentors, their components, and the operational principles that govern fermentation processes.

Fermentation is a type of redox metabolism carried out in the absence of oxygen. During fermentation, organic molecules (e.g., glucose) are catabolized and donate electrons to other organic molecules. In the process, ATP and organic end products (e.g., lactate) are formed.

Because oxygen is not required, it is an alternative to aerobic respiration. Over 25 % of bacteria and archaea carry out fermentation. They live in the gut, sediments, food, and other environments. Eukaryotes, including humans and other animals, also carry out fermentation.

Fermentation is important in several areas of human society. Humans have used fermentation in production of food for 13,000 years. Humans and their livestock have microbes in the gut that carry out fermentation, releasing products used by the host for energy. Fermentation is used at an industrial level to produce commodity chemicals, such as ethanol and lactate. In total, fermentation forms more than 50 metabolic end products with a wide range of uses.

9.1 HISTORY OF THE USE OF FERMENTATION

The use of fermentation, particularly for beverages, has existed since the Neolithic and has been documented dating from 7000 to 6600 BCE in Jiahu, China, 5000 BCE in India, Ayurveda mentions many Medicated Wines, 6000 BCE in Georgia, 3150 BCE in ancient Egypt, 3000 BCE in Babylon, 2000 BCE in pre-Hispanic Mexico, and 1500 BC in Sudan. Fermented foods have a religious significance in Judaism and

Christianity. The Baltic god Rugutis was worshiped as the agent of fermentation.

In 1837, Charles Cagniard de la Tour, Theodor Schwann and Friedrich Traugott Kützing independently published papers concluding, as a result of microscopic investigations, that yeast is a living organism that reproduces by budding. Schwann boiled grape juice to kill the yeast and found that no fermentation would occur until new yeast was added. However, a lot of chemists, including Antoine Lavoisier, continued to view fermentation as a simple chemical reaction and rejected the notion that living organisms could be involved. This was seen as a reversion to vitalism and was lampooned in an anonymous publication by Justus von Liebig and Friedrich Wöhler.

The turning point came when Louis Pasteur (1822–1895), during the 1850s and 1860s, repeated Schwann's experiments and showed fermentation is initiated by living organisms in a series of investigations. In 1857, Pasteur showed lactic acid fermentation is caused by living organisms. In 1860, he demonstrated how bacteria cause souring in milk, a process formerly thought to be merely a chemical change. His work in identifying the role of microorganisms in food spoilage led to the process of pasteurization.

In 1877, working to improve the French brewing industry, Pasteur published his famous paper on fermentation, "*Etudes sur la Bière*", which was translated into English in 1879 as "Studies on fermentation". He defined fermentation (incorrectly) as "Life without air", yet he correctly showed how specific types of microorganisms cause specific types of fermentations and specific end-products.

Although showing fermentation resulted from the action of living microorganisms was a breakthrough, it did not explain the basic nature of fermentation; nor did it prove it is caused by microorganisms which appear to be always present. Many scientists, including Pasteur, had unsuccessfully attempted to extract the fermentation enzyme from yeast.

Success came in 1897 when the German chemist Eduard Buechner ground up yeast, extracted a juice from them, then found to his amazement this "dead" liquid would ferment a sugar solution, forming carbon dioxide and alcohol much like living yeasts.

Buechner's results are considered to mark the birth of biochemistry. The "unorganized ferments" behaved just like the organized ones. From that time on, the term enzyme came to be applied to all ferments. It was then understood fermentation is caused by enzymes produced by microorganisms. In 1907, Buechner won the Nobel Prize in chemistry for his work.

Advances in microbiology and fermentation technology have continued steadily up until the present. For example, in the 1930s, it was discovered microorganisms could be mutated with physical and chemical treatments to be higher-yielding, faster-growing, tolerant of less oxygen, and able to use a more concentrated medium. Strain selection and hybridization developed as well, affecting most modern food fermentations.

9.1.1 Post 1930s

The field of fermentation has been critical to the production of a wide range of consumer goods, from food and drink to industrial chemicals and pharmaceuticals. Since its early beginnings in ancient civilizations, the use of fermentation has continued to evolve and expand, with new techniques and technologies driving advances in product quality, yield, and efficiency. The period from the 1930s onward saw a number of significant advancements in fermentation technology, including the development of new processes for producing high-value products like antibiotics and enzymes, the increasing importance of fermentation in the production of bulk chemicals, and a growing interest in the use of fermentation for the production of functional foods and nutraceuticals.

The 1950s and 1960s saw the development of new fermentation technologies, such as the use of immobilized cells and enzymes, which allowed for more precise control over fermentation processes and increased the production of high-value products like antibiotics and enzymes. In the 1970s and 1980s, fermentation became increasingly important in the production of bulk chemicals like ethanol, lactic acid, and citric acid.

This led to the development of new fermentation techniques and the use of genetically engineered microorganisms to improve yields and reduce production costs. In the 1990s and 2000s, there was a growing interest in the use of fermentation for the production of functional foods and nutraceuticals, which have potential health benefits beyond basic nutrition. This led to the development of new fermentation processes and the use of probiotics and other functional ingredients.

Overall, the period from 1930 onward saw significant advancements in the use of fermentation for industrial purposes, leading to the production of a wide range of fermented products that are now consumed around the world.

9.1.2 Definition

The definition of fermentation has evolved over the years. The most modern definition is catabolism where organic compounds are both the electron donor and acceptor. A common electron donor is glucose, and pyruvate is a common electron acceptor. This definition distinguishes fermentation from aerobic respiration, where oxygen is the acceptor, and types of anaerobic respiration where inorganic compound is the acceptor.

Fermentation had been defined differently in the past. In 1876, Louis Pasteur defined it as "la vie sans air" (life without air). This definition came before the discovery of anaerobic respiration. Later, it had been defined as catabolism that forms ATP through only substrate-level phosphorylation. However, several pathways of fermentation have been discovered to form ATP through an electron transport chain and ATP synthase, also.

9.1.3 Biological Role and Prevalence

Fermentation is used by organisms to generate ATP energy for metabolism. One advantage is that it requires no oxygen or other external electron acceptors, and thus it can be carried when those electron acceptors are absent. A disadvantage is that it produces relatively little ATP, yielding only between 2 to 4.5 per glucose compared to 32 for aerobic respiration.

Over 25% of bacteria and archaea carry out fermentation. This type of metabolism is most common in the phylum Bacillota, and it is least common in Actinomycetota. Their most common habitat is host-associated ones, such as the gut.

Animals, including humans, also carry out fermentation. The product of fermentation in humans is lactate, and it is formed during anaerobic exercise or in cancerous cells. No animal is known to survive on fermentation alone, even as one parasitic animal (Henneguya zschokkei) is known to survive without oxygen.

9.1.4 Substrates and Products of Fermentation

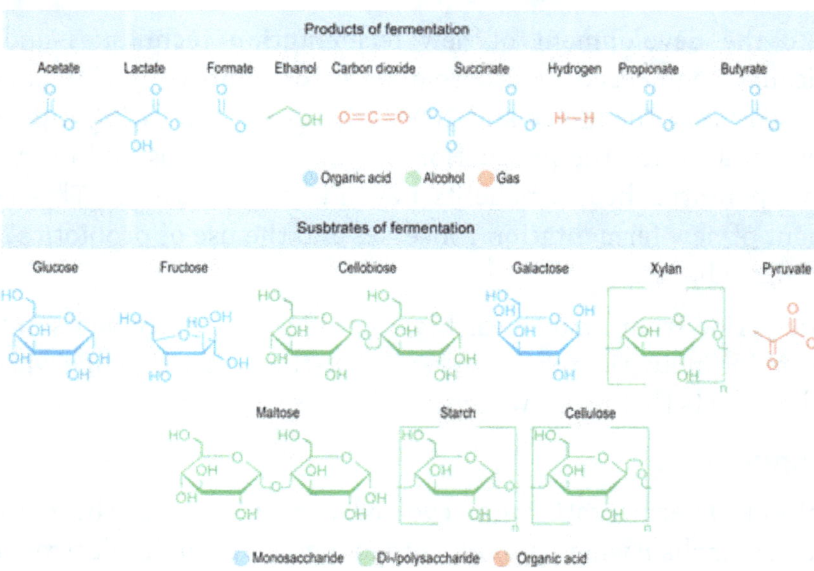

Fig 1: The Most Common Substrates and Products of Fermentation

Fermentation uses a range of substrates and forms a variety of metabolic end products. Of the 55 end products formed, the most common are acetate and lactate.

Of the 46 chemically-defined substrates that have been reported, the most common are glucose and other sugars.

9.1.5 Biochemical Overview

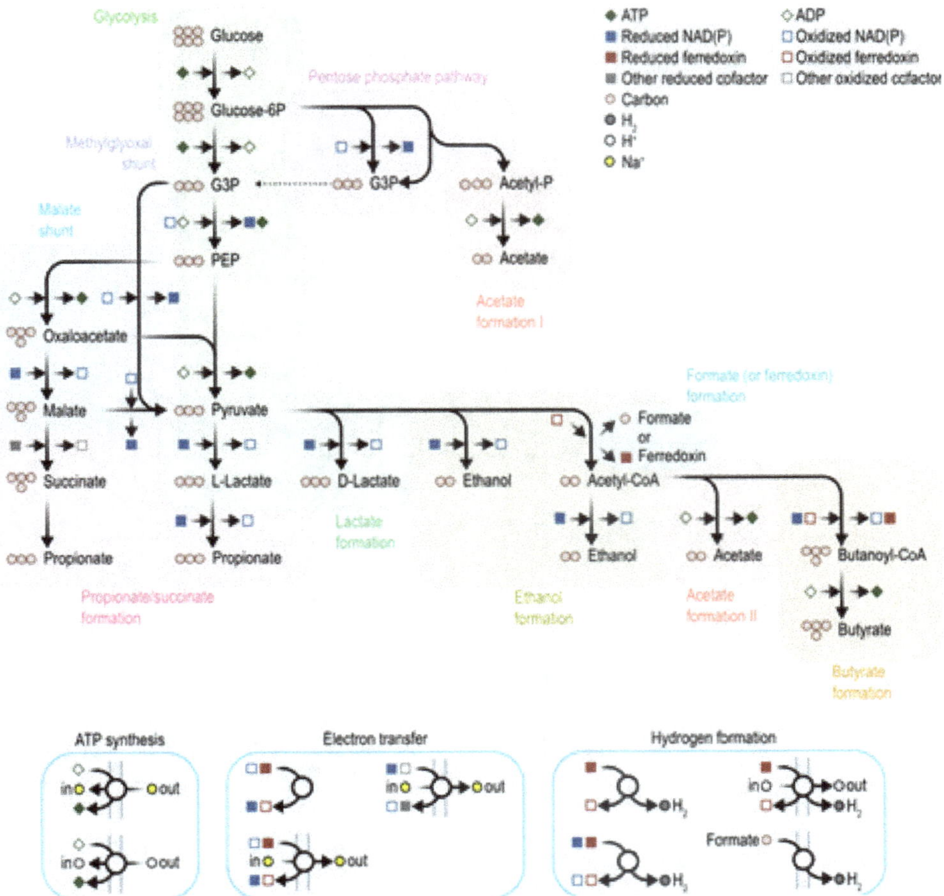

Fig 2: Overview of the Biochemical Pathways for Fermentation of Glucose

When an organic compound is fermented, it is broken down to a simpler molecule and releases electrons.

The electrons are transferred to a redox cofactor, which in turn transfers them to an organic compound. ATP is generated in the process, and it can be formed by substrate-level phosphorylation or by ATP synthase.

When glucose is fermented, it enters glycolysis or the pentose phosphate pathway and is converted to pyruvate. From pyruvate, pathways branch out to form a number of end products (e.g. lactate). At several points, electrons are released and accepted by redox cofactors (NAD and ferredoxin). At later points, these cofactors donate electrons to their final acceptor and become oxidized. ATP is also formed at several points in the pathway.

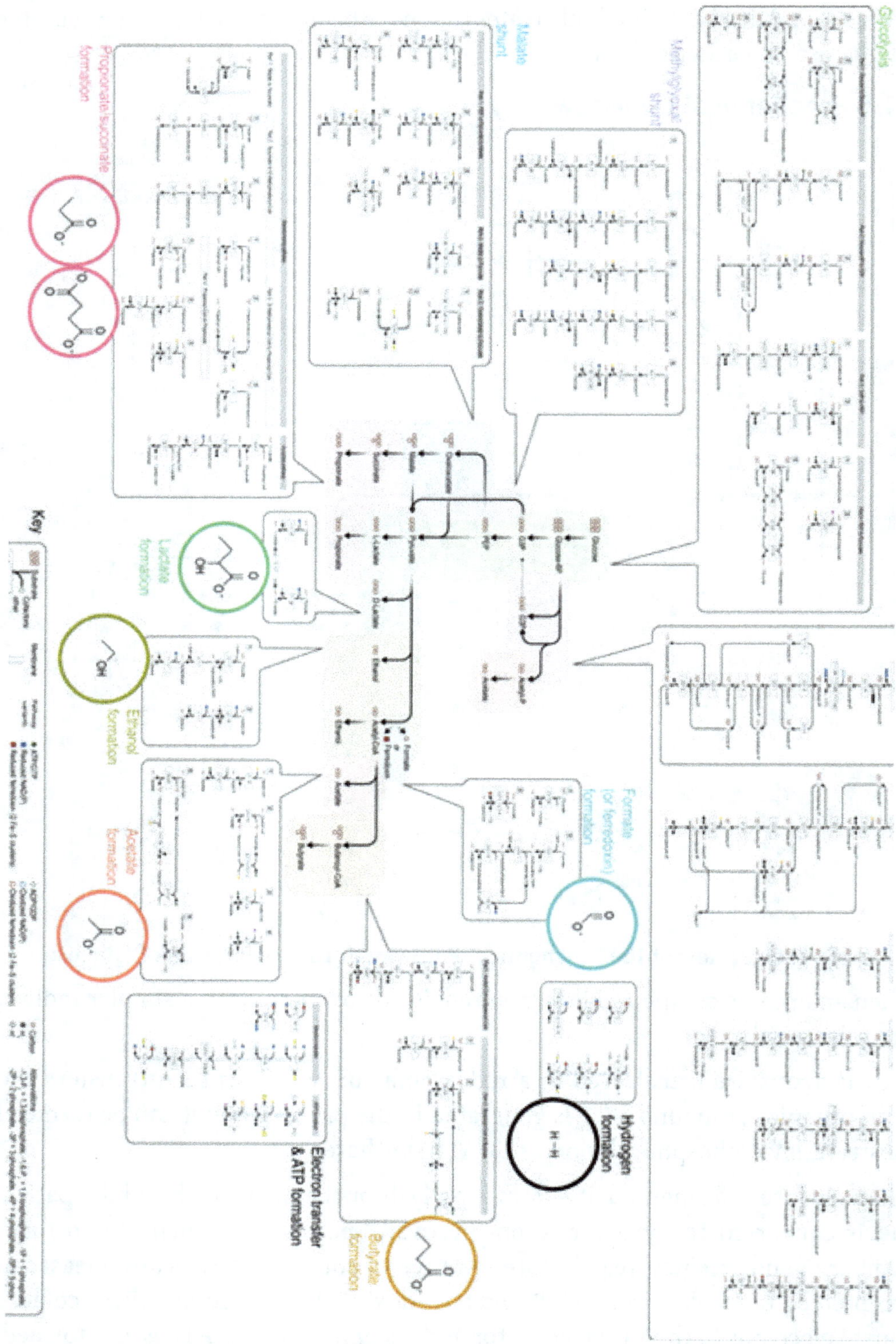

Fig 3: The Biochemical Pathways of Fermentation of Glucose in Poster Format

While fermentation is simple in overview, its details are more complex. Across organisms, fermentation of glucose involves over 120 different biochemical reactions. Further, multiple pathways can be responsible for forming the same product. For forming acetate from its immediate precursor (pyruvate or acetyl-CoA), six separate pathways have been found.

9.1.6 Biochemistry of Individual Products

Ethanol

In ethanol fermentation, one glucose molecule is converted into two ethanol molecules and two carbon dioxide (CO_2) molecules. It is used to make bread dough rise: the carbon dioxide forms bubbles, expanding the dough into a foam. The ethanol is the intoxicating agent in alcoholic beverages such as wine, beer and liquor. Fermentation of feedstocks, including sugarcane, maize, and sugar beets, produces ethanol that is added to gasoline. In some species of fish, including goldfish and carp, it provides energy when oxygen is scarce (along with lactic acid fermentation). Before fermentation, a glucose molecule breaks down into two pyruvate molecules (glycolysis). The energy from this exothermic reaction is used to bind inorganic phosphates to ADP, which converts it to ATP, and convert NAD^+ to NADH. The pyruvates break down into two acetaldehyde molecules and give off two carbon dioxide molecules as waste products. The acetaldehyde is reduced into ethanol using the energy and hydrogen from NADH, and the NADH is oxidized into NAD^+ so that the cycle may repeat. The reaction is catalyzed by the enzymes pyruvate decarboxylase and alcohol dehydrogenase.

History of Bioethanol Fermentation

The history of ethanol as a fuel spans several centuries and is marked by a series of significant milestones. Samuel Morey, an American inventor, was the first to produce ethanol by fermenting corn in 1826. However, it was not until the California Gold Rush in the 1850s that ethanol was first used as a fuel in the United States. Rudolf Diesel demonstrated his engine, which could run on vegetable oils and ethanol, in 1895, but the widespread use of petroleum-based diesel engines made ethanol less popular as a fuel. In the 1970s, the oil crisis reignited interest in ethanol, and Brazil became a leader in ethanol production and use. The United States began producing ethanol on a large scale in the 1980s and 1990s as a fuel additive to gasoline, due to government regulations. Today, ethanol continues to be explored as a sustainable and renewable fuel source, with researchers developing new technologies and biomass sources for its production.

- **1826:** Samuel Morey, an American inventor, was the first to produce ethanol by fermenting corn. However, ethanol was not widely used as a fuel until many years later.

- **1850s:** Ethanol was first used as a fuel in the United States during the California Gold Rush. Miners used ethanol as a fuel for lamps and stoves because it was cheaper than whale oil.
- **1895:** German engineer Rudolf Diesel demonstrated his engine, which was designed to run on vegetable oils, including ethanol. However, the widespread use of diesel engines fueled by petroleum made ethanol less popular as a fuel.
- **1970s:** The oil crisis of the 1970s led to renewed interest in ethanol as a fuel. Brazil became a leader in ethanol production and use, due in part to government policies that encouraged the use of biofuels.
- **1980s–1990s:** The United States began to produce ethanol on a large scale as a fuel additive to gasoline. This was due to the passage of the Clean Air Act in 1990, which required the use of oxygenates, such as ethanol, to reduce emissions.
- **2000s–Present:** There has been continued interest in ethanol as a renewable and sustainable fuel. Researchers are exploring new sources of biomass for ethanol production, such as switchgrass and algae, and developing new technologies to improve the efficiency of the fermentation process.

Lactic Acid

Homolactic fermentation (producing only lactic acid) is the simplest type of fermentation. Pyruvate from glycolysis undergoes a simple redox reaction, forming lactic acid. Overall, one molecule of glucose (or any six-carbon sugar) is converted to two molecules of lactic acid:

$$C_6H_{12}O_6 \rightarrow 2\ CH_3CHOHCOOH$$

It occurs in the muscles of animals when they need energy faster than the blood can supply oxygen. It also occurs in some kinds of bacteria (such as lactobacilli) and some fungi. It is the type of bacteria that convert lactose into lactic acid in yogurt, giving it its sour taste. These lactic acid bacteria can carry out either homolactic fermentation, where the end-product is mostly lactic acid, or *heterolactic fermentation*, where some lactate is further metabolized to ethanol and carbon dioxide (via the phosphoketolase pathway), acetate, or other metabolic products, e.g.:

$$C_6H_{12}O_6 \rightarrow CH_3CHOHCOOH + C_2H_5OH + CO_2$$

If lactose is fermented (as in yogurts and cheeses), it is first converted into glucose and galactose (both six-carbon sugars with the same atomic formula):

$$C_{12}H_{22}O_{11} + H_2O \rightarrow 2\ C_6H_{12}O_6$$

Heterolactic fermentation is in a sense intermediate between lactic acid fermentation and other types, e.g., alcoholic fermentation. Reasons to go further and convert lactic acid into something else include:

- **The Acidity of Lactic Acid Impedes Biological Processes:** This can be beneficial to the fermenting organism as it drives out competitors that are unadapted to the acidity. As a result, the food will have a longer shelf life (one reason foods are purposely fermented in the first place); however, beyond a certain point, the acidity starts affecting the organism that produces it.
- **The High Concentration of Lactic Acid:** The final product of fermentation drives the equilibrium backwards (Le Chatelier's principle), decreasing the rate at which fermentation can occur and slowing down growth.
- Ethanol, into which lactic acid can be easily converted, is volatile and will readily escape, allowing the reaction to proceed easily. CO_2 is also produced, but it is only weakly acidic and even more volatile than ethanol.
- Acetic acid (another conversion product) is acidic and not as volatile as ethanol; however, in the presence of limited oxygen, its creation from lactic acid releases additional energy. It is a lighter molecule than lactic acid, forming fewer hydrogen bonds with its surroundings (due to having fewer groups that can form such bonds), thus is more volatile and will also allow the reaction to proceed more quickly.
- If propionic acid, butyric acid, and longer monocarboxylic acids are produced, the amount of acidity produced per glucose consumed will decrease, as with ethanol, allowing faster growth.

Hydrogen Gas

Hydrogen gas is produced in many types of fermentation as a way to regenerate NAD^+ from NADH. Electrons are transferred to ferredoxin, which in turn is oxidized by hydrogenase, producing H_2.

Hydrogen gas is a substrate for methanogens and sulfate reducers, which keep the concentration of hydrogen low and favor the production of such an energy-rich compound, but hydrogen gas at a fairly high concentration can nevertheless be formed, as in flatus. For example, *Clostridium pasteurianum* ferments glucose to butyrate, acetate, carbon dioxide, and hydrogen gas: The reaction leading to acetate is:

$$C_6H_{12}O_6 + 4\ H_2O \rightarrow 2\ CH_3COO^- + 2\ HCO_3^- + 4\ H^+ + 4\ H_2$$

Glyoxylate

Glyoxylate fermentation is a type of fermentation used by microbes that are able to utilize glyoxylate as a nitrogen source.

Other

Other types of fermentation include:

- Mixed acid fermentation,
- Butanediol fermentation,
- Butyrate fermentation,
- Caproate fermentation, and
- Acetone–butanol–ethanol fermentation.

In the Broader Sense

In food and industrial contexts, any chemical modification performed by a living being in a controlled container can be termed "fermentation".

Alternative Protein

Fermentation can be used to make alternative protein sources. It is commonly used to modify existing protein foods, including plant-based ones such as soy, into more flavorful forms such as tempeh and fermented tofu. More modern "fermentation" makes recombinant protein to help produce meat analogue, milk substitute, cheese analogues, and egg substitutes. Some examples are:

- Recombinant myoglobin for faux meat (Motif Foodworks).
- Recombinant leghemoglobin for faux meat (Impossible Foods).
- Recombinant whey for dairy replacement (Perfect Day).
- Recombinant egg white (EVERY).

Heme proteins such as myoglobin and hemoglobin give meat its characteristic texture, flavor, color, and aroma. The myoglobin and leghemoglobin ingredients can be used to replicate this property, despite them coming from a vat instead of meat.

Enzymes

Industrial fermentation can be used for enzyme production, where proteins with catalytic activity are produced and secreted by microorganisms. The development of fermentation processes, microbial strain engineering and recombinant gene technologies has enabled the commercialization of a wide range of enzymes. Enzymes are used in all kinds of industrial segments, such as food (lactose removal, cheese flavor), beverage (juice treatment), baking (bread softness, dough conditioning), animal feed, detergents (protein, starch and lipid stain removal), textile, personal care and pulp and paper industries.

9.1.7 Modes of Industrial Operation

Most industrial fermentation uses batch or fed-batch procedures, although continuous fermentation can be more economical if various challenges, particularly the difficulty of maintaining sterility, can be met.

Batch

In a batch process, all the ingredients are combined and the reactions proceed without any further input. Batch fermentation has been used for millennia to make bread and alcoholic beverages, and it is still a common method, especially when the process is not well understood. However, it can be expensive because the fermentor must be sterilized using high pressure steam between batches. Strictly speaking, there is often addition of small quantities of chemicals to control the pH or suppress foaming.

Batch fermentation goes through a series of phases. There is a lag phase in which cells adjust to their environment; then a phase in which exponential growth occurs. Once many of the nutrients have been consumed, the growth slows and becomes non-exponential, but production of *secondary metabolites* (including commercially important antibiotics and enzymes) accelerates. This continues through a stationary phase after most of the nutrients have been consumed, and then the cells die.

Fed-Batch

Fed-batch fermentation is a variation of batch fermentation where some of the ingredients are added during the fermentation. This allows greater control over the stages of the process. In particular, production of secondary metabolites can be increased by adding a limited quantity of nutrients during the non-exponential growth phase. Fed-batch operations are often sandwiched between batch operations.

Open

The high cost of sterilizing the fermentor between batches can be avoided using various open fermentation approaches that are able to resist contamination. One is to use a naturally evolved mixed culture. This is particularly favored in wastewater treatment, since mixed populations can adapt to a wide variety of wastes. Thermophilic bacteria can produce lactic acid at temperatures of around 50 °Celsius, sufficient to discourage microbial contamination; and ethanol has been produced at a temperature of 70 °C. This is just below its boiling point (78 °C), making it easy to extract. Halophilic bacteria can produce bioplastics in hypersaline conditions. Solid-state fermentation adds a small amount of water to a solid substrate; it is widely used in the food industry to produce flavors, enzymes and organic acids.

Continuous

In continuous fermentation, substrates are added and final products removed continuously. There are three varieties: chemostats, which hold nutrient levels constant; turbidostats, which keep cell mass constant; and plug flow reactors in which the culture medium flows steadily through a tube while the cells are recycled from the outlet to the inlet.

If the process works well, there is a steady flow of feed and effluent and the costs of repeatedly setting up a batch are avoided. Also, it can prolong the exponential growth phase and avoid byproducts that inhibit the reactions by continuously removing them.

However, it is difficult to maintain a steady state and avoid contamination, and the design tends to be complex. Typically the fermentor must run for over 500 hours to be more economical than batch processors.

9.2 TYPES OF FERMENTORS

Fermentors come in various designs tailored to specific applications, microbial types, and desired products. The primary classifications of fermentors include batch fermentors, continuous fermentors, and fed-batch fermentors.

9.2.1 Batch Fermentors

Batch fermentors are the simplest form of fermentation equipment. In this setup, all ingredients, including the culture, nutrients, and substrates, are added to the fermentor at the beginning of the process. The fermentation occurs over a defined period, after which the product is harvested, and the vessel is cleaned for the next batch. Batch fermentors are widely used for producing products like beer, wine, and certain pharmaceuticals. They allow for easier control of variables and are well-suited for smaller production scales. However, batch fermentation may not be as efficient for large-scale production due to downtime during cleaning and setup.

9.2.2 Continuous Fermentors

In continuous fermentors, substrates are continuously fed into the system while products and byproducts are simultaneously removed. This setup allows for the constant operation of the fermentation process, leading to higher overall productivity and reduced downtime. Continuous fermentors are particularly advantageous for large-scale operations, such as the production of ethanol or lactic acid. However, they require careful control of growth conditions, as any fluctuation can lead to contamination or reduced yield.

9.2.3 Fed-Batch Fermentors

Fed-batch fermentors represent a hybrid approach, combining elements of batch and continuous systems. In this setup, the initial substrate is added to the fermentor, but additional nutrients or substrates are fed into the system during the fermentation process. This method allows for better control over the growth environment and product formation. Fed-batch systems are often used for producing high-value products, such as antibiotics and recombinant proteins, as they can optimize microbial growth conditions without the risk of washout that can occur in continuous systems.

9.3 KEY COMPONENTS OF FERMENTORS

Understanding the key components of fermentors is essential for grasping their operation and functionality. Fermentors typically consist of several critical elements that work together to create optimal fermentation conditions.

9.3.1 Vessel

The fermentor vessel is the primary container where fermentation occurs. It is usually constructed from stainless steel or glass, allowing for easy cleaning and sterilization. The size of the vessel can vary widely, from small laboratory fermentors to large industrial tanks capable of holding thousands of liters. The design of the vessel often includes features such as curved bottoms to facilitate mixing and drainage.

9.3.2 Agitation System

Agitation is crucial in fermentation processes to ensure uniform mixing of the culture, nutrients, and gases. An effective agitation system enhances mass transfer, oxygen transfer, and heat transfer, promoting optimal microbial growth. Mechanical agitators, such as impellers or stirrers, are commonly used to provide adequate mixing. The speed of agitation must be carefully controlled, as excessive agitation can damage delicate microbial cells.

9.3.3 Aeration System

For aerobic fermentation processes, an aeration system is essential to supply the necessary oxygen for microbial growth. This system typically includes air compressors or spargers that introduce air into the fermentor. The design of the aeration system must ensure that oxygen is distributed evenly throughout the vessel, preventing localized anaerobic conditions that could inhibit growth or produce unwanted byproducts.

9.3.4 Temperature Control

Temperature is a critical factor in fermentation, influencing microbial metabolism and product formation. Fermentors are equipped with temperature control systems, including jackets or coils that circulate heating or cooling fluids. Automated temperature monitoring and control systems allow operators to maintain the desired temperature range throughout the fermentation process, ensuring optimal growth conditions.

9.3.5 pH Control

Maintaining the appropriate pH is vital for successful fermentation, as pH fluctuations can affect microbial activity and product yield. Fermentors are equipped with pH sensors and controllers that monitor and adjust pH levels by adding acids or bases as needed. This automated control ensures that the fermentation environment remains conducive to optimal microbial activity.

9.3.6 Sampling and Harvesting Ports

Fermentors typically include ports for sampling and harvesting the product. Sampling ports allow operators to periodically test the fermentation broth for parameters such as cell density, pH, and product concentration. Harvesting ports enable the efficient removal of the final product or biomass without disrupting the fermentation process.

9.4 OPERATION OF FERMENTATION EQUIPMENT

The operation of fermentation equipment involves several critical steps, each contributing to the overall success of the fermentation process.

9.4.1 Preparation and Sterilization

Before fermentation can begin, the fermentor and its components must be thoroughly cleaned and sterilized to prevent contamination by unwanted microorganisms. This is typically achieved using steam sterilization, where the fermentor is filled with steam to kill any existing microorganisms. The media, nutrients, and inoculum (the microbial culture) are also sterilized to ensure that only the desired organisms are introduced into the fermentor.

9.4.2 Inoculation

Once sterilization is complete, the inoculum is introduced into the fermentor along with the sterile growth medium. The inoculum is usually prepared in advance to ensure that it is healthy and viable. The inoculation step is critical, as the initial concentration of microbial cells can significantly influence the fermentation's success.

9.4.3 Monitoring and Control

During fermentation, continuous monitoring of critical parameters is essential for maintaining optimal conditions.

Automated systems and sensors track temperature, pH, dissolved oxygen, and agitation speed. This data is used to make real-time adjustments to the fermentation process, ensuring that the environment remains conducive to microbial growth and product formation.

9.4.4 Product Formation

As fermentation progresses, microorganisms metabolize the available substrates, leading to the production of desired metabolites, such as alcohol, acids, or proteins.

The rate of product formation is influenced by factors such as substrate concentration, microbial strain, and environmental conditions. Operators must carefully monitor these factors to optimize yields.

9.4.5 Harvesting

At the end of the fermentation process, the product is harvested through the designated ports. This may involve separating biomass from the liquid product, which can be accomplished using centrifugation or filtration techniques. The harvested product may then undergo further purification or processing, depending on its intended use.

9.5 CHALLENGES IN FERMENTATION PROCESSES

While fermentation is a powerful biotechnological process, it is not without its challenges. Several factors can impact the efficiency and effectiveness of fermentation equipment and processes.

9.5.1 Contamination

Contamination by unwanted microorganisms can lead to reduced yields and compromised product quality. Contamination can occur at various stages, including during inoculation, media preparation, and through equipment handling. Strict aseptic techniques and regular monitoring are necessary to mitigate contamination risks.

9.5.2 Process Variability

Fermentation processes can be influenced by numerous factors, including raw material quality, environmental conditions, and microbial health. Variability in these factors can lead to inconsistent product yields and quality. Implementing robust process control strategies and monitoring techniques can help minimize variability and improve reproducibility.

9.5.3 Scale-Up Challenges

Scaling up fermentation processes from laboratory to industrial scales presents several challenges. Factors such as heat transfer, mixing, and mass transfer can behave differently at larger scales, affecting microbial growth and product formation. Careful consideration of these factors during the design phase is crucial for successful scale-up.

9.6 FUTURE DIRECTIONS IN FERMENTATION TECHNOLOGY

As biotechnological applications continue to expand, so too do the advancements in fermentation technology. Emerging trends and innovations are shaping the future of fermentation processes and equipment.

9.6.1 Automation and Control Systems

The increasing complexity of fermentation processes has led to the development of advanced automation and control systems. These systems utilize data analytics, artificial intelligence, and machine learning to optimize fermentation parameters in real-time, enhancing productivity and reducing human error.

9.6.2 Bioreactor Design Innovations

New bioreactor designs are being developed to improve the efficiency of fermentation processes. Innovations such as single-use bioreactors, which minimize cleaning and sterilization requirements, and multi-phase bioreactors, which facilitate the simultaneous growth of different microorganisms, are gaining traction in the industry.

9.6.3 Integration of Biotechnology and Fermentation

The integration of biotechnology with fermentation processes is opening new avenues for product development. Techniques such as metabolic engineering and synthetic biology are being applied to enhance microbial strains, enabling the production of novel compounds and improving existing fermentation processes.

9.7 CONCLUSION

Fermentors and the operation of fermentation equipment are fundamental to the success of fermentation processes in various industries. Understanding the different types of fermentors, their components, and operational principles is essential for optimizing microbial growth and product formation. As the field of fermentation technology continues to evolve, advancements in automation, bioreactor design, and biotechnology integration will play a critical role in shaping the future of fermentation processes. By addressing the challenges associated with fermentation and embracing innovative solutions, the industry can enhance productivity, product quality, and sustainability in biotechnological applications.

CHAPTER 10

DOWNSTREAM PROCESSING: EXTRACTION OF FERMENTATION PRODUCTS

Downstream processing refers to the recovery and the purification of biosynthetic products, particularly pharmaceuticals, from natural sources such as animal tissue, plant tissue or fermentation broth, including the recycling of salvageable components as well as the proper treatment and disposal of waste. It is an essential step in the manufacture of pharmaceuticals such as antibiotics, hormones (e.g. insulin and human growth hormone), antibodies (e.g. infliximab and abciximab) and vaccines; antibodies and enzymes used in diagnostics; industrial enzymes; and natural fragrance and flavor compounds. Downstream processing is usually considered a specialized field in biochemical engineering, which is itself a specialization within chemical engineering. Many of the key technologies were developed by chemists and biologists for laboratory-scale separation of biological and synthetic products, whilst the role of biochemical and chemical engineers is to develop the technologies towards larger production capacities.

Downstream processing and analytical bioseparation both refer to the separation or purification of biological products, but at different scales of operation and for different purposes. Downstream processing implies manufacture of a purified product fit for a specific use, generally in marketable quantities, while analytical bioseparation refers to purification for the sole purpose of measuring a component or components of a mixture, and may deal with sample sizes as small as a single cell.

10.1 STAGES

A widely recognized heuristic for categorizing downstream processing operations divides them into four groups which are applied in order to bring a product from its natural state as a component of a tissue, cell or fermentation broth through progressive improvements in purity and concentration.

Removal of insolubles is the first step and involves the capture of the product as a solute in a particulate-free liquid, for example the separation of cells, cell debris or other particulate matter from fermentation broth containing an antibiotic.

Typical operations to achieve this are filtration, centrifugation, sedimentation, precipitation, flocculation, electro-precipitation, and gravity settling. Additional operations such as grinding, homogenization, or leaching, required to recover products from solid sources such as plant and animal tissues, are usually included in this group.

Product isolation is the removal of those components whose properties vary considerably from that of the desired product. For most products, water is the chief impurity and isolation steps are designed to remove most of it, reducing the volume of material to be handled and concentrating the product. Solvent extraction, adsorption, ultrafiltration, and precipitation are some of the unit operations involved.

Product purification is done to separate those contaminants that resemble the product very closely in physical and chemical properties. Consequently, steps in this stage are expensive to carry out and require sensitive and sophisticated equipment. This stage contributes a significant fraction of the entire downstream processing expenditure. Examples of operations include affinity, size exclusion, reversed phase chromatography, ion-exchange chromatography, crystallization and fractional precipitation.

Product polishing describes the final processing steps which end with packaging of the product in a form that is stable, easily transportable and convenient. Crystallization, desiccation, lyophilization and spray drying are typical unit operations. Depending on the product and its intended use, polishing may also include operations to sterilize the product and remove or deactivate trace contaminants which might compromise product safety. Such operations might include the removal of viruses or depyrogenation. A few product recovery methods may be considered to combine two or more stages. For example, expanded bed adsorption (Vennapusa *et al.* 2008) accomplishes removal of insolubles and product isolation in a single step. Affinity chromatography often isolates and purifies in a single step.

10.2 IMPORTANCE OF DOWNSTREAM PROCESSING

Downstream processing is vital to the overall success and profitability of fermentation-based industries. It often accounts for a significant portion of the production costs, especially in biopharmaceuticals, where purity and quality are critical. The main objectives of downstream processing are:

- Efficient recovery of the desired product.
- Purification to meet the required quality and safety standards.
- Concentration of the product for storage or transport.
- Reduction of contaminants such as waste and byproducts.

The effectiveness of downstream processing can greatly influence product yield, production costs, and the overall scalability of a fermentation process.

10.3 STAGES OF DOWNSTREAM PROCESSING

The downstream processing of fermentation products can be divided into several key stages. Each stage serves a specific purpose in the separation, concentration, and purification of the product from the fermentation broth.

10.3.1 Cell Separation

The first step in downstream processing is the removal of microbial cells and other solids from the fermentation broth. This step is necessary to obtain a clear solution that contains the target product. Various methods are used for cell separation, depending on the size and nature of the microorganisms, as well as the type of fermentation product.

- **Centrifugation:** This method uses high-speed rotation to create a centrifugal force, causing solid particles (such as microbial cells) to sediment at the bottom of a centrifuge tube or vessel. Centrifugation is highly effective for separating large cells or particles but can be energy-intensive and costly on a large scale.
- **Filtration:** Filtration is another commonly used method for cell separation, where the fermentation broth is passed through a membrane or filter that retains the solid particles while allowing the liquid (containing the product) to pass through. This method is often employed for smaller-scale operations or for products that are sensitive to shear forces.
- **Flocculation:** In some cases, chemical agents called flocculants are added to the fermentation broth to aggregate cells into larger clumps, which can then be more easily separated by filtration or sedimentation.

10.3.2 Product Isolation

Once the cells have been removed, the next stage involves isolating the desired fermentation product from the remaining liquid. This is often one of the most challenging aspects of downstream processing, as fermentation broths typically contain a complex mixture of proteins, metabolites, and other molecules. Several techniques are used for product isolation:

- **Solvent Extraction:** In this process, an organic solvent is added to the broth to selectively dissolve the target product. The solvent phase containing the product is then separated from the aqueous phase. Solvent extraction is particularly useful for the recovery of hydrophobic compounds, such as certain antibiotics and biofuels. However, the choice of solvent must be carefully considered to avoid denaturing the product or introducing toxicity.

- **Precipitation**: Precipitation involves the addition of chemicals (such as salts or organic solvents) to the broth, which causes the product to precipitate out of solution. The precipitated product can then be recovered by centrifugation or filtration. This method is commonly used for protein-based products, as proteins can be selectively precipitated using ammonium sulfate or ethanol.
- **Membrane Separation**: Membrane separation techniques, such as ultrafiltration and dialysis, use semi-permeable membranes to concentrate and purify products based on their molecular size. Ultrafiltration is particularly useful for concentrating proteins, enzymes, and other large molecules.

10.3.3 Product Purification

After the product has been isolated, further purification is often necessary to remove contaminants such as impurities, byproducts, and residual solvents. The level of purification required depends on the intended use of the product—pharmaceuticals, for example, require extremely high purity levels. Several advanced purification techniques are used in downstream processing:

- **Chromatography**: Chromatography is one of the most powerful and versatile purification techniques in downstream processing. It separates components based on their interactions with a stationary phase and a mobile phase. Different types of chromatography, such as ion-exchange chromatography, affinity chromatography, and size-exclusion chromatography, are used to achieve high levels of product purity. This method is particularly effective for purifying proteins, peptides, and other complex molecules.
- **Crystallization**: For certain products, crystallization can be an effective method of purification. By carefully controlling the temperature and solvent conditions, the product can be induced to form crystals, which can then be separated from the liquid phase. Crystallization is commonly used in the purification of organic acids, antibiotics, and certain small-molecule drugs.
- **Dialysis:** Dialysis is a membrane-based technique used to remove small molecules, salts, and other impurities from larger molecules, such as proteins. In dialysis, the product is placed in a semi-permeable membrane bag, which is then submerged in a buffer solution. The small molecules diffuse out of the bag, leaving the purified product inside.

10.3.4 Product Concentration

Once the product has been purified, it often needs to be concentrated for further use, storage, or transportation. Concentration can be achieved by removing water or other solvents from the product solution. Common methods for product concentration include:

- **Evaporation:** Evaporation involves heating the product solution to remove water or solvents, thereby increasing the concentration of the target product. Care must be taken to avoid excessive heat, which can denature or degrade heat-sensitive products, such as proteins and enzymes.
- **Freeze-Drying (Lyophilization):** Freeze-drying is a gentle method of concentration used for heat-sensitive products. In this process, the product is first frozen, and then the ice is removed by sublimation under vacuum conditions. Freeze-drying is particularly useful for pharmaceuticals and other products that require long-term stability and preservation.
- **Ultrafiltration:** Ultrafiltration can also be used for product concentration, especially for large molecules like proteins and enzymes. By applying pressure to the product solution, water and small impurities are forced through a semi-permeable membrane, leaving behind a concentrated product.

10.4 PRODUCT FORMULATION AND FINAL PROCESSING

Once the product has been concentrated, it may undergo further processing to prepare it for its final use. This stage, known as product formulation, involves adding stabilizers, preservatives, or other excipients to ensure the product remains stable during storage and transport.

The formulation must be carefully designed to meet the specific requirements of the product, such as its shelf life, solubility, and bioavailability. For pharmaceuticals, this step may also involve converting the product into its final dosage form, such as tablets, capsules, or injectable solutions. In the case of biofuels or industrial enzymes, the product may be blended with other substances or packaged for large-scale distribution.

10.5 CHALLENGES IN DOWNSTREAM PROCESSING

Downstream processing presents several challenges, especially in large-scale industrial applications. Some of the most common challenges include:

- **High Costs:** Downstream processing can be expensive, particularly for products that require high levels of purity, such as pharmaceuticals. The cost of equipment, solvents, and energy consumption can significantly impact the overall economics of the production process.

- **Product Loss:** During each stage of downstream processing, some loss of the target product is inevitable. Minimizing product loss while maintaining high purity levels is a constant challenge in the industry.
- **Contamination Risks:** Contamination by unwanted microorganisms or chemicals can occur at any stage of downstream processing. Maintaining strict aseptic conditions and monitoring for contamination is essential to ensure product quality.
- **Scalability:** Scaling up downstream processes from laboratory to industrial scale can be difficult, as factors such as mixing, mass transfer, and heat transfer may behave differently at larger scales. Ensuring that laboratory processes can be efficiently translated to industrial production is a critical challenge in downstream processing.

10.6 CONCLUSION

Downstream processing is a vital aspect of industrial microbiology and biotechnology, responsible for recovering, purifying, and concentrating products from fermentation processes. From the initial separation of microbial cells to the final product formulation, each stage of downstream processing requires careful consideration and optimization. Advances in technology, such as improved membrane filtration techniques, automation, and novel chromatography methods, are continuously enhancing the efficiency and cost-effectiveness of downstream processing. However, challenges such as high costs, product loss, and contamination risks remain critical areas of focus for further innovation and improvement in the field.

CHAPTER 11

SIGNIFICANCE AND PROCESSES OF STERILITY IN INDUSTRIAL MICROBIOLOGY

Industrial microbiology plays a critical role in various sectors such as pharmaceuticals, food and beverage, biotechnology, and agriculture. In these industries, sterility is essential to ensure the safety and efficacy of microbial processes and products. The importance of sterility lies in preventing contamination by unwanted microorganisms, which can compromise the quality, potency, and shelf life of products, while also posing health risks. This essay explores the significance of sterility in industrial microbiology and delves into the processes used to achieve and maintain sterile conditions.

11.1 STERILIZATION

Sterilization refers to any process that removes, kills, or deactivates all forms of life (particularly microorganisms such as fungi, bacteria, spores, and unicellular eukaryotic organisms) and other biological agents (such as prions or viruses) present in or on a specific surface, object, or fluid. Sterilization can be achieved through various means, including heat, chemicals, irradiation, high pressure, and filtration. Sterilization is distinct from disinfection, sanitization, and pasteurization, in that those methods reduce rather than eliminate all forms of life and biological agents present. After sterilization, an object is referred to as being sterile or aseptic.

11.1.1 Applications

Foods

One of the first steps toward modernized sterilization was made by Nicolas Appert, who discovered that application of heat over a suitable period slowed the decay of foods and various liquids, preserving them for safe consumption for a longer time than was typical.

Canning of foods is an extension of the same principle and has helped to reduce food borne illness ("food poisoning"). Other methods of sterilizing foods include ultra-high temperature processing (which uses a shorter duration of heating), food irradiation and high pressure (pascalization).

In the context of food, sterility typically refers to *commercial sterility*, "the absence of microorganisms capable of growing in the food at normal non-refrigerated conditions at which the food is likely to be held during distribution and storage" according to the *Codex Allimentarius*.

Medicine and Surgery

In general, surgical instruments and medications that enter an already aseptic part of the body (such as the bloodstream, or penetrating the skin) must be sterile. Examples of such instruments include scalpels, hypodermic needles, and artificial pacemakers. This is also essential in the manufacture of parenteral pharmaceuticals.

Preparation of injectable medications and intravenous solutions for fluid replacement therapy requires not only sterility but also well-designed containers to prevent entry of adventitious agents after initial product sterilization.

Most medical and surgical devices used in healthcare facilities are made of materials that are able to go under steam sterilization. However, since 1950, there has been an increase in medical devices and instruments made of materials (e.g., plastics) that require low-temperature sterilization.

Ethylene oxide gas has been used since the 1950s for heat- and moisture-sensitive medical devices. Within the past 15 years, a number of new, low-temperature sterilization systems (e.g., vaporized hydrogen peroxide, peracetic acid immersion, ozone) have been developed and are being used to sterilize medical devices.

Spacecraft

There are strict international rules to protect the contamination of Solar System bodies from biological material from Earth. Standards vary depending on both the type of mission and its destination; the more likely a planet is considered to be habitable, the stricter the requirements are.

Many components of instruments used on spacecraft cannot withstand very high temperatures, so techniques not requiring excessive temperatures are used as tolerated, including heating to at least 120 °C (248 °F), chemical sterilization, oxidization, ultraviolet, and irradiation.

11.1.2 Quantification

The aim of sterilization is the reduction of initially present microorganisms or other potential pathogens.

The degree of sterilization is commonly expressed by multiples of the decimal reduction time, or D-value, denoting the time needed to reduce the initial number N_0 to one tenth (10^{-1}) of its original value. Then the number of microorganisms N after sterilization time t is given by

$$\frac{N}{N_0} = 10^{\left(-\frac{t}{D}\right)}$$

The D-value is a function of sterilization conditions and varies with the type of microorganism, temperature, water activity, pH etc.. For steam sterilization typically the temperature, in degrees Celsius, is given as an index.

Theoretically, the likelihood of the survival of an individual microorganism is never zero. To compensate for this, the overkill method is often used. Using the overkill method, sterilization is performed by sterilizing for longer than is required to kill the bioburden present on or in the item being sterilized. This provides a sterility assurance level (SAL) equal to the probability of a non-sterile unit.

For high-risk applications, such as medical devices and injections, a sterility assurance level of at least 10^{-6} is required by the United States Food and Drug Administration (FDA).

11.1.3 Heat

Steam

Steam sterilization, also known as moist heat sterilization, uses heated saturated steam under pressure to inactivate or kill microorganisms via denaturation of macromolecules, primarily proteins. This method is a faster process than dry heat sterilization. Steam sterilization is performed using an autoclave, sometimes called a converter or steam sterilizer. Steam sterilization cycles can be categorized as either pre-vacuum or gravity displacement.

Typical steam sterilization cycles are between 3 and 30 minutes at 121–134 °C (250–273 °F) at 100 kPa (15 psi), but adjustments may be made depending on the bioburden of the article being sterilized, its resistance (D-value) to steam sterilization, the article's heat tolerance, and the required sterility assurance level. Following the completion of a cycle, liquids in a pressurized autoclave must be cooled slowly to avoid boiling over when the pressure is released. This may be achieved by gradually depressurizing the sterilization chamber and allowing liquids to evaporate under a negative pressure, while cooling the contents.

Proper autoclave treatment will inactivate all resistant bacterial spores in addition to fungi, bacteria, and viruses, but is not expected to eliminate all prions, which vary in their resistance. For prion elimination, various recommendations state 121–132 °C (250–270 °F) for 60 minutes or 134 °C (273 °F) for at least 18 minutes. The 263K scrapie prion is inactivated relatively quickly by such sterilization procedures; however, other strains of scrapie, and strains of Creutzfeldt-Jakob disease (CKD) and bovine spongiform encephalopathy (BSE) are more resistant. Using mice as test animals, one experiment showed that heating BSE positive brain tissue at 134–138 °C (273–280 °F) for 18 minutes resulted in only a 2.5 log decrease in prion infectivity.

Most autoclaves have meters and charts that record or display information, particularly temperature and pressure as a function of time. The information is checked to ensure that the conditions required for sterilization have been met. Indicator tape is often placed on the packages of products prior to autoclaving, and some packaging incorporates indicators. The indicator changes color when exposed to steam, providing a visual confirmation.

Biological indicators can also be used to independently confirm autoclave performance. Simple biological indicator devices are commercially available, based on microbial spores. Most contain spores of the heat-resistant microbe *Geobacillus stearothermophilus* (formerly *Bacillus stearothermophilus*), which is extremely resistant to steam sterilization. Biological indicators may take the form of glass vials of spores and liquid media, or as spores on strips of paper inside glassine envelopes. These indicators are placed in locations where it is difficult for steam to reach to verify that steam is penetrating there.

For autoclaving, cleaning is critical. Extraneous biological matter or grime may shield organisms from steam penetration. Proper cleaning can be achieved through physical scrubbing, sonication, ultrasound, or pulsed air.

Pressure cooking and canning is analogous to autoclaving, and when performed correctly renders food sterile.

To sterilize waste materials that are chiefly composed of liquid, a purpose-built effluent decontamination system can be utilized. These devices can function using a variety of sterilants, although using heat via steam is most common.

Dry

Dry heat was the first method of sterilization and is a longer process than moist heat sterilization. The destruction of microorganisms through the use of dry heat is a gradual phenomenon. With longer exposure to lethal temperatures, the number of killed microorganisms increases. Forced ventilation of hot air can be used to increase the rate at which heat is transferred to an organism and reduce the temperature and amount of time needed to achieve sterility. At higher temperatures, shorter exposure times are required to kill organisms. This can reduce heat-induced damage to food products. The standard setting for a hot air oven is at least two hours at 160 °C (320 °F). A rapid method heats air to 463.15 K (190.00 °C; 374.00 °F) for 6 minutes for unwrapped objects and 12 minutes for wrapped objects. Dry heat has the advantage that it can be used on powders and other heat-stable items that are adversely affected by steam (e.g. it does not cause rusting of steel objects).

Flaming

Flaming is done to inoculation loops and straight-wires in microbiology labs for streaking. Leaving the loop in the flame of a Bunsen burner or alcohol burner until it glows red ensures that any infectious agent is inactivated. This is commonly used for small metal or glass objects, but not for large objects. However, during the initial heating, infectious material may be sprayed from the wire surface before it is killed, contaminating nearby surfaces and objects. Therefore, special heaters have been developed that surround the inoculating loop with a heated cage, ensuring that such sprayed material does not further contaminate the area. Another problem is that gas flames may leave carbon or other residues on the object if the object is not heated enough. A variation on flaming is to dip the object in a 70% or more concentrated solution of ethanol, then briefly touch the object to a Bunsen burner flame. The ethanol will ignite and burn off rapidly, leaving less residue than a gas flame

Incineration

Incineration is a waste treatment process that involves the combustion of organic substances contained in waste materials. This method also burns any organism to ash. It is used to sterilize medical and other biohazardous waste before it is discarded with non-hazardous waste. Bacteria incinerators are mini furnaces that incinerate and kill off any microorganisms that may be on an inoculating loop or wire.

Tyndallization

Named after John Tyndall, tyndallization is an obsolete and lengthy process designed to reduce the level of activity of sporulating microbes that are left by a simple boiling water method. The process involves boiling for a period (typically 20 minutes) at atmospheric pressure, cooling, incubating for a day, and then repeating the process a total of three to four times. The incubation periods are to allow heat-resistant spores surviving the previous boiling period to germinate to form the heat-sensitive vegetative (growing) stage, which can be killed by the next boiling step. This is effective because many spores are stimulated to grow by the heat shock. The procedure only works for media that can support bacterial growth, and will not sterilize non-nutritive substrates like water. Tyndallization is also ineffective against prions.

Glass Bead Sterilizers

Glass bead sterilizers work by heating glass beads to 250 °C (482 °F). Instruments are then quickly doused in these glass beads, which heat the object while physically scraping contaminants off their surface. Glass bead sterilizers were once a common sterilization method employed in dental offices as well as biological laboratories, but are not approved by the U.S. Food and Drug Administration (FDA) and Centers for Disease Control and Prevention (CDC) to be used as a sterilizers since 1997. They are still popular in European and Israeli dental practices, although there are no current evidence-based guidelines for using this sterilizer.

11.1.4 Chemical Sterilization

Chemicals are also used for sterilization. Heating provides a reliable way to rid objects of all transmissible agents, but it is not always appropriate if it will damage heat-sensitive materials such as biological materials, fiber optics, electronics, and many plastics. In these situations chemicals, either in a gaseous or liquid form, can be used as sterilants. While the use of gas and liquid chemical sterilants avoids the problem of heat damage, users must ensure that the article to be sterilized is chemically compatible with the sterilant being used and that the sterilant is able to reach all surfaces that must be sterilized (typically cannot penetrate packaging). In addition, the use of chemical sterilants poses new challenges for workplace safety, as the properties that make chemicals effective sterilants usually make them harmful to humans. The procedure for removing sterilant residue from the sterilized materials varies depending on the chemical and process that is used.

Ethylene Oxide

Ethylene oxide (EO, EtO) gas treatment is one of the common methods used to sterilize, pasteurize, or disinfect items because of its wide range of material compatibility. It is also used to process items that are sensitive to processing with other methods, such as radiation (gamma, electron beam, X-ray), heat (moist or dry), or other chemicals. Ethylene oxide treatment is the most common chemical sterilization method, used for approximately 70% of total sterilizations, and for over 50% of all disposable medical devices.

Ethylene oxide treatment is generally carried out between 30 and 60 °C (86 and 140 °F) with relative humidity above 30% and a gas concentration between 200 and 800 mg/L. Typically, the process lasts for several hours. Ethylene oxide is highly effective, as it penetrates all porous materials, and it can penetrate through some plastic materials and films. Ethylene oxide kills all known microorganisms, such as bacteria (including spores), viruses, and fungi (including yeasts and moulds), and is compatible with almost all materials even when repeatedly applied. It is flammable, toxic, and carcinogenic; however, only with a reported potential for some adverse health effects when not used in compliance with published requirements. Ethylene oxide sterilizers and processes require biological validation after sterilizer installation, significant repairs or process changes.

The traditional process consists of a preconditioning phase (in a separate room or cell), a processing phase (more commonly in a vacuum vessel and sometimes in a pressure rated vessel), and an aeration phase (in a separate room or cell) to remove EO residues and lower by-products such as ethylene chlorohydrin (EC or ECH) and, of lesser importance, ethylene glycol (EG). An alternative process, known as all-in-one processing, also exists for some products whereby all three phases are performed in the vacuum or pressure rated vessel. This latter option can facilitate faster overall processing time and residue dissipation.

The most common EO processing method is the gas chamber method. To benefit from economies of scale, EO has traditionally been delivered by filling a large chamber with a combination of gaseous EO either as pure EO, or with other gases used as diluents; diluents include chlorofluorocarbons (CFCs), hydrochlorofluorocarbons (HCFCs), and carbon dioxide.

Ethylene oxide is still widely used by medical device manufacturers. Since EO is explosive at concentrations above 3%, EO was traditionally supplied with an inert carrier gas, such as a CFC or HCFC. The use of CFCs or HCFCs as the carrier gas was banned because of concerns of ozone depletion. These halogenated hydrocarbons are being replaced by systems using 100% EO, because of regulations and the high cost of the blends. In hospitals, most EO sterilizers use single-use cartridges because of the convenience and ease of use compared to the former plumbed gas cylinders of EO blends.

It is important to adhere to patient and healthcare personnel government specified limits of EO residues in and/or on processed products, operator exposure after processing, during storage and handling of EO gas cylinders, and environmental emissions produced when using EO.

The U.S. Occupational Safety and Health Administration (OSHA) has set the permissible exposure limit (PEL) at 1 ppm - calculated as an eight-hour time-weighted average (TWA) - and 5 ppm as a 15-minute excursion limit (EL). The National Institute for Occupational Safety and Health's (NIOSH) immediately dangerous to life and health limit (IDLH) for EO is 800 ppm. The odor threshold is around 500 ppm, so EO is imperceptible until concentrations are well above the OSHA PEL. Therefore, OSHA recommends that continuous gas monitoring systems be used to protect workers using EO for processing.

Nitrogen Dioxide

Nitrogen dioxide (NO_2) gas is a rapid and effective sterilant for use against a wide range of microorganisms, including common bacteria, viruses, and spores. The unique physical properties of NO_2 gas allow for sterilant dispersion in an enclosed environment at room temperature and atmospheric pressure. The mechanism for lethality is the degradation of DNA in the spore core through nitration of the phosphate backbone, which kills the exposed organism as it absorbs NO_2. This degradations occurs at even very low concentrations of the gas. NO_2 has a boiling point of 21 °C (70 °F) at sea level, which results in a relatively highly saturated vapour pressure at ambient temperature. Because of this, liquid NO_2 may be used as a convenient source for the sterilant gas. Liquid NO_2 is often referred to by the name of its dimer, dinitrogen tetroxide (N_2O_4). Additionally, the low levels of concentration required, coupled with the high vapour pressure, assures that no condensation occurs on the devices being sterilized. This means that no aeration of the devices is required immediately following the sterilization cycle. NO_2 is also less corrosive than other sterilant gases, and is compatible with most medical materials and adhesives.

The most-resistant organism (MRO) to sterilization with NO_2 gas is the spore of *Geobacillus stearothermophilus*, which is the same MRO for both steam and hydrogen peroxide sterilization processes. The spore form of *G. stearothermophilus* has been well characterized over the years as a biological indicator in sterilization applications. Microbial inactivation of *G. stearothermophilus* with NO_2 gas proceeds rapidly in a log-linear fashion, as is typical of other sterilization processes. Noxilizer, Inc. has commercialized this technology to offer contract sterilization services for medical devices at its Baltimore, Maryland (U.S.) facility. This has been demonstrated in Noxilizer's lab in multiple studies and is supported by published reports from other labs. These same properties also allow for quicker removal of the sterilant and residual gases through aeration of the enclosed environment. The combination of rapid lethality and easy removal of the gas allows for shorter overall cycle times during the sterilization (or decontamination) process and a lower level of sterilant residuals than are found with other sterilization methods. Eniware, LLC has developed a portable, power-free sterilizer that uses no electricity, heat or water. The 25 liter unit makes sterilization of surgical instruments possible for austere forward surgical teams, in health centers throughout the world with intermittent or no electricity and in disaster relief and humanitarian crisis situations. The four hour cycle uses a single use gas generation ampoule and a disposable scrubber to remove nitrogen dioxide gas.

Ozone

Ozone is used in industrial settings to sterilize water and air, as well as a disinfectant for surfaces. It has the benefit of being able to oxidize most organic matter. On the other hand, it is a toxic and unstable gas that must be produced on-site, so it is not practical to use in many settings.

Ozone offers many advantages as a sterilant gas; ozone is a very efficient sterilant because of its strong oxidizing properties (E=2.076 vs SHE) capable of destroying a wide range of pathogens, including prions, without the need for handling hazardous chemicals since the ozone is generated within the sterilizer from medical-grade oxygen. The high reactivity of ozone means that waste ozone can be destroyed by passing over a simple catalyst that reverts it to oxygen and ensures that the cycle time is relatively short. The disadvantage of using ozone is that the gas is very reactive and very hazardous. The NIOSH's immediately dangerous to life and health limit (IDLH) for ozone is 5 ppm, 160 times smaller than the 800 ppm IDLH for ethylene oxide. NIOSH and OSHA has set the PEL for ozone at 0.1 ppm, calculated as an eight-hour time-weighted average. The sterilant gas manufacturers include many safety features in their products but prudent practice is to provide continuous monitoring of exposure to ozone, in order to provide a rapid warning in the event of a leak. Monitors for determining workplace exposure to ozone are commercially available.

Glutaraldehyde and Formaldehyde

Glutaraldehyde and formaldehyde solutions (also used as fixatives) are accepted liquid sterilizing agents, provided that the immersion time is sufficiently long. To kill all spores in a clear liquid can take up to 22 hours with glutaraldehyde and even longer with formaldehyde. The presence of solid particles may lengthen the required period or render the treatment ineffective. Sterilization of blocks of tissue can take much longer, due to the time required for the fixative to penetrate. Glutaraldehyde and formaldehyde are volatile, and toxic by both skin contact and inhalation. Glutaraldehyde has a short shelf-life (<2 weeks), and is expensive. Formaldehyde is less expensive and has a much longer shelf-life if some methanol is added to inhibit polymerization to paraformaldehyde, but is much more volatile. Formaldehyde is also used as a gaseous sterilizing agent; in this case, it is prepared on-site by depolymerization of solid paraformaldehyde. Many vaccines, such as the original Salk polio vaccine, are sterilized with formaldehyde.

Hydrogen Peroxide

Hydrogen peroxide, in both liquid and as vaporized hydrogen peroxide (VHP), is another chemical sterilizing agent. Hydrogen peroxide is a strong oxidant, which allows it to destroy a wide range of pathogens. Hydrogen peroxide is used to sterilize heat- or temperature-sensitive articles, such as rigid endoscopes. In medical sterilization, hydrogen peroxide is used at higher concentrations, ranging from around 35% up to 90%. The biggest advantage of hydrogen peroxide as a sterilant is the short cycle time. Whereas the cycle time for ethylene oxide may be 10 to 15 hours, some modern hydrogen peroxide sterilizers have a cycle time as short as 28 minutes.

Drawbacks of hydrogen peroxide include material compatibility, a lower capability for penetration and operator health risks. Products containing cellulose, such as paper, cannot be sterilized using VHP and products containing nylon may become brittle. The penetrating ability of hydrogen peroxide is not as good as ethylene oxide and so there are limitations on the length and diameter of the lumen of objects that can be effectively sterilized. Hydrogen peroxide is a primary irritant and the contact of the liquid solution with skin will cause bleaching or ulceration depending on the concentration and contact time. It is relatively non-toxic when diluted to low concentrations, but is a dangerous oxidizer at high concentrations (> 10% w/w).

The vapour is also hazardous, primarily affecting the eyes and respiratory system. Even short term exposures can be hazardous and NIOSH has set the IDLH at 75 ppm, less than one tenth the IDLH for ethylene oxide (800 ppm). Prolonged exposure to lower concentrations can cause permanent lung damage and consequently, OSHA has set the permissible exposure limit to 1.0 ppm, calculated as an eight-hour time-weighted average. Sterilizer manufacturers go to great lengths to make their products safe through careful design and incorporation of many safety features, though there are still workplace exposures of hydrogen peroxide from gas sterilizers documented in the FDA MAUDE database. When using any type of gas sterilizer, prudent work practices should include good ventilation, a continuous gas monitor for hydrogen peroxide and good work practices and training.

Vaporized hydrogen peroxide (VHP) is used to sterilize large enclosed and sealed areas, such as entire rooms and aircraft interiors. Although toxic, VHP breaks down in a short time to water and oxygen.

Peracetic Acid

Peracetic acid (0.2%) is a recognized sterilant by the FDA for use in sterilizing medical devices such as endoscopes. Peracetic acid which is also known as peroxyacetic acid is a chemical compound often used in disinfectants such as sanitizers. It is most commonly produced by the reaction of acetic acid and hydrogen peroxide with each other by using an acid catalyst. Peracetic acid is never sold in unstabilized solutions which is why it is considered to be environmentally friendly. Peracetic acid is a colorless liquid and the molecular formula of peracetic acid is $C_2H_4O_3$ or CH3COOOH. More recently, peracetic acid is being used throughout the world as more people are using fumigation to decontaminate surfaces to reduce the risk of COVID-19 and other diseases.

Potential for Chemical Sterilization of Prions

Prions are highly resistant to chemical sterilization. Treatment with aldehydes, such as formaldehyde, have actually been shown to increase prion resistance. Hydrogen peroxide (3%) for one hour was shown to be ineffective, providing less than 3 logs (10^{-3}) reduction in contamination. Iodine, formaldehyde, glutaraldehyde, and peracetic acid also fail this test (one hour treatment). Only chlorine, phenolic compounds, guanidinium thiocyanate, and sodium hydroxide reduce prion levels by more than 4 logs; chlorine (too corrosive to use on certain objects) and sodium hydroxide are the most consistent. Many studies have shown the effectiveness of sodium hydroxide.

11.1.5 Radiation Sterilization

Sterilization can be achieved using electromagnetic radiation, such as ultraviolet light, X-rays and gamma rays, or irradiation by subatomic particles such as by electron beams. Electromagnetic or particulate radiation can be energetic enough to ionize atoms or molecules (ionizing radiation), or less energetic (non-ionizing radiation).

Non-Ionizing Radiation Sterilization

Ultraviolet light irradiation (UV, from a germicidal lamp) is useful for sterilization of surfaces and some transparent objects. Many objects that are transparent to visible light absorb UV. UV irradiation is routinely used to sterilize the interiors of biological safety cabinets between uses, but is ineffective in shaded areas, including areas under dirt (which may become polymerized after prolonged irradiation, so that it is very difficult to remove). It also damages some plastics, such as polystyrene foam if exposed for prolonged periods of time.

Ionizing Radiation Sterilization

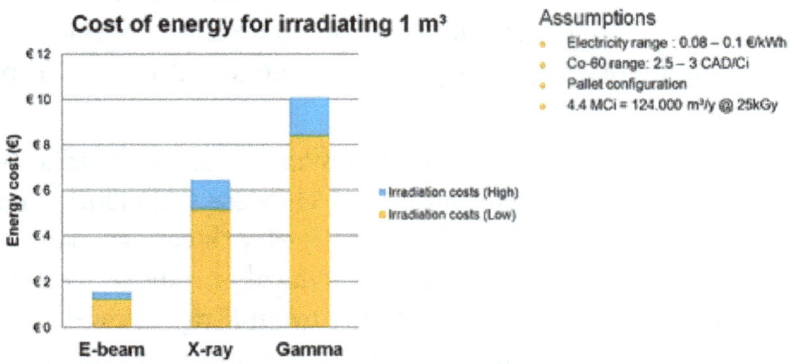

Fig 1: Efficiency Illustration of the Different Radiation Technologies (Electron Beam, X-ray, Gamma Rays)

The safety of irradiation facilities is regulated by the International Atomic Energy Agency of the United Nations and monitored by the different national Nuclear Regulatory Commissions (NRC).

The radiation exposure accidents that have occurred in the past are documented by the agency and thoroughly analyzed to determine the cause and improvement potential. Such improvements are then mandated to retrofit existing facilities and future design.

Gamma radiation is very penetrating, and is commonly used for sterilization of disposable medical equipment, such as syringes, needles, cannulas and IV sets, and food. It is emitted by a radioisotope, usually cobalt-60 (^{60}Co) or caesium-137 (^{137}Cs), which have photon energies of up to 1.3 and 0.66 MeV, respectively.

Use of a radioisotope requires shielding for the safety of the operators while in use and in storage. With most designs, the radioisotope is lowered into a water-filled source storage pool, which absorbs radiation and allows maintenance personnel to enter the radiation shield. One variant keeps the radioisotope under water at all times and lowers the product to be irradiated in the water in hermetically sealed bells; no further shielding is required for such designs. Other uncommonly used designs use dry storage, providing movable shields that reduce radiation levels in areas of the irradiation chamber. An incident in Decatur, Georgia, US, where water-soluble caesium-137 leaked into the source storage pool, requiring NRC intervention has led to use of this radioisotope being almost entirely discontinued in favour of the more costly, non-water-soluble cobalt-60. Cobalt-60 gamma photons have about twice the energy, and hence greater penetrating range, of caesium-137-produced radiation.

Electron beam processing is also commonly used for sterilization. Electron beams use an on-off technology and provide a much higher dosing rate than gamma or X-rays. Due to the higher dose rate, less exposure time is needed and thereby any potential degradation to polymers is reduced. Because electrons carry a charge, electron beams are less penetrating than both gamma and X-rays. Facilities rely on substantial concrete shields to protect workers and the environment from radiation exposure.

High-energy X-rays (produced by *bremsstrahlung*) allow irradiation of large packages and pallet loads of medical devices. They are sufficiently penetrating to treat multiple pallet loads of low-density packages with very good dose uniformity ratios. X-ray sterilization does not require chemical or radioactive material: high-energy X-rays are generated at high intensity by an X-ray generator that does not require shielding when not in use. X-rays are generated by bombarding a dense material (target) such as tantalum or tungsten with high-energy electrons, in a process known as *bremsstrahlung* conversion. These systems are energy-inefficient, requiring much more electrical energy than other systems for the same result.

Irradiation with X-rays, gamma rays, or electrons does not make materials radioactive, because the energy used is too low. Generally an energy of at least 10 MeV is needed to induce radioactivity in a material. Neutrons and very high-energy particles can make materials radioactive, but have good penetration, whereas lower energy particles (other than neutrons) cannot make materials radioactive, but have poorer penetration.

Sterilization by irradiation with gamma rays may however affect material properties. Irradiation is used by the United States Postal Service to sterilize mail in the Washington, D.C. area. Some foods (e.g. spices and ground meats) are sterilized by irradiation. Subatomic particles may be more or less penetrating and may be generated by a radioisotope or a device, depending upon the type of particle.

11.1.6 Sterile Filtration

Fluids that would be damaged by heat, irradiation or chemical sterilization, such as drug solution, can be sterilized by microfiltration using membrane filters. This method is commonly used for heat labile pharmaceuticals and protein solutions in medicinal drug processing. A microfilter with pore size of usually 0.22 μm will effectively remove microorganisms. Some staphylococcal species have, however, been shown to be flexible enough to pass through 0.22 μm filters. In the processing of biologics, viruses must be removed or inactivated, requiring the use of nanofilters with a smaller pore size (20–50 nm). Smaller pore sizes lower the flow rate, so in order to achieve higher total throughput or to avoid premature blockage, pre-filters might be used to protect small pore membrane filters. Tangential flow filtration (TFF) and alternating tangential flow (ATF) systems also reduce particulate accumulation and blockage.

Membrane filters used in production processes are commonly made from materials such as mixed cellulose ester or polyethersulfone (PES). The filtration equipment and the filters themselves may be purchased as pre-sterilized disposable units in sealed packaging or must be sterilized by the user, generally by autoclaving at a temperature that does not damage the fragile filter membranes. To ensure proper functioning of the filter, the membrane filters are integrity tested post-use and sometimes before use.

The nondestructive integrity test assures the filter is undamaged and is a regulatory requirement. Typically, terminal pharmaceutical sterile filtration is performed inside of a cleanroom to prevent contamination.

11.1.7 Preservation of Sterility

Instruments that have undergone sterilization can be maintained in such condition by containment in sealed packaging until use. Aseptic technique is the act of maintaining sterility during procedures.

11.2 IMPORTANCE OF STERILITY IN INDUSTRIAL MICROBIOLOGY

Sterility is crucial in industrial microbiology to avoid contamination that could disrupt microbial fermentation or biochemical reactions. In the pharmaceutical industry, where microbial cells or their metabolites are used to produce antibiotics, vaccines, and enzymes, maintaining sterility is essential to prevent the growth of pathogens or spoilage organisms. Similarly, in the food and beverage industry, contamination by harmful microorganisms can lead to spoilage, reduced product quality, and the potential for foodborne illnesses. Sterility ensures the safety of food products and prolongs shelf life, minimizing economic losses.

In biotechnological applications, sterility is equally important. Genetically modified organisms (GMOs) are often engineered for specific purposes, such as producing insulin or other therapeutic proteins.

The presence of non-sterile conditions could introduce foreign microbes that interfere with the intended microbial process or lead to mutations in the GMOs, reducing their efficacy. Additionally, in the field of agriculture, where microbial inoculants are used to improve soil fertility or control plant diseases, sterility ensures that only the desired strains are introduced into the environment.

Ensuring sterility also has significant implications for human health. In the manufacture of sterile pharmaceuticals, contamination by harmful bacteria, viruses, or fungi can lead to life-threatening infections, particularly in immunocompromised patients.

This underscores the need for stringent sterility protocols in industrial microbiology to maintain the integrity of products and protect public health.

11.3 KEY PROCESSES TO ACHIEVE STERILITY

To achieve and maintain sterility in industrial microbiology, various sterilization techniques and protocols are employed. These processes can be broadly classified into physical, chemical, and biological methods.

11.3.1 Physical Methods of Sterilization

- **Heat Sterilization:** Heat is one of the most common and effective methods for sterilizing equipment, culture media, and other materials in industrial microbiology. There are two primary forms of heat sterilization: moist heat and dry heat.
- **Moist Heat Sterilization (Autoclaving):** This method involves using steam under pressure to destroy microorganisms, including their spores. Autoclaving is highly effective and widely used in industries to sterilize culture media, glassware, and certain instruments. Typically, materials are exposed to temperatures of 121°C for 15-20 minutes, depending on the load size. The high temperature and pressure kill microbes by denaturing their proteins and disrupting their cell membranes.
- **Dry Heat Sterilization:** This method uses high temperatures (160-170°C) for extended periods (1-2 hours) to achieve sterility. Dry heat is suitable for materials that cannot withstand moisture, such as metal

instruments or glass containers. The heat oxidizes cellular components, leading to the death of microorganisms.

- **Filtration:** Filtration is a physical method used to remove microorganisms from liquids or gases. Membrane filters with pore sizes ranging from 0.2 to 0.45 micrometers are commonly used to trap bacteria and larger microbes. Filtration is particularly useful for sterilizing heat-sensitive liquids, such as culture media, vitamins, and antibiotics, which cannot be autoclaved.
- **Radiation:** Ionizing radiation, such as gamma rays or electron beams, can be used to sterilize materials by damaging the DNA and cellular structures of microorganisms. This method is often employed for sterilizing medical devices, pharmaceuticals, and packaging materials. Ultraviolet (UV) radiation is another option, though it is generally less effective than ionizing radiation and is used primarily for surface sterilization in laboratories and cleanrooms.

11.3.2 Chemical Methods of Sterilization

Ethylene Oxide Gas

Ethylene oxide (EO) gas is a widely used chemical sterilant, particularly for medical devices and plastic materials that cannot tolerate high heat or moisture. EO gas penetrates the material and reacts with proteins and DNA to kill microorganisms. However, it requires careful handling due to its toxicity and the need for extensive aeration to remove residual gas after sterilization.

Hydrogen Peroxide Vapor (HPV)

Hydrogen peroxide vapor is another chemical sterilant used in industrial microbiology. It is effective at killing a broad spectrum of microorganisms, including bacteria, viruses, fungi, and spores. HPV is commonly used to decontaminate enclosed spaces, such as isolators, cleanrooms, and production equipment.

Peracetic Acid

Peracetic acid is an oxidizing agent used for the sterilization of medical and pharmaceutical equipment. It is highly effective against a wide range of microorganisms and is often used in combination with hydrogen peroxide in sterilizing solutions. This method is commonly used for sterilizing surgical instruments, endoscopes, and other reusable equipment in industries.

11.3.3 Biological Methods of Sterilization

Biological methods of sterilization rely on the use of biological indicators, which contain known quantities of resistant spores. These indicators are used to validate the effectiveness of sterilization processes. After exposure to the sterilization process, the biological indicators are incubated to determine whether any spores survived. If no growth is detected, the process is considered effective.

This method is particularly important for industries that require stringent validation of their sterilization protocols, such as pharmaceutical manufacturing.

11.4 MAINTAINING STERILITY DURING PRODUCTION

Achieving sterility is only one part of the challenge in industrial microbiology; maintaining sterile conditions throughout production is equally crucial. To this end, industrial facilities employ a range of procedures and technologies designed to prevent contamination during manufacturing.

11.4.1 Cleanrooms and Controlled Environments

Cleanrooms are specially designed environments where air quality, temperature, humidity, and pressure are carefully controlled to minimize the presence of airborne microorganisms and particulate matter.

Personnel working in cleanrooms must wear protective clothing, including gowns, gloves, masks, and shoe covers, to reduce the risk of introducing contaminants into the sterile environment. The air in cleanrooms is continuously filtered through high-efficiency particulate air (HEPA) filters, which remove particles as small as 0.3 micrometers.

11.4.2 Aseptic Techniques

Aseptic techniques are employed to prevent contamination during the handling of sterile materials.

This includes practices such as sterilizing work surfaces, using sterile instruments, and minimizing exposure to the environment. In biotechnological and pharmaceutical industries, aseptic processing is critical for the production of sterile products, including injectable drugs and vaccines.

11.4.3 Sterile Filtration of Air and Gases

In some industries, sterile filtration of air and gases is necessary to maintain a contamination-free environment. HEPA filters are commonly used in air-handling systems to remove microbial contaminants. Similarly, compressed gases used in industrial processes, such as nitrogen or oxygen, may be filtered to ensure that they are sterile before coming into contact with sensitive materials or equipment.

11.5 CONCLUSION

Sterility is a cornerstone of industrial microbiology, ensuring the safety, quality, and efficacy of products across diverse industries.

The processes used to achieve and maintain sterility—whether through physical, chemical, or biological methods—are designed to eliminate or control microbial contamination at every stage of production.

From the manufacture of pharmaceuticals to the production of food and biotechnology products, sterility is essential for preventing contamination, safeguarding public health, and ensuring the success of microbial processes.

As technology continues to evolve, the methods used to ensure sterility will likely become even more sophisticated, further enhancing the capabilities of industrial microbiology.

SECTION 4:

PRODUCTION OF METABOLITES AS BULK CHEMICALS OR AS INPUTS IN OTHER PROCESSES

CHAPTER 12

BIOCATALYSTS, IMMOBILIZED ENZYMES AND IMMOBILIZED CELLS

12.1 BIOCATALYSTS

Biocatalysis refers to the use of living (biological) systems or their parts to speed up (catalyze) chemical reactions. In biocatalytic processes, natural catalysts, such as enzymes, perform chemical transformations on organic compounds. Both enzymes that have been more or less isolated and enzymes still residing inside living cells are employed for this task. Modern biotechnology, specifically directed evolution, has made the production of modified or non-natural enzymes possible.

This has enabled the development of enzymes that can catalyze novel small molecule transformations that may be difficult or impossible using classical synthetic organic chemistry. Utilizing natural or modified enzymes to perform organic synthesis is termed chemoenzymatic synthesis; the reactions performed by the enzyme are classified as chemoenzymatic reactions.

12.1.1 History

Biocatalysis underpins some of the oldest chemical transformations known to humans, for brewing predates recorded history. The oldest records of brewing are about 6000 years old and refer to the Sumerians. The employment of enzymes and whole cells have been important for many industries for centuries.

The most obvious uses have been in the food and drink businesses where the production of wine, beer, cheese etc. is dependent on the effects of the microorganisms. More than one hundred years ago, biocatalysis was employed to do chemical transformations on non-natural man-made organic compounds, with the last 30 years seeing a substantial increase in the application of biocatalysis to produce fine chemicals, especially for the pharmaceutical industry.

Since biocatalysis deals with enzymes and microorganisms, it is historically classified separately from "homogeneous catalysis" and "heterogeneous catalysis". However, mechanistically speaking, biocatalysis is simply a special case of heterogeneous catalysis.

12.1.2 Advantages of Chemoenzymatic Synthesis

- Enzymes are environmentally benign, being completely degraded in the environment.
- Most enzymes typically function under mild or biological conditions, which minimizes problems of undesired side-reactions such as decomposition, isomerization, racemization and rearrangement, which often plague traditional methodology.
- Enzymes selected for chemoenzymatic synthesis can be immobilized on a solid support. These immobilized enzymes demonstrate improved stability and re-usability.
- Through the development of protein engineering, specifically site-directed mutagenesis and directed evolution, enzymes can be modified to enable non-natural reactivity. Modifications may also allow for a broader substrate range, enhance reaction rate or catalyst turnover.
- Enzymes exhibit extreme selectivity towards their substrates. Typically enzymes display three major types of selectivity:
  - *Chemoselectivity:* Since the purpose of an enzyme is to act on a single type of functional group, other sensitive functionalities, which would normally react to a certain extent under chemical catalysis, survive. As a result, biocatalytic reactions tend to be "cleaner" and laborious purification of product(s) from impurities emerging through side-reactions can largely be omitted.
  - *Regioselectivity and Diastereoselectivity:* Due to their complex three-dimensional structure, enzymes may distinguish between functional groups which are chemically situated in different regions of the substrate molecule.
  - *Enantioselectivity:* Since almost all enzymes are made from L-amino acids, enzymes are chiral catalysts. As a consequence, any type of chirality present in the substrate molecule is "recognized" upon the formation of the enzyme-substrate complex. Thus a prochiral substrate may be transformed into an optically active product and both enantiomers of a racemic substrate may react at different rates.

These reasons, and especially the latter, are the major reasons why synthetic chemists have become interested in biocatalysis.

This interest in turn is mainly due to the need to synthesize enantiopure compounds as chiral building blocks for Pharmaceutical drugs and agrochemicals.

12.1.3 Asymmetric Biocatalysis

The use of biocatalysis to obtain enantiopure compounds can be divided into two different methods:

- Kinetic resolution of a racemic mixture.
- Biocatalyzed asymmetric synthesis.

In kinetic resolution of a racemic mixture, the presence of a chiral object (the enzyme) converts one of the stereoisomers of the reactant into its product at a greater reaction rate than for the other reactant stereoisomer. The stereochemical mixture has now been transformed into a mixture of two different compounds, making them separable by normal methodology.

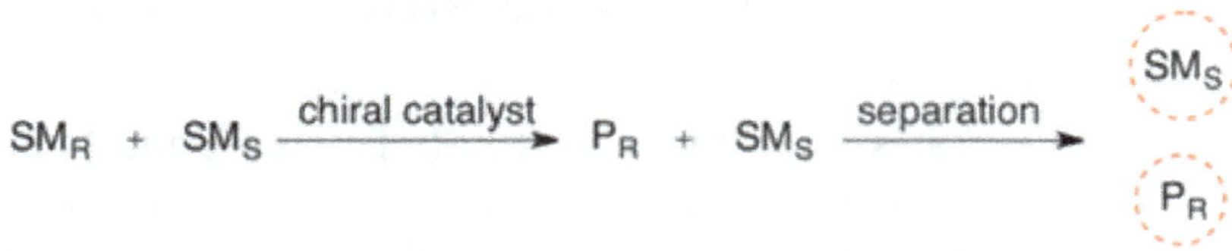

Biocatalyzed kinetic resolution is utilized extensively in the purification of racemic mixtures of synthetic amino acids. Many popular amino acid synthesis routes, such as the Strecker Synthesis, result in a mixture of R and S enantiomers. This mixture can be purified by (I) acylating the amine using an anhydride and then (II) selectively deacylating only the L enantiomer using hog kidney acylase. These enzymes are typically extremely selective for one enantiomer leading to very large differences in rate, allowing for selective deacylation. Finally the two products are now separable by classical techniques, such as chromatography.

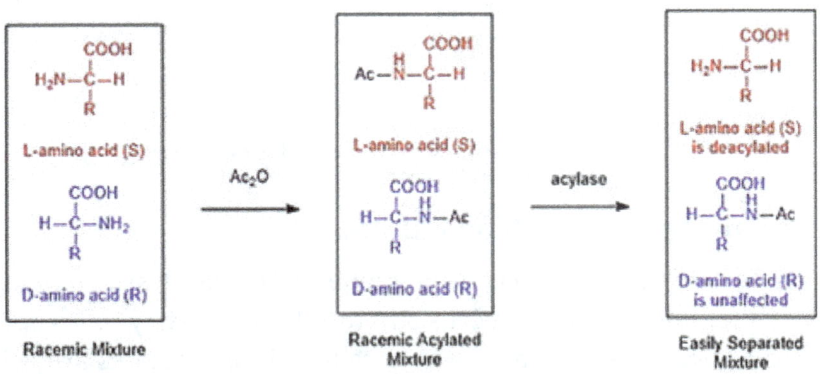

The maximum yield in such kinetic resolutions is 50%, since a yield of more than 50% means that some of wrong isomer also has reacted, giving a lower enantiomeric excess. Such reactions must therefore be terminated before equilibrium is reached. If it is possible to perform such resolutions under conditions where the two substrate- enantiomers are racemizing continuously, all substrate may in theory be converted into enantiopure product. This is called dynamic resolution. In biocatalyzed asymmetric synthesis, a non-chiral unit becomes chiral in such a way that the different possible stereoisomers are formed in different quantities. The chirality is introduced into the substrate by influence of enzyme, which is chiral. Yeast is a biocatalyst for the enantioselective reduction of ketones.

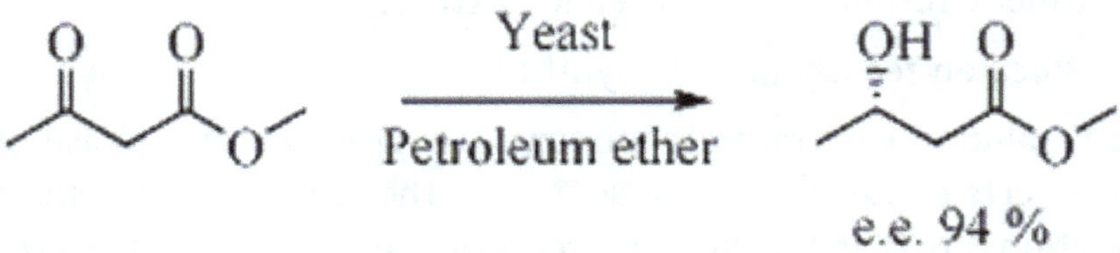

The Baeyer–Villiger oxidation is another example of a biocatalytic reaction. In one study a specially designed mutant of *Candida antarctica* was found to be an effective catalyst for the Michael addition of acrolein with acetylacetone at 20 °C in absence of additional solvent.

Another study demonstrates how racemic nicotine (mixture of S and R-enantiomers 1 in *scheme* 3) can be deracemized in a one-pot procedure involving a monoamine oxidase isolated from Aspergillus niger which is able to oxidize only the amine S-enantiomer to the imine 2 and involving an ammonia–borane reducing couple which can reduce the imine 2 back to the amine 1. In this way the S-enantiomer will continuously be consumed by the enzyme while the R-enantiomer accumulates. It is even possible to stereoinvert pure S to pure R.

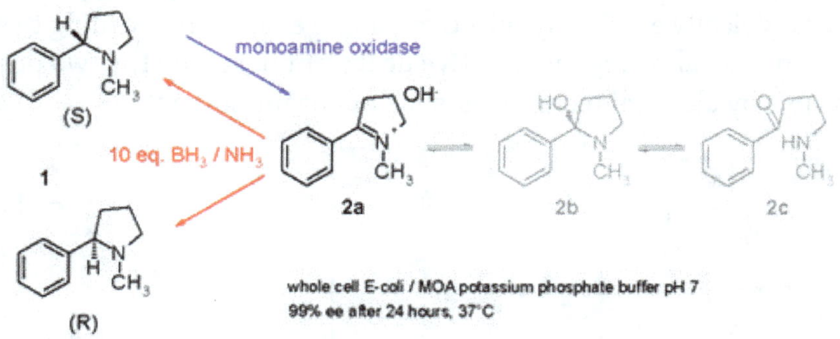

12.1.4 Photoredox Enabled Biocatalysis

Recently, photoredox catalysis has been applied to biocatalysis, enabling unique, previously inaccessible transformations. Photoredox chemistry relies upon light to generate free radical intermediates. These radical intermediates are achiral thus racemic mixtures of product are obtained when no external chiral environment is provided. Enzymes can provide this chiral environment within the active site and stabilize a particular conformation and favoring formation of one, enantiopure product. Photoredox enabled biocatalysis reactions fall into two categories:

- Internal coenzyme/cofactor photocatalyst.
- External photocatalyst.

Certain common hydrogen atom transfer (HAT) cofactors (NADPH and Flavin) can operate as single electron transfer (SET) reagents. Although these species are capable of HAT without irradiation, their redox potentials are enhance by nearly

2.0 V upon visible light irradiation. When paired with their respective enzymes (typically ene-reductases) This phenomenon has been utilized by chemists to develop enantioselective reduction methodologies. For example medium sized lactams can be synthesized in the chiral environment of an ene-reductase through a reductive, baldwin favored, radical cyclization terminated by enantioselective HAT from NADPH.

The second category of photoredox enabled biocatalytic reactions use an external photocatalyst (PC). Many types of PCs with a large range of redox potentials can be utilized, allowing for greater tunability of reactive compared to using a cofactor. Rose bengal, and external PC, was utilized in tandem with an oxidoreductase to enantioselectively deacylate medium sized alpha-acyl-ketones.

Using an external PC has some downsides. For example, external PCs typically complicate reaction design because the PC may react with both the bound and unbound substrate. If a reaction occurs between the unbound substrate and the PC, enantioselectivity is lost and other side reactions may occur.

12.1.5 Agricultural Uses

Bioenzymes are also bio catalyst. They are prepared by fermentation of organic waste, jaggery and water in ratio 3:1:10 for three months. It increases the soil microbe population and speeds up composting and decomposition and so is included in catalyts. It heals the soil. It is one of the best best organic liquid fertilizer. It is diluted with water.

12.2 IMMOBILIZED ENZYMES

An immobilized enzyme is an enzyme, with restricted mobility, attached to an inert, insoluble material—such as calcium alginate (produced by reacting a mixture of sodium alginate solution and enzyme solution with calcium chloride).

This can provide increased resistance to changes in conditions such as pH or temperature. It also lets enzymes be held in place throughout the reaction, following which they are easily separated from the products and may be used again - a far more efficient process and so is widely used in industry for enzyme catalysed reactions. An alternative to enzyme immobilization is whole cell immobilization. Immobilized enzymes are easily to be handled, simply separated from their products, and can be reused.

Enzymes are bio-catalysts which play an essential role in the enhancement of chemical reactions in cells without being persistently modified, wasted, nor resulting in the loss of equilibrium of chemical reactions. Although the characteristics of enzymes are extremely unique, their utility in the industry is limited due to the lack of re-usability, stability, and high-cost of production.

12.2.1 History

The first synthetic immobilized enzyme was made in the 1950s, performed by the inclusion of enzyme into polymeric matrices or binding onto carrier substances. Also cross-linking procedure was applied by cross-linking of protein alone or along with the addition of inert materials. Over the last decade various immobilization methods have been developed. Binding the enzyme to previously synthesized carrier materials for example is the mostly preferred method so far. Newly, the procedure of cross-linking of crystals of enzyme is also considered as an exciting substitute. Utilization rate of immobilized enzymes is growing constantly.

12.2.2 Considerations

Before performing any kind of immobilization techniques, some factors should be in mind. It is necessary to understand the chemical and physical effects on an enzyme following immobilization. Enzyme stability and kinetic characteristics can be altered due to changes in the microenvironment conditions of the enzyme after entrapment, support material attachment, or products of enzymatic actions for instance. Additionally, it is important to consider maintaining the tertiary structure of an enzyme prior to immobilizing to have a functional enzyme. Similarly, another crucial site for the functionality of an enzyme is the active-site, which should also be maintained while enzyme is being attached to a surface for immobilization, it is a must to have a selective method for the attachment of surface/material to not end up with an immobilized, but dysfunctional enzyme. Consequently, there are three foundational factors to be thought of for the production of functional immobilized enzymes: immobilization supports selection, conditions and methods of immobilization.

Support selection

For a support material to be ideal, it must be hydrophilic, inert towards enzymes, biocompatible, microbial attack and compression resistant, and must be affordable. Support materials can be organic or inorganic, synthetic or natural (depending on the composition), since they are biomaterial types at the end. There is no universal type of a support material to be used for the immobilization of all enzymes. However, there are some commonly used supports such as silica-based carriers, acrylic resins, synthetic polymers, active membranes and exchange resins. One of the hardest processes before the immobilization process itself, is the selection of support material since it relies on the enzyme type, reaction of media, safety policy of hydrodynamic and reaction conditions. As different types of support give different physical and chemical characteristics and properties, which would effect enzyme function, such as: Hydrophilicity/hydrophobicity, surface chemistry, and pore size.

12.2.3 Methods

Enzymes can be immobilized by physical, or chemical methods including:

Physical

- A straightforward method for reversible immobilization, involving the enzymes being adsorbed or attached physically onto a support substance. Adsorption can take place through weak non-specific forces, such as van der waals, hydrogen bonds, and hydrophobic interactions, whereas in ionic bonding the enzymes are bound through salt linkages.
- Adsorption on glass, alginate beads or matrix: Enzyme is attached to the outside of an inert material. In general, this method is the slowest among those listed here. As adsorption is not a chemical reaction, the active site of the immobilized enzyme may be blocked by the matrix or bead, greatly reducing the activity of the enzyme.

Entrapment

- This is an irreversible physical immobilization technique which can be considered as a physical restriction of enzyme in a specified area/ space. It can be used for raising mechanical stability and can be also used for the reduction of leaching events of enzymes. Since the enzyme in this process does not interact chemically with the polymer/ material of the support fibers/lattice, it remains protected from denaturation with time.
- Basically, the enzyme is trapped in insoluble beads or microspheres, such as calcium alginate beads. However, these insoluble substances hinder the arrival of the substrate, and the exit of products.

Chemical

Cross-Linking

- **Cross-Linkage:** Another irreversible method that does not require a support material for the attachment of enzyme molecules. In this technique, the molecules of enzymes are covalently bonded to each other to create a matrix consisting of almost only enzyme. The reaction ensures that the binding site does not cover the enzyme's active site, the activity of the enzyme is only affected by immobility. However, the inflexibility of the covalent bonds precludes the self-healing properties exhibited by chemo-adsorbed self-assembled monolayers. Use of a spacer molecule like poly(ethylene glycol) helps in the reduction of steric hindrance by the substrate in this case.

Covalent Bonding

- The enzyme is bound covalently to an insoluble support (such as silica gel or macroporous polymer beads with epoxide groups). This approach provides the strongest enzyme/support interaction, and so the lowest protein leakage during catalysis.
- The activity of the enzyme being covalently bound is dependent on several factors including: shape, and size of carrier material, coupling method type, the composition and coupling special conditions of carrier material.

Affinity-tag binding: is an immobilization method combining physical, and chemical methods where enzymes may be immobilized to a surface, e.g. in a porous material, using non-covalent or covalent Protein tags. This technology has been established for protein purification purposes. This technique is the generally applicable, and can be performed without prior enzyme purification with a pure preparation as the result. Porous glass and derivatives thereof are used, where the porous surface can be adapted in terms of hydrophobicity to suit the enzyme in question.

12.2.4 Random versus Site-Directed

Numerous enzymes of biotechnological importance have been immobilized on various supports (inorganic, organic, composite and nanomaterials) via random multipoint attachment. However, immobilization via random chemical modification results in a heterogeneous protein population where more than one side chains (amino, carboxyl, thiol etc) present in proteins are linked with the support with potential reduction in activity due to restriction of substrate access to the active site.

In contrast, in site-directed enzyme immobilization, the support can be linked to a single specific amino acid (generally N- or C-termini) in a protein molecule away from the active-site. This way maximal enzyme activity is retained due to the free access of the substrate to the active-site. These strategies are mainly chemical but may additionally require genetic and enzymatic methods to generate functional groups (that are absent in protein) on the support and enzyme.

The choice of SDCM method depends on many factors, such as the type of enzyme (less stable psychrophilic, or more stable thermophilic homologue), pH stability of enzyme, the availability of N- or C-termini to the reagent, non-interference of the enzyme terminus with the enzyme activity, type of catalytic amino acid residue, the availability, price and the ease of preparation of reagents. For example, the generation of complementary clickable functionalities (alkyne and azide) on the support and enzyme is one of the most convenient way for immobilizing enzymes via site-directed chemical modification.

12.2.5 Substrate Immobilization

Another widely used application of the immobilization approach together with enzymes has been the enzymatic reactions on immobilized substrates. This approach facilitates the analysis of enzyme activities and mimics the performance of enzymes on e.g. cell walls.

12.2.6 Commercial Use

Immobilized enzymes have important application uses as they reduce costs and improve the outcome of the reaction they catalyze. Advantages include:

- **Convenience:** Minuscule amounts of protein dissolve in the reaction, so workup can be much easier. Upon completion, reaction mixtures typically contain only solvent and reaction products.
- **Economy:** The immobilized enzyme is easily removed from the reaction making it easy to recycle the biocatalyst. This is particularly useful in processes such as the production of Lactose Free Milk, as the milk can be drained from a container leaving the enzyme (Lactase) inside ready for the next batch.
- **Stability:** Immobilized enzymes typically have greater thermal and operational stability than the soluble form of the enzyme.

In the past, biological washing powders and detergents contained many proteases and lipases that broke down dirt. However, when the cleaning products contacted human skin, they created allergic reactions. This is why immobilization of enzymes is important, for many application fields.

Immobilized enzymes are used in various applications including: food, chemical, pharmaceutical, and medical industry. In the food industry for example, Immobilized enzymes are used for the manufacturing of several types of zero-calorie sweetners, Allulose for instance is an epimer of fructose, which is different structurally, resulting in it not being absorbable by human bodies when ingested. Another example of immobilized-enzyme-based sweetners include: Tagatose (immobilized β-galactosidase).

In the chemical (cosmetics) industry as well, immobilized enzymes are used for the production of emollient esters by utilizing immobilized CalB enzyme. The first company to use such method is Evonik company in 2000. The enzyme Lipase-CalB in its immobilized state is actually used in other pharmaceutical applications for the production of Odanacatib, and Sofosbuvir.

12.3 IMMOBILIZED CELLS

Immobilized cell technology is a method of air filtration and purification that uses whole cell immobilization. It is a process whereby microfine particulate matter is removed from the air by attracting charged particulates in the air to a bio-reactive mass, or bioreactor, which enzymatically renders them inert.

Almost all airborne particulate matter carries a charge, either negative or positive. As air moves through an immobilized cell technology system, those charged particles are attracted to a water cascade, that functions as neutral ground, and that pulls the particulates into the bioreactor. (These airborne particulates consist of inert materials [dust, pollen, etc.], volatile organic compounds, and hazardous air pollutants.)

Once inside the bioreactor, the particulates are oxidized in a solution of water, enzymes, and bacteria breaking them down into base elements. Oxidation through immobilized cell technology is 12 times more efficient than natural oxidation is under the same conditions.

The key difference between traditional air filtration systems and those that employ immobilized cell technology is that the latter does not employ mesh filters. It requires water, electrical power, and biomass for the bioreactor. The base, inert, elements it produces need to be removed periodically.

CHAPTER 13
PRODUCTION OF BIOFUEL AND INDUSTRIAL ALCOHOL

The increasing global demand for energy, coupled with concerns about the environmental impact of fossil fuels, has led to a growing interest in renewable energy sources such as biofuels and industrial alcohols. Biofuels, including bioethanol, biodiesel, and biogas, are derived from biological materials and offer a cleaner, more sustainable alternative to traditional petroleum-based fuels. Industrial alcohols, such as ethanol and methanol, are crucial not only as fuels but also as solvents and raw materials for various industrial processes.

13.1 BIOFUEL: AN OVERVIEW

Biofuel is a fuel that is produced over a short time span from biomass, rather than by the very slow natural processes involved in the formation of fossil fuels such as oil. Biofuel can be produced from plants or from agricultural, domestic or industrial biowaste. Biofuels are mostly used for transportation, but can also be used for heating and electricity. Biofuels (and bioenergy in general) are regarded as a renewable energy source. The use of biofuel has been subject to criticism regarding the "food vs fuel" debate, varied assessments of their sustainability, and possible deforestation and biodiversity loss as a result of biofuel production.

In general, biofuels emit fewer greenhouse gas emissions when burned in an engine and are generally considered carbon-neutral fuels as the carbon emitted has been captured from the atmosphere by the crops used in production. However, life-cycle assessments of biofuels have shown large emissions associated with the potential land-use change required to produce additional biofuel feedstocks.

The outcomes of lifecycle assessments (LCAs) for biofuels are highly situational and dependent on many factors including the type of feedstock, production routes, data variations, and methodological choices. This could be added to emphasize the complexity and variability in assessing the environmental impacts of biofuels. Estimates about the climate impact from biofuels vary widely based on the methodology and exact situation examined. Therefore, the climate change mitigation potential of biofuel varies considerably: in some scenarios emission levels are comparable to fossil fuels, and in other scenarios the biofuel emissions result in negative emissions.

Global demand for biofuels is predicted to increase by 56% over 2022–2027. By 2027 worldwide biofuel production is expected to supply 5.4% of the world's fuels for transport including 1% of aviation fuel.

Demand for aviation biofuel is forecast to increase. However some policy has been criticised for favoring ground transportation over aviation.

The two most common types of biofuel are bioethanol and biodiesel. Brazil is the largest producer of bioethanol, while the EU is the largest producer of biodiesel. The energy content in the global production of bioethanol and biodiesel is 2.2 and 1.8 EJ per year, respectively.

Bioethanol is an alcohol made by fermentation, mostly from carbohydrates produced in sugar or starch crops such as maize, sugarcane, or sweet sorghum. Cellulosic biomass, derived from non-food sources, such as trees and grasses, is also being developed as a feedstock for ethanol production.

Ethanol can be used as a fuel for vehicles in its pure form (E100), but it is usually used as a gasoline additive to increase octane ratings and improve vehicle emissions.

Biodiesel is produced from oils or fats using transesterification. It can be used as a fuel for vehicles in its pure form (B100), but it is usually used as a diesel additive to reduce levels of particulates, carbon monoxide, and hydrocarbons from diesel-powered vehicles.

13.1.1 Conventional Biofuels (First Generation)

First-generation biofuels (also denoted as "conventional biofuels") are made from food crops grown on arable land.

The crop's sugar, starch, or oil content is converted into biodiesel or ethanol, using transesterification, or yeast fermentation.

13.1.2 Advanced Biofuels

To avoid a "food versus fuel" dilemma, second-generation biofuels and third-generation biofuels (also called advanced biofuels or sustainable biofuels or drop-in biofuels) are made from feedstocks which do not directly compete with food or feed crop such as waste products and energy crops. A wide range of waste feedstocks such as those derived from agriculture and forestry activities like rice straw, rice husk, wood chips, and sawdust can be used to produce advanced biofuels through biochemical and thermochemical processes.

The feedstock used to make the fuels either grow on arable land but are byproducts of the main crop, or they are grown on marginal land. Second-generation feedstocks also include straw, bagasse, perennial grasses, jatropha, waste vegetable oil, municipal solid waste and so forth.

13.1.3 Types

Liquid

Ethanol

Biologically produced alcohols, most commonly ethanol, and less commonly propanol and butanol, are produced by the action of microorganisms and enzymes through the fermentation of sugars or starches (easiest to produce) or cellulose (more difficult to produce).The IEA estimates that ethanol production used 20% of sugar supplies and 13% of corn supplies in 2021.

Ethanol fuel is the most common biofuel worldwide, particularly in Brazil. Alcohol fuels are produced by fermentation of sugars derived from wheat, corn, sugar beets, sugar cane, molasses and any sugar or starch from which alcoholic beverages such as whiskey, can be made (such as potato and fruit waste, etc.). Production methods used are enzyme digestion (to release sugars from stored starches), fermentation of the sugars, distillation and drying. The distillation process requires significant energy input to generate heat. Heat is sometimes generated with unsustainable natural gas fossil fuel, but cellulosic biomass such as bagasse is the most common fuel in Brazil, while pellets, wood chips and also waste heat are more common in Europe. Corn-to-ethanol and other food stocks has led to the development of cellulosic ethanol.

Other Bioalcohols

Methanol is currently produced from natural gas, a non-renewable fossil fuel. In the future it is hoped to be produced from biomass as biomethanol. This is technically feasible, but the production is currently being postponed for concerns that the economic viability is still pending.

The methanol economy is an alternative to the hydrogen economy to be contrasted with today's hydrogen production from natural gas.

Butanol (C_{4H9OH}) is formed by ABE fermentation (acetone, butanol, ethanol) and experimental modifications of the process show potentially high net energy gains with biobutanol as the only liquid product. Biobutanol is often claimed to provide a direct replacement for gasoline, because it will produce more energy than ethanol and allegedly can be burned "straight" in existing gasoline engines (without modification to the engine or car), is less corrosive and less water-soluble than ethanol, and could be distributed via existing infrastructures. *Escherichia coli* strains have also been successfully engineered to produce butanol by modifying their amino acid metabolism.

One drawback to butanol production in *E. coli* remains the high cost of nutrient rich media, however, recent work has demonstrated *E. coli* can produce butanol with minimal nutritional supplementation. Biobutanol is sometimes called biogasoline, which is incorrect as it is chemically different, being an alcohol and not a hydrocarbon like gasoline.

Biodiesel

Biodiesel is the most common biofuel in Europe. It is produced from oils or fats using transesterification and is a liquid similar in composition to fossil/mineral diesel. Chemically, it consists mostly of fatty acid methyl (or ethyl) esters (FAMEs). Feedstocks for biodiesel include animal fats, vegetable oils, soy, rapeseed, jatropha, mahua, mustard, flax, sunflower, palm oil, hemp, field pennycress, *Pongamia pinnata* and algae. Pure biodiesel (B100, also known as "neat" biodiesel) currently reduces emissions with up to 60% compared to diesel Second generation B100. As of 2020, researchers at Australia's CSIRO have been studying safflower oil as an engine lubricant, and researchers at Montana State University's Advanced Fuels Center in the US have been studying the oil's performance in a large diesel engine, with results described as a "breakthrough".

Biodiesel can be used in any diesel engine and modified equipment when mixed with mineral diesel. It can also be used in its pure form (B100) in diesel engines, but some maintenance and performance problems may occur during wintertime utilization, since the fuel becomes somewhat more viscous at lower temperatures, depending on the feedstock used.

Electronically controlled 'common rail' and 'Unit Injector' type systems from the late 1990s onwards can only use biodiesel blended with conventional diesel fuel. These engines have finely metered and atomized multiple-stage injection systems that are very sensitive to the viscosity of the fuel.

Many current-generation diesel engines are designed to run on B100 without altering the engine itself, although this depends on the fuel rail design.

Since biodiesel is an effective solvent and cleans residues deposited by mineral diesel, engine filters may need to be replaced more often, as the biofuel dissolves old deposits in the fuel tank and pipes. It also effectively cleans the engine combustion chamber of carbon deposits, helping to maintain efficiency.

Biodiesel is an oxygenated fuel, meaning it contains a reduced amount of carbon and higher hydrogen and oxygen content than fossil diesel. This improves the combustion of biodiesel and reduces the particulate emissions from unburnt carbon. However, using pure biodiesel may increase NO_x-emissions Biodiesel is also safe to handle and transport because it is non-toxic and biodegradable, and has a high flash point of about 300 °F (148 °C) compared to petroleum diesel fuel, which has a flash point of 125 °F (52 °C). In many European countries, a 5% biodiesel blend is widely used and is available at thousands of gas stations. In France, biodiesel is incorporated at a rate of 8% in the fuel used by all French diesel vehicles. Avril Group produces under the brand Diester, a fifth of 11 million tons of biodiesel consumed annually by the European Union. It is the leading European producer of biodiesel.

Green Diesel

Green diesel can be produced from a combination of biochemical and thermochemical processes. Conventional green diesel is produced through hydroprocessing biological oil feedstocks, such as vegetable oils and animal fats. Recently, it is produced using series of thermochemical processes such as pyrolysis and hydroprocessing. In the thermochemical route, syngas produced from gasification, bio-oil produced from pyrolysis or biocrude produced from hydrothermal liquefaction is upgraded to green diesel using hydroprocessing. Hydroprocessing is the process of using hydrogen to reform a molecular structure. For example, hydrocracking which is a widely used hydroprocessing technique in refineries is used at elevated temperatures and pressure in the presence of a catalyst to break down larger molecules, such as those found in vegetable oils, into shorter hydrocarbon chains used in diesel engines. Green diesel may also be called renewable diesel, drop-in biodiesel, hydrotreated vegetable oil (HVO fuel) or hydrogen-derived renewable diesel. Unlike biodiesel, green diesel has exactly the same chemical properties as petroleum-based diesel. It does not require new engines, pipelines or infrastructure to distribute and use, but has not been produced at a cost that is competitive with petroleum. Gasoline versions are also being developed. Green diesel is being developed in Louisiana and Singapore by ConocoPhillips, Neste Oil, Valero, Dynamic Fuels, and Honeywell UOP as well as Preem in Gothenburg, Sweden, creating what is known as Evolution Diesel.

Straight Vegetable Oil

Straight unmodified edible vegetable oil is generally not used as fuel, but lower-quality oil has been used for this purpose. Used vegetable oil is increasingly being processed into biodiesel, or (more rarely) cleaned of water and particulates and then used as a fuel.

The IEA estimates that biodiesel production used 17% of global vegetable oil supplies in 2021. Oils and fats reacted with 10 pounds of a short-chain alcohol (usually methanol) in the presence of a catalyst (usually sodium hydroxide [NaOH] can be hydrogenated to give a diesel substitute. The resulting product is a straight-chain hydrocarbon with a high cetane number, low in aromatics and sulfur and does not contain oxygen. Hydrogenated oils can be blended with diesel in all proportions. They have several advantages over biodiesel, including good performance at low temperatures, no storage stability problems and no susceptibility to microbial attack.

Biogasoline

Biogasoline can be produced biologically and thermochemically. Using biological methods, a study led by Professor Lee Sang-yup at the Korea Advanced Institute of Science and Technology (KAIST) and published in the international science journal *Nature* used modified *E. coli* fed with glucose found in plants or other non-food crops to produce biogasoline with the produced enzymes. The enzymes converted the sugar into fatty acids and then turned these into hydrocarbons that were chemically and structurally identical to those found in commercial gasoline fuel. The thermochemical approach of producing biogasoline are similar to those used to produced biodiesel. Biogasoline may also be called drop-in gasoline or renewable gasoline.

Bioethers

Bioethers (also referred to as fuel ethers or oxygenated fuels) are cost-effective compounds that act as octane rating enhancers. "Bioethers are produced by the reaction of reactive iso-olefins, such as iso-butylene, with bioethanol." Bioethers are created from wheat or sugar beets, and also be produced from the waste glycerol that results from the production of biodiesel.

They also enhance engine performance, while significantly reducing engine wear and toxic exhaust emissions. By greatly reducing the amount of ground-level ozone emissions, they contribute to improved air quality. In transportation fuel there are six ether additives: dimethyl ether (DME), diethyl ether (DEE), methyl *tert*-butyl ether (MTBE), ethyl *tert*-butyl ether (ETBE), *tert*-amyl methyl ether (TAME), and *tert*-amyl ethyl ether (TAEE).

The European Fuel Oxygenates Association identifies MTBE and ETBE as the most commonly used ethers in fuel to replace lead. Ethers were introduced in Europe in the 1970s to replace the highly toxic compound. Although Europeans still use bioether additives, the U.S. Energy Policy Act of 2005 lifted a requirement for reformulated gasoline to include an oxygenate, leading to less MTBE being added to fuel. Although bioethers are likely to replace ethers produced from petroleum in the UK, it is highly unlikely they will become a fuel in and of itself due to the low energy density.

Aviation Biofuel

An aviation biofuel (also known as bio-jet fuel or bio-aviation fuel (BAF)) is a biofuel used to power aircraft and is a sustainable aviation fuel (SAF). The International Air Transport Association (IATA) considers it a key element in reducing the environmental impact of aviation. Aviation biofuel is used to decarbonize medium and long-haul air travel. These types of travel generate the most emissions, and could extend the life of older aircraft types by lowering their carbon footprint.

Synthetic paraffinic kerosene (SPK) refers to any non-petroleum-based fuel designed to replace kerosene jet fuel, which is often, but not always, made from biomass.

Biofuels are biomass-derived fuels from plants, animals, or waste; depending on which type of biomass is used, they could lower CO_2 emissions by 20–98% compared to conventional jet fuel. The first test flight using blended biofuel was in 2008, and in 2011, blended fuels with 50% biofuels were allowed on commercial flights. In 2023 SAF production was 600 million liters, representing 0.2% of global jet fuel use.

Aviation biofuel can be produced from plant or animal sources such as *Jatropha*, algae, tallows, waste oils, palm oil, Babassu, and *Camelina* (bio-SPK); from solid biomass using pyrolysis processed with a Fischer–Tropsch process (FT-SPK); with an alcohol-to-jet (ATJ) process from waste fermentation; or from synthetic biology through a solar reactor. Small piston engines can be modified to burn ethanol.

Sustainable biofuels are an alternative to electrofuels. Sustainable aviation fuel is certified as being sustainable by a third-party organisation.

SAF technology faces significant challenges due to feedstock constraints. The oils and fats known as hydrotreated esters and fatty acids (Hefa), crucial for SAF production, are in limited supply as demand increases. Although advanced e-fuels technology, which combines waste CO_2 with clean hydrogen, presents a promising solution, it is still under development and comes with high costs. To overcome these issues, SAF developers are exploring more readily available feedstocks such as woody biomass and agricultural and municipal waste, aiming to produce lower-carbon jet fuel more sustainably and efficiently.

Gaseous

Biogas and Biomethane

Biogas is a mixture composed primarily of methane and carbon dioxide produced by the process of anaerobic digestion of organic material by micro-organisms. Other trace components of this mixture includes water vapor, hydrogen sulfide, siloxanes, hydrocarbons, ammonia, oxygen, carbon monoxide, and nitrogen. It can be produced either from biodegradable waste materials or by the use of energy crops fed into anaerobic digesters to supplement gas yields.

The solid byproduct, digestate, can be used as a biofuel or a fertilizer. When CO_2 and other impurities are removed from biogas, it is called biomethane. The CO_2 can also be combined with hydrogen in methanation to form more methane.

Biogas can be recovered from mechanical biological treatment waste processing systems. Landfill gas, a less clean form of biogas, is produced in landfills through naturally occurring anaerobic digestion. If it escapes into the atmosphere, it acts as a greenhouse gas. In Sweden, "waste-to-energy" power plants capture methane biogas from garbage and use it to power transport systems. Farmers can produce biogas from cattle manure via anaerobic digesters.

Syngas

Syngas, a mixture of carbon monoxide, hydrogen and various hydrocarbons, is produced by partial combustion of biomass (combustion with an amount of oxygen that is not sufficient to convert the biomass completely to carbon dioxide and water). Before partial combustion the biomass is dried and sometimes pyrolysed. Syngas is more efficient than direct combustion of the original biofuel; more of the energy contained in the fuel is extracted.

Syngas may be burned directly in internal combustion engines, turbines or high-temperature fuel cells. The wood gas generator, a wood-fueled gasification reactor, can be connected to an internal combustion engine.

Syngas can be used to produce methanol, dimethyl ether and hydrogen, or converted via the Fischer–Tropsch process to produce a diesel substitute, or a mixture of alcohols that can be blended into gasoline. Gasification normally relies on temperatures greater than 700 °C. Lower-temperature gasification is desirable when co-producing biochar, but results in syngas polluted with tar.

Solid

The term "biofuels" is also used for solid fuels that are made from biomass, even though this is less common.

13.1.4 Research into other Types

Algae-Based Biofuels

Algae can be produced in ponds or tanks on land, and out at sea. Algal fuels have high yields, a high ignition point, can be grown with minimal impact on fresh water resources, can be produced using saline water and wastewater, and are biodegradable and relatively harmless to the environment if spilled. However, production requires large amounts of energy and fertilizer, the produced fuel degrades faster than other biofuels, and it does not flow well in cold temperatures.

By 2017, due to economic considerations, most efforts to produce fuel from algae have been abandoned or changed to other applications. Third and fourth-generation biofuels also include biofuels that are produced by bioengineered organisms i.e. algae and cyanobacteria. Algae and cyanobacteria will use water, carbon dioxide, and solar energy to produce biofuels. This method of biofuel production is still at the research level. The biofuels that are secreted by the bioengineered organisms are expected to have higher photon-to-fuel conversion efficiency, compared to older generations of biofuels. One of the advantages of this class of biofuels is that the cultivation of the organisms that produce the biofuels does not require the use of arable land. The disadvantages include the cost of cultivating the biofuel-producing organisms being very high.

Electrofuels and Solar Fuels

Electrofuels and solar fuels may or may not be biofuels, depending on whether they contain biological elements. Electrofuels are made by storing electrical energy in the chemical bonds of liquids and gases. The primary targets are butanol, biodiesel, and hydrogen, but include other alcohols and carbon-containing gases such as methane and butane. A solar fuel is a synthetic chemical fuel produced from solar energy. Light is converted to chemical energy, typically by reducing protons to hydrogen, or carbon dioxide to organic compounds.

13.1.5 Bio-Digesters

A bio-digester is a mechanized toilet that uses decomposition and sedimentation to turn human waste into a renewable fuel called biogas. Biogas can be made from substances like agricultural waste and sewage.

The bio-digester uses a process called anaerobic digestion to produce biogas. Anaerobic digestion uses a chemical process to break down organic matter with the use of microorganisms in the absence of oxygen to produce biogas. The processes involved in anaerobic respiration are hydrolysis, acidogenesis, acetogenesis, and methanogenesis.

13.1.6 Extent of Production and Use

Global biofuel production was 81 Mtoe in 2017 which represented an annual increase of about 3% compared to 2010. In 2017, the US was the largest biofuel producer in the world producing 37 Mtoe, followed by Brazil and South America at 23 Mtoe and Europe (mainly Germany) at 12 Mtoe.

An assessment from 2017 found that: “Biofuels will never be a major transport fuel as there is just not enough land in the world to grow plants to make biofuel for all vehicles. It can however, be part of an energy mix to take us into a future of renewable energy.”

In 2021, worldwide biofuel production provided 4.3% of the world’s fuels for transport, including a very small amount of aviation biofuel.

By 2027, worldwide biofuel production is expected to supply 5.4% of the world’s fuels for transport including 1% of aviation fuel.

The US, Europe, Brazil and Indonesia are driving the majority of biofuel consumption growth. This demand for biodiesel, renewable diesel and biojet fuel is projected to increase by 44% (21 billion litres) over 2022-2027.

13.1.7 Issues

Issues relating to biofuel are social, economic, environmental and technical problems that may arise from biofuel production and use. Social and economic issues include the "food vs fuel" debate and the need to develop responsible policies and economic instruments to ensure sustainable biofuel production. Farming for biofuels feedstock can be detrimental to the environment if not done sustainably. Environmental concerns include deforestation, biodiversity loss and soil erosion as a result of land clearing for biofuels agriculture. While biofuels can contribute to reduction in global carbon emissions, indirect land use change for biofuel production can have the inverse effect. Technical issues include possible modifications necessary to run the engine on biofuel, as well as energy balance and efficiency. The International Resource Panel outlined the wider and interrelated factors that need to be considered when deciding on the relative merits of pursuing one biofuel over another. The IRP concluded that not all biofuels perform equally in terms of their effect on climate, energy security and ecosystems, and suggested that environmental and social effects need to be assessed throughout the entire life-cycle.

Environmental Impacts

Estimates about the climate impact from biofuels vary widely based on the methodology and exact situation examined. In general, biofuels emit fewer greenhouse gas emissions when burned in an engine and are generally considered carbon-neutral fuels as the carbon they emit has been captured from the atmosphere by the crops used in biofuel production. They can have greenhouse gas emissions ranging from as low as -127.1 gCO_2eq per MJ when carbon capture is incorporated into their production to those exceeding 95 gCO_2eq per MJ when land-use change is significant. Several factors are responsible for the variation in emission numbers of biofuel, such as feedstock and its origin, fuel production technique, system boundary definitions, and energy sources. However, many government policies, such as those by the European Union and the UK, require that biofuels have at least 65% greenhouse gas emissions savings (or 70% if it is renewable fuels of non-biological origins) relative to fossil fuels.

Life-cycle assessments of first-generation biofuels have shown large emissions associated with the potential land-use change required to produce additional biofuel feedstocks. If no land-use change is involved, first-generation biofuels can—on average—have lower emissions than fossil fuels. However, biofuel production can compete with food crop production. Up to 40% of corn produced in the United States is used to make ethanol and worldwide 10% of all grain is turned into biofuel. A 50% reduction in grain used for biofuels in the US and Europe would replace all of Ukraine's grain exports. Several studies have shown that reductions in emissions from biofuels are achieved at the expense of other impacts, such as acidification, eutrophication, water footprint and biodiversity loss.

Second-generation biofuels are thought to increase environmental sustainability since the non-food part of plants is being used to produce second-generation biofuels instead of being disposed of. But the use of second-generation biofuels increases the competition for lignocellulosic biomass, increasing the cost of these biofuels.

In theory, third-generation biofuels, produced from algae, shouldn't harm the environment more than first- or second-generation biofuels due to lower changes in land use and the fact that they do not require pesticide use for production. When looking at the data however, it has been shown that the environmental cost to produce the infrastructure and energy required for third generation biofuel production, are higher than the benefits provided from the biofuels use.

The European Commission has officially approved a measure to phase out palm oil-based biofuels by 2030. Unsustainable palm oil agriculture has caused significant environmental and social problems, including deforestation and pollution.

The production of biofuels can be very energy intensive, which, if generated from non-renewable sources, can heavily mitigate the benefits gained through biofuel use.

A solution proposed to solve this issue is to supply biofuel production facilities with excess nuclear energy, which can supplement the power provided by fossil fuels. This can provide a carbon inexpensive solution to help reduce the environmental impacts of biofuel production.

Indirect Land Use Change Impacts of Biofuels

The indirect land use change impacts of biofuels, also known as ILUC or iLUC (pronounced as i-luck), relates to the unintended consequence of releasing more carbon emissions due to land-use changes around the world induced by the expansion of croplands for ethanol or biodiesel production in response to the increased global demand for biofuels.

As farmers worldwide respond to higher crop prices in order to maintain the global food supply-and-demand balance, pristine lands are cleared to replace the food crops that were diverted elsewhere to biofuels' production.

Because natural lands, such as rainforests and grasslands, store carbon in their soil and biomass as plants grow each year, clearance of wilderness for new farms translates to a net increase in greenhouse gas emissions. Due to this off-site change in the carbon stock of the soil and the biomass, indirect land use change has consequences in the greenhouse gas (GHG) balance of a biofuel.

13.2 IMPORTANCE OF BIOFUELS AND INDUSTRIAL ALCOHOLS

Biofuels and industrial alcohols have gained significance due to their potential to reduce greenhouse gas emissions, enhance energy security, and create economic opportunities through the use of renewable resources.

Unlike fossil fuels, biofuels are derived from renewable feedstocks such as agricultural crops, waste materials, and algae, making them a more sustainable energy option. The combustion of biofuels emits significantly lower levels of carbon dioxide (CO_2) and other pollutants, contributing to efforts to mitigate climate change.

Industrial alcohols, particularly ethanol, play a crucial role in a variety of sectors, including transportation, pharmaceuticals, and cosmetics. Ethanol is widely used as a biofuel additive to gasoline (often referred to as gasohol) to reduce emissions and improve fuel efficiency. Methanol, another type of industrial alcohol, is used as a feedstock for producing chemicals, plastics, and formaldehyde. The production of biofuels and industrial alcohols, therefore, has far-reaching implications for both the energy sector and broader industrial applications.

13.3 FEEDSTOCKS FOR BIOFUEL AND INDUSTRIAL ALCOHOL PRODUCTION

The production of biofuels and industrial alcohols depends on the availability and type of feedstocks. These feedstocks are broadly classified into three categories: first-generation, second-generation, and third-generation.

13.3.1 First-Generation Feedstocks

First-generation feedstocks refer to food crops such as corn, sugarcane, wheat, and soybean. These crops are rich in carbohydrates (starches or sugars) or lipids, making them suitable for biofuel production. The most common biofuel produced from first-generation feedstocks is bioethanol, which is derived through the fermentation of sugars in crops like corn or sugarcane. For instance, in Brazil, sugarcane is the primary feedstock for bioethanol production, while in the United States, corn is predominantly used.

Biodiesel, another form of biofuel, is produced from vegetable oils derived from soybean, rapeseed, or palm oil. These oils undergo a transesterification process, where they are reacted with alcohol (usually methanol) to produce biodiesel and glycerol as a byproduct.

However, the use of first-generation feedstocks has raised ethical concerns regarding food security. Since these feedstocks are also used for food production, diverting them for biofuel production can lead to higher food prices and exacerbate food shortages in some regions.

13.3.2 Second-Generation Feedstocks

Second-generation feedstocks, also known as lignocellulosic biomass, are derived from non-food sources such as agricultural residues (e.g., corn stover, wheat straw), forest residues, and dedicated energy crops (e.g., switchgrass, miscanthus). These feedstocks consist primarily of cellulose, hemicellulose, and lignin, which must be broken down into fermentable sugars before biofuels can be produced.

The production of bioethanol from lignocellulosic biomass involves several steps: pretreatment, enzymatic hydrolysis, fermentation, and distillation. Pretreatment is necessary to break down the tough cell walls of the biomass and make the cellulose more accessible to enzymes. Enzymatic hydrolysis then converts the cellulose into glucose, which can be fermented by microorganisms such as *Saccharomyces cerevisiae* to produce ethanol.

The advantage of second-generation feedstocks is that they do not compete directly with food crops, making them a more sustainable option for biofuel production. However, the conversion processes are more complex and costly compared to first-generation biofuels, which has limited their large-scale commercial deployment.

13.3.3 Third-Generation Feedstocks

Third-generation feedstocks primarily refer to algae, which are considered one of the most promising sources of biofuel. Algae have a high oil content, grow rapidly, and can be cultivated on non-arable land, making them an attractive option for biodiesel production. Unlike terrestrial crops, algae can produce biomass year-round and have a much higher yield per unit area.

There are two main types of algae used for biofuel production: microalgae and macroalgae (seaweed). Microalgae are particularly suited for biodiesel production because of their high lipid content, which can be extracted and converted into biodiesel through transesterification.

Additionally, algae can be used to produce bioethanol and biogas through fermentation and anaerobic digestion, respectively.

The challenges associated with third-generation feedstocks include the high cost of cultivation, harvesting, and oil extraction. Despite these challenges, ongoing research and technological advancements are expected to make algae-based biofuels more economically viable in the future.

13.4 PROCESSES INVOLVED IN BIOFUEL AND INDUSTRIAL ALCOHOL PRODUCTION

The production of biofuels and industrial alcohols involves a series of biochemical, thermochemical, and mechanical processes. These processes vary depending on the type of feedstock and the desired end product.

13.4.1 Fermentation for Bioethanol Production

Fermentation is the most widely used process for producing bioethanol from carbohydrate-rich feedstocks. During fermentation, microorganisms such as yeast (e.g., *Saccharomyces cerevisiae*) or bacteria metabolize sugars into ethanol and carbon dioxide. The fermentation process typically takes place in large fermentation tanks under controlled conditions of temperature, pH, and oxygen levels to optimize ethanol yield.

For first-generation bioethanol production, the feedstock (e.g., corn, sugarcane) is first processed to extract the fermentable sugars. In the case of corn, the starch is broken down into glucose through enzymatic hydrolysis. For sugarcane, the sugars are readily available in the form of sucrose, which can be directly fermented by yeast.

In the case of second-generation bioethanol production, lignocellulosic biomass must undergo pretreatment and enzymatic hydrolysis to release the sugars, which are then fermented. After fermentation, the ethanol is separated from the fermentation broth through distillation.

13.4.2 Transesterification for Biodiesel Production

Biodiesel is produced through a chemical process known as transesterification, where vegetable oils or animal fats react with an alcohol (usually methanol) in the presence of a catalyst (e.g., sodium hydroxide or potassium hydroxide). The transesterification reaction breaks down the triglycerides in the oil into fatty acid methyl esters (FAME), which are the chemical components of biodiesel. Glycerol is produced as a byproduct of the reaction and can be used in other industries, such as cosmetics and pharmaceuticals. The production of biodiesel from algae involves similar processes, with the added step of oil extraction. The harvested algae are dried, and the oil is extracted using solvents or mechanical pressing. The extracted oil is then converted into biodiesel through transesterification.

13.4.3 Anaerobic Digestion for Biogas Production

Biogas is produced through anaerobic digestion, a biological process in which organic matter is broken down by microorganisms in the absence of oxygen. This process occurs naturally in environments such as landfills and wetlands, but it can also be controlled in industrial settings to produce biogas from agricultural waste, sewage sludge, or food waste.

Anaerobic digestion takes place in large reactors known as digesters, where the organic material is decomposed into biogas (a mixture of methane and carbon dioxide) and digestate (a nutrient-rich byproduct). The biogas can be used as a renewable energy source for heating, electricity generation, or as a vehicle fuel.

13.5 INDUSTRIAL APPLICATIONS OF ALCOHOLS

In addition to their use as biofuels, industrial alcohols such as ethanol and methanol have a wide range of applications. Ethanol is used as a solvent in the production of pharmaceuticals, cosmetics, and personal care products. It is also a key ingredient in the production of alcoholic beverages and is used as a disinfectant in hand sanitizers and cleaning products. Methanol, on the other hand, is used as a feedstock in the production of formaldehyde, which is an important raw material for manufacturing plastics, resins, and adhesives. Methanol is also used in the production of synthetic fuels, as an antifreeze agent, and in the synthesis of acetic acid.

13.6 CONCLUSION

The production of biofuels and industrial alcohols holds great promise for reducing dependence on fossil fuels, lowering greenhouse gas emissions, and promoting sustainable energy solutions. The use of renewable feedstocks such as agricultural crops, waste materials, and algae presents opportunities to harness biological resources for energy and industrial purposes. While challenges remain, particularly in terms of cost and technological feasibility, continued research and development are expected to drive advancements in biofuel production and expand the role of industrial alcohols in diverse sectors. As the world moves toward a more sustainable future, biofuels and industrial alcohols will play an increasingly important role in the global energy landscape.

CHAPTER 14
PRODUCTION OF ORGANIC ACIDS

Organic acids are a group of important chemical compounds used in various industries, including food and beverage, pharmaceuticals, cosmetics, textiles, and agriculture. They are characterized by the presence of one or more carboxyl (-COOH) groups and are typically produced either through chemical synthesis or microbial fermentation. The global demand for organic acids continues to rise due to their versatility and eco-friendliness, making the biological production of these acids, particularly through fermentation, an increasingly attractive method.

14.1 CHARACTERISTICS

In general, organic acids are weak acids and do not dissociate completely in water, whereas the strong mineral acids do. Lower molecular mass organic acids such as formic and lactic acids are miscible in water, but higher molecular mass organic acids, such as benzoic acid, are insoluble in molecular (neutral) form. On the other hand, most organic acids are very soluble in organic solvents. *p*-Toluenesulfonic acid is a comparatively strong acid used in organic chemistry often because it is able to dissolve in the organic reaction solvent. Exceptions to these solubility characteristics exist in the presence of other substituents that affect the polarity of the compound.

14.2 APPLICATIONS

Simple organic acids like formic or acetic acids are used for oil and gas well stimulation treatments. These organic acids are much less reactive with metals than are strong mineral acids like hydrochloric acid (HCl) or mixtures of HCl and hydrofluoric acid (HF). For this reason, organic acids are used at high temperatures or when long contact times between acid and pipe are needed.

The conjugate bases of organic acids such as citrate and lactate are often used in biologically compatible buffer solutions. Citric and oxalic acids are used as rust removal. As acids, they can dissolve the iron oxides, but without damaging the base metal as do stronger mineral acids. In the dissociated form, they may be able to chelate the metal ions, helping to speed removal.

Biological systems create many more complex organic acids such as L-lactic, citric, and D-glucuronic acids that contain hydroxyl or carboxyl groups. Human blood and urine contain these plus organic acid degradation products of amino acids, neurotransmitters, and intestinal bacterial action on food components. Examples of these categories are alpha-ketoisocaproic, vanilmandelic, and D-lactic acids, derived from catabolism of L-leucine and epinephrine (adrenaline) by human tissues and catabolism of dietary carbohydrate by intestinal bacteria, respectively. Organic acids (C_1–C_7) are widely distributed in nature as normal constituents of plants or animal tissues. They are also formed through microbial fermentation of carbohydrates mainly in the large intestine. They are sometimes found in their sodium, potassium, or calcium salts, or even stronger double salts.

14.2.1 In Food

Organic acids are used in food preservation because of their effects on bacteria. The key basic principle on the mode of action of organic acids on bacteria is that non-dissociated (non-ionized) organic acids can penetrate the bacteria cell wall and disrupt the normal physiology of certain types of bacteria that we call *pH-sensitive*, meaning that they cannot tolerate a wide internal and external pH gradient. Among those bacteria are *Escherichia coli*, *Salmonella* spp., *C. perfringens*, *Listeria monocytogenes*, and *Campylobacter* species.

Upon passive diffusion of organic acids into the bacteria, where the pH is near or above neutrality, the acids will dissociate and raise the bacteria internal pH, leading to situations that will not impair nor stop the growth of bacteria. On the other hand, the anionic part of the organic acids that can escape the bacteria in its dissociated form will accumulate within the bacteria and disrupt few metabolic functions, leading to osmotic pressure increase, incompatible with the survival of the bacteria.

It has been well demonstrated that the state of the organic acids (undissociated or dissociated) is not important to define their capacity to inhibit the growth of bacteria, compared to undissociated acids. Lactic acid and its salts sodium lactate and potassium lactate are widely used as antimicrobials in food products, in particular, dairy and poultry such as ham and sausages.

14.2.2 In Nutrition and Animal Feeds

Organic acids have been used successfully in pig production for more than 25 years. Although less research has been done in poultry, organic acids have also been found to be effective in poultry production. Organic acids added to feeds should be protected to avoid their dissociation in the crop and in the intestine (high pH segments) and reach far into the gastrointestinal tract, where the bulk of the bacteria population is located. From the use of organic acids in poultry and pigs, one can expect an improvement in performance similar to or better than that of antibiotic growth promoters, without the public health concern, a preventive effect on the intestinal problems like necrotic enteritis in chickens and *Escherichia coli* infection in young pigs. Also one can expect a reduction of the carrier state for *Salmonella* species and *Campylobacter* species.

14.2.3 Ongoing Research

In addition to the end uses previously seen, organic acids have been tested for the following applications:

- Barbero-López and colleagues tested at the University of Eastern Finland the potential use of three organic acids, acetic, formic and propionic acids, in wood preservation. They showed a high antifungal potential against the decaying fungi tested (brown rotting fungi *Coniophora puteana, Rhodonia placenta* and *Gloeophyllum trabeum;* White rotting fungus *Trametes versicolor)* in Petri dish. However, when they treated wood with organic acids, the acids leached out from wood and did not prevent degradation. Additionally, the organic acids' acidity may have caused chemical degradation on wood. Additionally, in a more recent study, the ecotoxicity of several natural wood preservatives was compared, and the results indicated a very low toxicity of propionic acid.

14.3 IMPORTANCE OF ORGANIC ACIDS

Organic acids are vital in multiple sectors due to their wide range of properties. In the food and beverage industry, they serve as preservatives, flavor enhancers, acidulants, and pH regulators. Citric acid, for instance, is widely used to enhance flavor and act as a preservative in soft drinks, candies, and sauces. Similarly, lactic acid is employed in the fermentation of dairy products such as yogurt and cheese, contributing to texture and flavor.

In the pharmaceutical industry, organic acids are used as excipients, intermediates for drug production, and active ingredients. For example, salicylic acid is a precursor for the synthesis of aspirin, while ascorbic acid (vitamin C) is used for its antioxidant properties in various health supplements.

Industrial applications of organic acids extend to biodegradable polymers, detergents, and bio-based materials. With increasing demand for sustainable products, the production of bio-based organic acids via microbial fermentation is becoming an essential part of the green chemistry movement.

14.4 METHODS OF ORGANIC ACID PRODUCTION

The production of organic acids can be categorized into two major methods: chemical synthesis and microbial fermentation. While chemical synthesis remains the traditional approach, microbial fermentation has gained prominence due to its sustainability and the availability of renewable feedstocks.

14.4.1 Chemical Synthesis

Chemical synthesis of organic acids involves the reaction of specific chemicals, often derived from petroleum-based sources, under controlled conditions to form the desired organic acid. For example, acetic acid is traditionally produced through the oxidation of ethanol or hydrocarbons. Chemical synthesis has long been the primary method due to its established infrastructure and relatively high yield. However, chemical synthesis has several drawbacks, including the reliance on non-renewable fossil fuels and the generation of harmful byproducts. As environmental concerns grow and industries seek sustainable alternatives, chemical synthesis is gradually being replaced by biotechnological methods such as microbial fermentation.

14.4.2 Microbial Fermentation

Microbial fermentation is a biological process that involves the use of microorganisms, such as bacteria, fungi, and yeast, to convert renewable biomass into organic acids. This method is more eco-friendly compared to chemical synthesis, as it utilizes renewable feedstocks, generates fewer byproducts, and operates under milder conditions. The microorganisms used in fermentation metabolize sugars, starches, or other organic materials into acids as a byproduct of their metabolic processes. There are two main types of fermentation processes used in the production of organic acids: batch fermentation and continuous fermentation.

- **Batch Fermentation:** In this method, the microbial culture is grown in a closed system with a fixed amount of substrate. The process continues until the substrate is consumed or the organic acid concentration reaches a desired level, after which the product is harvested, and the process is restarted with a fresh culture. This is the traditional method used for producing organic acids like lactic acid and citric acid.
- **Continuous Fermentation:** This process allows for the constant feeding of substrate into the system, while the product and waste materials are continuously removed. Continuous fermentation is more efficient for large-scale production and ensures a steady supply of organic acids. However, it requires precise control of environmental conditions and microbial growth rates.

The microorganisms used for organic acid production are often genetically modified to optimize their yield and tolerance to the acidic environment they create during fermentation.

Genetic engineering can also enhance the metabolic pathways that lead to the production of specific organic acids, making microbial fermentation a flexible and efficient method for industrial production.

14.5 TYPES OF ORGANIC ACIDS AND THEIR PRODUCTION

Several organic acids are produced on an industrial scale, with each acid having its own unique production process and applications. Below are some of the most important organic acids, along with a brief overview of how they are produced.

14.5.1 Citric Acid

Citric acid is one of the most widely produced organic acids globally, with applications in the food and beverage, pharmaceutical, and cosmetic industries. It is commonly produced through microbial fermentation using the filamentous fungus *Aspergillus niger*. The fermentation process involves the use of glucose or molasses as a carbon source, which is metabolized by the fungus to produce citric acid. In industrial settings, the fermentation process typically occurs in large bioreactors under controlled conditions of pH and temperature to optimize citric acid production. After fermentation, the citric acid is recovered from the fermentation broth through filtration, precipitation, or crystallization processes.

14.5.2 Lactic Acid

Lactic acid is an important organic acid used in food preservation, biodegradable plastics, and pharmaceuticals. It is produced through microbial fermentation using lactic acid bacteria (e.g., *Lactobacillus* species) or certain fungi. The process involves the fermentation of carbohydrates, such as glucose or lactose, into lactic acid. Lactic acid can be produced in two forms: L(+)-lactic acid and D(-)-lactic acid, depending on the microorganism used. Lactic acid fermentation can be carried out in either batch or continuous systems, and the product is recovered through filtration and purification processes. Lactic acid has gained significant attention due to its use in the production of polylactic acid (PLA), a biodegradable plastic that serves as a sustainable alternative to conventional petroleum-based plastics.

14.5.3 Acetic Acid

Acetic acid is another widely produced organic acid, used primarily as a chemical reagent and in the production of vinegar, which is its dilute form. Industrially, acetic acid is produced both chemically, through the oxidation of ethanol, and biologically, through the fermentation of ethanol by acetic acid bacteria (e.g., *Acetobacter* species). The microbial production of acetic acid is particularly significant in the food and beverage industry, where it is used as a preservative and flavoring agent in products such as pickles, sauces, and salad dressings.

14.5.4 Succinic Acid

Succinic acid is an emerging bio-based organic acid with applications in biodegradable plastics, food additives, and pharmaceuticals. It is produced through microbial fermentation using bacteria such as *Escherichia coli* or fungi like *Saccharomyces cerevisiae*. The fermentation of glucose or other sugars by these microorganisms results in the production of succinic acid, which is then recovered through precipitation or crystallization. Bio-based succinic acid has the potential to replace petroleum-derived chemicals in various industries, particularly in the production of bioplastics and biodegradable polymers.

14.5.5 Other Organic Acids

Other notable organic acids produced through microbial fermentation include fumaric acid, malic acid, propionic acid, and itaconic acid. Each of these acids has its own unique production process, depending on the microorganism used and the substrate provided. For example, fumaric acid is produced using the fungus *Rhizopus oryzae*, while propionic acid is produced by propionibacteria through the fermentation of sugars.

14.6 INDUSTRIAL APPLICATIONS OF ORGANIC ACIDS

Organic acids have a diverse range of applications across various industries. Some of the key uses of these acids are outlined below:

- **Food and Beverage Industry:** Organic acids such as citric acid, lactic acid, and acetic acid are commonly used as acidulants, preservatives, and flavor enhancers in foods and beverages. They help extend shelf life, improve flavor, and maintain the desired pH levels in products like soft drinks, dairy products, and condiments.
- **Pharmaceutical Industry:** Organic acids play an important role in the pharmaceutical industry as excipients, stabilizers, and active ingredients. For example, citric acid is used as a buffering agent in medications, while salicylic acid is used in acne treatments and anti-inflammatory drugs.
- **Biodegradable Plastics:** Lactic acid is used in the production of polylactic acid (PLA), a biodegradable polymer that serves as an environmentally friendly alternative to petroleum-based plastics. Similarly, succinic acid is used to produce biodegradable plastics and resins.
- **Agriculture:** Organic acids are used in animal feed to improve digestion and promote growth. Acetic acid and propionic acid are also used as preservatives in silage and feedstock to prevent spoilage by microorganisms.

14.7 CONCLUSION

The production of organic acids is a crucial part of many industries due to their wide range of applications and growing importance in sustainable industrial practices. While chemical synthesis remains a method of production, microbial fermentation is becoming increasingly important as a more eco-friendly and cost-effective alternative. With the rising demand for bio-based products, the microbial production of organic acids is poised to play a pivotal role in the future of green chemistry and industrial biotechnology.

Through advances in fermentation technologies and genetic engineering, the production of organic acids will continue to expand, offering innovative solutions to global challenges in food security, healthcare, and environmental sustainability.

CHAPTER 15
PRODUCTION OF AMINO ACIDS BY FERMENTATION

Amino acids are the building blocks of proteins and play critical roles in various biological processes, from metabolism to cellular repair. Beyond their biological significance, amino acids have a broad range of industrial applications. They are used in the food and feed industries as nutritional supplements, in the pharmaceutical industry as precursors for drug synthesis, and in cosmetics for skin care products. Due to the growing demand for amino acids in various sectors, fermentation-based production methods have gained prominence.

15.1 AMINO ACID: AN OVERVIEW

Amino acids are organic compounds that contain both amino and carboxylic acid functional groups. Although over 500 amino acids exist in nature, by far the most important are the 22 α-amino acids incorporated into proteins. Only these 22 appear in the genetic code of life.

Amino acids can be classified according to the locations of the core structural functional groups (alpha- (α-), beta- (β-), gamma- (γ-) amino acids, etc.); other categories relate to polarity, ionization, and side-chain group type (aliphatic, acyclic, aromatic, polar, etc.). In the form of proteins, amino-acid *residues* form the second-largest component (water being the largest) of human muscles and other tissues. Beyond their role as residues in proteins, amino acids participate in a number of processes such as neurotransmitter transport and biosynthesis. It is thought that they played a key role in enabling life on Earth and its emergence.

Amino acids are formally named by the IUPAC-IUBMB Joint Commission on Biochemical Nomenclature in terms of the fictitious "neutral" structure shown in the illustration. For example, the systematic name of alanine is 2-aminopropanoic acid, based on the formula CH_3-CH(NH_2)-COOH. The Commission justified this approach as follows:

- The systematic names and formulas given refer to hypothetical forms in which amino groups are unprotonated and carboxyl groups are undissociated. This convention is useful to avoid various nomenclatural problems but should not be taken to imply that these structures represent an appreciable fraction of the amino-acid molecules.

15.1.1 History

The first few amino acids were discovered in the early 1800s. In 1806, French chemists Louis-Nicolas Vauquelin and Pierre Jean Robiquet isolated a compound from asparagus that was subsequently named asparagine, the first amino acid to be discovered. Cystine was discovered in 1810, although its monomer, cysteine, remained undiscovered until 1884. Glycine and leucine were discovered in 1820. The last of the 20 common amino acids to be discovered was threonine in 1935 by William Cumming Rose, who also determined the essential amino acids and established the minimum daily requirements of all amino acids for optimal growth.

The unity of the chemical category was recognized by Wurtz in 1865, but he gave no particular name to it. The first use of the term "amino acid" in the English language dates from 1898, while the German term, *Aminosäure*, was used earlier. Proteins were found to yield amino acids after enzymatic digestion or acid hydrolysis. In 1902, Emil Fischer and Franz Hofmeister independently proposed that proteins are formed from many amino acids, whereby bonds are formed between the amino group of one amino acid with the carboxyl group of another, resulting in a linear structure that Fischer termed "peptide".

15.1.2 General Structure

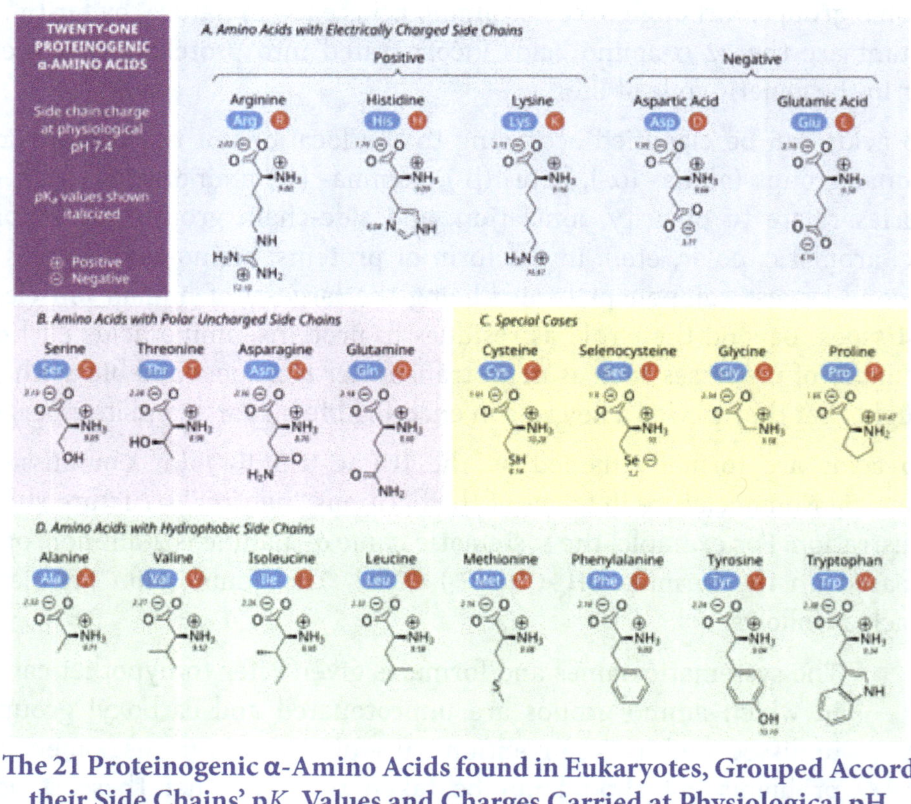

Fig 1: The 21 Proteinogenic α-Amino Acids found in Eukaryotes, Grouped According to their Side Chains' pK_a Values and Charges Carried at Physiological pH

2-, alpha-, or α-**amino acids** have the generic formula $H_2NCHRCOOH$ in most cases, where R is an organic substituent known as a "side chain".

Of the many hundreds of described amino acids, 22 are proteinogenic ("protein-building"). It is these 22 compounds that combine to give a vast array of peptides and proteins assembled by ribosomes. Non-proteinogenic or modified amino acids may arise from post-translational modification or during nonribosomal peptide synthesis.

Chirality

The carbon atom next to the carboxyl group is called the α–carbon. In proteinogenic amino acids, it bears the amine and the R group or side chain specific to each amino acid. With four distinct substituents, the α–carbon is stereogenic in all α-amino acids except glycine. All chiral proteogenic amino acids have the L configuration. They are "left-handed" enantiomers, which refers to the stereoisomers of the alpha carbon. A few D-amino acids ("right-handed") have been found in nature, e.g., in bacterial envelopes, as a neuromodulator (D-serine), and in some antibiotics. Rarely, D-amino acid residues are found in proteins, and are converted from the L-amino acid as a post-translational modification.

Side Chains

Polar Charged Side Chains

Five amino acids possess a charge at neutral pH. Often these side chains appear at the surfaces on proteins to enable their solubility in water, and side chains with opposite charges form important electrostatic contacts called salt bridges that maintain structures within a single protein or between interfacing proteins.

Many proteins bind metal into their structures specifically, and these interactions are commonly mediated by charged side chains such as aspartate, glutamate and histidine. Under certain conditions, each ion-forming group can be charged, forming double salts.

The two negatively charged amino acids at neutral pH are aspartate (Asp, D) and glutamate (Glu, E). The anionic carboxylate groups behave as Brønsted bases in most circumstances. Enzymes in very low pH environments, like the aspartic protease pepsin in mammalian stomachs, may have catalytic aspartate or glutamate residues that act as Brønsted acids.

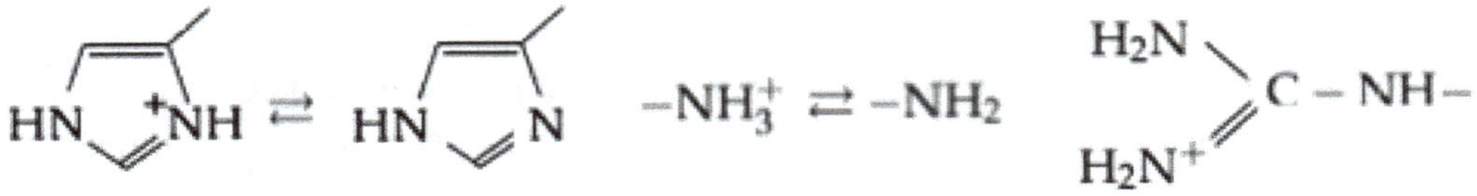

Fig 2: Functional Groups Found in Histidine (Left), Lysine (Middle) and Arginine (Right)

There are three amino acids with side chains that are cations at neutral pH: arginine (Arg, R), lysine (Lys, K) and histidine (His, H). Arginine has a charged guanidino group and lysine a charged alkyl amino group, and are fully protonated at pH 7. Histidine's imidazole group has a pK_a of 6.0, and is only around 10% protonated at neutral pH. Because histidine is easily found in its basic and conjugate acid forms it often participates in catalytic proton transfers in enzyme reactions.

Polar Uncharged Side Chains

The polar, uncharged amino acids serine (Ser, S), threonine (Thr, T), asparagine (Asn, N) and glutamine (Gln, Q) readily form hydrogen bonds with water and other amino acids. They do not ionize in normal conditions, a prominent exception being the catalytic serine in serine proteases. This is an example of severe perturbation, and is not characteristic of serine residues in general. Threonine has two chiral centers, not only the L (2*S*) chiral center at the α-carbon shared by all amino acids apart from achiral glycine, but also (3*R*) at the β-carbon. The full stereochemical specification is (2*S*,3*R*)-L-threonine.

Hydrophobic Side Chains

Nonpolar amino acid interactions are the primary driving force behind the processes that fold proteins into their functional three dimensional structures. None of these amino acids' side chains ionize easily, and therefore do not have pK_as, with the exception of tyrosine (Tyr, Y). The hydroxyl of tyrosine can deprotonate at high pH forming the negatively charged phenolate. Because of this one could place tyrosine into the polar, uncharged amino acid category, but its very low solubility in water matches the characteristics of hydrophobic amino acids well.

Special Case Side Chains

Several side chains are not described well by the charged, polar and hydrophobic categories. Glycine (Gly, G) could be considered a polar amino acid since its small size means that its solubility is largely determined by the amino and carboxylate groups. However, the lack of any side chain provides glycine with a unique flexibility among amino acids with large ramifications to protein folding. Cysteine (Cys, C) can also form hydrogen bonds readily, which would place it in the polar amino acid category, though it can often be found in protein structures forming covalent bonds, called disulphide bonds, with other cysteines. These bonds influence the folding and stability of proteins, and are essential in the formation of antibodies. Proline (Pro, P) has an alkyl side chain and could be considered hydrophobic, but because the side chain joins back onto the alpha amino group it becomes particularly inflexible when incorporated into proteins. Similar to glycine this influences protein structure in a way unique among amino acids. Selenocysteine (Sec, U) is a rare amino acid not directly encoded by DNA, but is incorporated into proteins via the ribosome. Selenocysteine has a lower redox potential compared to the similar cysteine, and participates in several unique enzymatic reactions. Pyrrolysine (Pyl, O) is another amino acid not encoded in DNA, but synthesized into protein by ribosomes. It is found in archaeal species where it participates in the catalytic activity of several methyltransferases.

β- and γ-Amino Acids

Amino acids with the structure NH+3-CXY-CXY-CO-2, such as β-alanine, a component of carnosine and a few other peptides, are β-amino acids. Ones with the structure NH+3-CXY-CXY-CXY-CO-2 are γ-amino acids, and so on, where X and Y are two substituents (one of which is normally H).

Zwitterions

The common natural forms of amino acids have a zwitterionic structure, with -NH+3 (-NH+2- in the case of proline) and -CO-2 functional groups attached to the same C atom, and are thus α-amino acids, and are the only ones found in proteins during translation in the ribosome.

In aqueous solution at pH close to neutrality, amino acids exist as zwitterions, i.e. as dipolar ions with both NH+3 and CO-2 in charged states, so the overall structure is NH+3-CHR-CO-2. At physiological pH the so-called "neutral forms" -NH_2-CHR-CO_2H are not present to any measurable degree. Although the two charges in the zwitterion structure add up to zero it is misleading to call a species with a net charge of zero "uncharged".

In strongly acidic conditions (pH below 3), the carboxylate group becomes protonated and the structure becomes an ammonio carboxylic acid, NH+3-CHR-CO_2H.

This is relevant for enzymes like pepsin that are active in acidic environments such as the mammalian stomach and lysosomes, but does not significantly apply to intracellular enzymes. In highly basic conditions (pH greater than 10, not normally seen in physiological conditions), the ammonio group is deprotonated to give NH_2-CHR-CO-2.

Although various definitions of acids and bases are used in chemistry, the only one that is useful for chemistry in aqueous solution is that of Brønsted: an acid is a species that can donate a proton to another species, and a base is one that can accept a proton. This criterion is used to label the groups in the above illustration. The carboxylate side chains of aspartate and glutamate residues are the principal Brønsted bases in proteins. Likewise, lysine, tyrosine and cysteine will typically act as a Brønsted acid. Histidine under these conditions can act both as a Brønsted acid and a base.

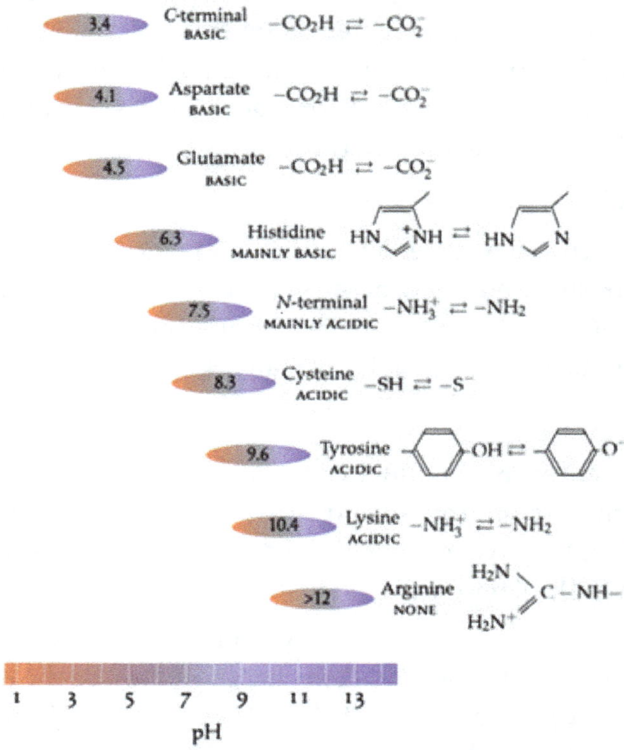

Fig 3: Ionization and Brønsted Character of N-Terminal Amino, C-Terminal Carboxylate, and Side Chains of Amino Acid Residues

Isoelectric Point

For amino acids with uncharged side-chains the zwitterion predominates at pH values between the two pK_a values, but coexists in equilibrium with small amounts of net negative and net positive ions. At the midpoint between the two pK_a values, the trace amount of net negative and trace of net positive ions balance, so that average net charge of all forms present is zero.

For amino acids with charged side chains, the pK_a of the side chain is involved. Thus for aspartate or glutamate with negative side chains, the terminal amino group is essentially entirely in the charged form -NH+3, but this positive charge needs to be balanced by the state with just one C-terminal carboxylate group is negatively charged.

Similar considerations apply to other amino acids with ionizable side-chains, including not only glutamate (similar to aspartate), but also cysteine, histidine, lysine, tyrosine and arginine with positive side chains.

Amino acids have zero mobility in electrophoresis at their isoelectric point, although this behaviour is more usually exploited for peptides and proteins than single amino acids. Zwitterions have minimum solubility at their isoelectric point, and some amino acids (in particular, with nonpolar side chains) can be isolated by precipitation from water by adjusting the pH to the required isoelectric point.

15.1.3 Physicochemical Properties

The 20 canonical amino acids can be classified according to their properties. Important factors are charge, hydrophilicity or hydrophobicity, size, and functional groups. These properties influence protein structure and protein–protein interactions. The water-soluble proteins tend to have their hydrophobic residues (Leu, Ile, Val, Phe, and Trp) buried in the middle of the protein, whereas hydrophilic side chains are exposed to the aqueous solvent. (In biochemistry, a residue refers to a specific monomer *within* the polymeric chain of a polysaccharide, protein or nucleic acid.) The integral membrane proteins tend to have outer rings of exposed hydrophobic amino acids that anchor them in the lipid bilayer. Some peripheral membrane proteins have a patch of hydrophobic amino acids on their surface that sticks to the membrane. In a similar fashion, proteins that have to bind to positively charged molecules have surfaces rich in negatively charged amino acids such as glutamate and aspartate, while proteins binding to negatively charged molecules have surfaces rich in positively charged amino acids like lysine and arginine. For example, lysine and arginine are present in large amounts in the low-complexity regions of nucleic-acid binding proteins. There are various hydrophobicity scales of amino acid residues.

Some amino acids have special properties. Cysteine can form covalent disulfide bonds to other cysteine residues. Proline forms a cycle to the polypeptide backbone, and glycine is more flexible than other amino acids.

Glycine and proline are strongly present within low complexity regions of both eukaryotic and prokaryotic proteins, whereas the opposite is the case with cysteine, phenylalanine, tryptophan, methionine, valine, leucine, isoleucine, which are highly reactive, or complex, or hydrophobic.

Many proteins undergo a range of posttranslational modifications, whereby additional chemical groups are attached to the amino acid residue side chains sometimes producing lipoproteins (that are hydrophobic), or glycoproteins (that are hydrophilic) allowing the protein to attach temporarily to a membrane. For example, a signaling protein can attach and then detach from a cell membrane, because it contains cysteine residues that can have the fatty acid palmitic acid added to them and subsequently removed.

Table of Standard Amino Acid Abbreviations and Properties

Although one-letter symbols are included in the table, IUPAC–IUBMB recommend that “Use of the one-letter symbols should be restricted to the comparison of long sequences". The one-letter notation was chosen by IUPAC-IUB based on the following rules:

- Initial letters are used where there is no ambuiguity: C cysteine, H histidine, I isoleucine, M methionine, S serine, V valine,
- Where arbitrary assignment is needed, the structurally simpler amino acids are given precedence: A Alanine, G glycine, L leucine, P proline, T threonine,
- F *PH*enylalanine and R a*R*ginine are assigned by being phonetically suggestive,
- W tryptophan is assigned based on the double ring being visually suggestive to the bulky letter W,
- K lysine and Y tyrosine are assigned as alphabetically nearest to their initials L and T (note that U was avoided for its similarity with V, while X was reserved for undetermined or atypical amino acids); for tyrosine the mnemonic t*Y*rosine was also proposed,
- D aspartate was assigned arbitrarily, with the proposed mnemonic aspar*D*ic acid; E glutamate was assigned in alphabetical sequence being larger by merely one methylene –CH2– group,
- N asparagine was assigned arbitrarily, with the proposed mnemonic asparagi*N*e; Q glutamine was assigned in alphabetical sequence of those still available (note again that O was avoided due to similarity with D), with the proposed mnemonic *Q*lutamine.

15.1.4 Occurrence and Functions in Biochemistry

Proteinogenic Amino Acids

Amino acids are the precursors to proteins. They join by condensation reactions to form short polymer chains called peptides or longer chains called either polypeptides or proteins.

These chains are linear and unbranched, with each amino acid residue within the chain attached to two neighboring amino acids. In nature, the process of making proteins encoded by RNA genetic material is called *translation* and involves the step-by-step addition of amino acids to a growing protein chain by a ribozyme that is called a ribosome. The order in which the amino acids are added is read through the genetic code from an mRNA template, which is an RNA derived from one of the organism's genes.

Twenty-two amino acids are naturally incorporated into polypeptides and are called proteinogenic or natural amino acids. Of these, 20 are encoded by the universal genetic code. The remaining 2, selenocysteine and pyrrolysine, are incorporated into proteins by unique synthetic mechanisms. Selenocysteine is incorporated when the mRNA being translated includes a SECIS element, which causes the UGA codon to encode selenocysteine instead of a stop codon. Pyrrolysine is used by some methanogenic archaea in enzymes that they use to produce methane. It is coded for with the codon UAG, which is normally a stop codon in other organisms.

Several independent evolutionary studies have suggested that Gly, Ala, Asp, Val, Ser, Pro, Glu, Leu, Thr may belong to a group of amino acids that constituted the early genetic code, whereas Cys, Met, Tyr, Trp, His, Phe may belong to a group of amino acids that constituted later additions of the genetic code.

Standard vs. Nonstandard Amino Acids

The 20 amino acids that are encoded directly by the codons of the universal genetic code are called *standard* or *canonical* amino acids. A modified form of methionine (*N*-formylmethionine) is often incorporated in place of methionine as the initial amino acid of proteins in bacteria, mitochondria and plastids (including chloroplasts). Other amino acids are called *nonstandard* or *non-canonical*. Most of the nonstandard amino acids are also non-proteinogenic (i.e. they cannot be incorporated into proteins during translation), but two of them are proteinogenic, as they can be incorporated translationally into proteins by exploiting information not encoded in the universal genetic code.

The two nonstandard proteinogenic amino acids are selenocysteine (present in many non-eukaryotes as well as most eukaryotes, but not coded directly by DNA) and pyrrolysine (found only in some archaea and at least one bacterium). The incorporation of these nonstandard amino acids is rare. For example, 25 human proteins include selenocysteine in their primary structure, and the structurally characterized enzymes (selenoenzymes) employ selenocysteine as the catalytic moiety in their active sites. Pyrrolysine and selenocysteine are encoded via variant codons. For example, selenocysteine is encoded by stop codon and SECIS element.

N-formylmethionine (which is often the initial amino acid of proteins in bacteria, mitochondria, and chloroplasts) is generally considered as a form of methionine rather than as a separate proteinogenic amino acid. Codon–tRNA combinations not found in nature can also be used to "expand" the genetic code and form novel proteins known as alloproteins incorporating non-proteinogenic amino acids.

Non-Proteinogenic Amino Acids

Aside from the 22 proteinogenic amino acids, many *non-proteinogenic* amino acids are known. Those either are not found in proteins (for example carnitine, GABA, levothyroxine) or are not produced directly and in isolation by standard cellular machinery. For example, hydroxyproline, is synthesised from proline. Another example is selenomethionine). Non-proteinogenic amino acids that are found in proteins are formed by post-translational modification. Such modifications can also determine the localization of the protein, e.g., the addition of long hydrophobic groups can cause a protein to bind to a phospholipid membrane. Examples:

- The carboxylation of glutamate allows for better binding of calcium cations,
- Hydroxyproline, generated by hydroxylation of proline, is a major component of the connective tissue collagen.
- Hypusine in the translation initiation factor EIF5A, contains a modification of lysine.

Some non-proteinogenic amino acids are not found in proteins. Examples include 2-aminoisobutyric acid and the neurotransmitter gamma-aminobutyric acid. Non-proteinogenic amino acids often occur as intermediates in the metabolic pathways for standard amino acids – for example, ornithine and citrulline occur in the urea cycle, part of amino acid catabolism. A rare exception to the dominance of α-amino acids in biology is the β-amino acid beta alanine (3-aminopropanoic acid), which is used in plants and microorganisms in the synthesis of pantothenic acid (vitamin B_5), a component of coenzyme A.

In Mammalian Nutrition

Amino acids are not typical component of food: animals eat proteins. The protein is broken down into amino acids in the process of digestion. They are then used to synthesize new proteins, other biomolecules, or are oxidized to urea and carbon dioxide as a source of energy.

The oxidation pathway starts with the removal of the amino group by a transaminase; the amino group is then fed into the urea cycle. The other product of transamidation is a keto acid that enters the citric acid cycle. Glucogenic amino acids can also be converted into glucose, through gluconeogenesis. Of the 20 standard amino acids, nine (His, Ile, Leu, Lys, Met, Phe, Thr, Trp and Val) are called essential amino acids because the human body cannot synthesize them from other compounds at the level needed for normal growth, so they must be obtained from food.

Semi-Essential and Conditionally Essential Amino Acids, and Juvenile Requirements

In addition, cysteine, tyrosine, and arginine are considered semiessential amino acids, and taurine a semi-essential aminosulfonic acid in children. Some amino acids are conditionally essential for certain ages or medical conditions. Essential amino acids may also vary from species to species. The metabolic pathways that synthesize these monomers are not fully developed.

Non-Protein Functions

Many proteinogenic and non-proteinogenic amino acids have biological functions beyond being precursors to proteins and peptides.In humans, amino acids also have important roles in diverse biosynthetic pathways. Defenses against herbivores in plants sometimes employ amino acids. Examples:

Standard Amino Acids

- Tryptophan is a precursor of the neurotransmitter serotonin.
- Tyrosine (and its precursor phenylalanine) are precursors of the catecholamine neurotransmitters dopamine, epinephrine and norepinephrine and various trace amines.
- Phenylalanine is a precursor of phenethylamine and tyrosine in humans. In plants, it is a precursor of various phenylpropanoids, which are important in plant metabolism.
- Glycine is a precursor of porphyrins such as heme.
- Arginine is a precursor of nitric oxide.
- Ornithine and *S*-adenosylmethionine are precursors of polyamines.
- Aspartate, glycine, and glutamine are precursors of nucleotides. However, not all of the functions of other abundant nonstandard amino acids are known.

Roles for Nonstandard Amino Acids

- Carnitine is used in lipid transport.
- gamma-aminobutyric acid is a neurotransmitter.
- 5-HTP (5-hydroxytryptophan) is used for experimental treatment of depression.
- L-DOPA (L-dihydroxyphenylalanine) for Parkinson's treatment,
- Eflornithine inhibits ornithine decarboxylase and used in the treatment of sleeping sickness.
- Canavanine, an analogue of arginine found in many legumes is an antifeedant, protecting the plant from predators.
- Mimosine found in some legumes, is another possible antifeedant. This compound is an analogue of tyrosine and can poison animals that graze on these plants.

15.1.5 Uses in Industry

Animal Feed

Amino acids are sometimes added to animal feed because some of the components of these feeds, such as soybeans, have low levels of some of the essential amino acids, especially of lysine, methionine, threonine, and tryptophan. Likewise amino acids are used to chelate metal cations in order to improve the absorption of minerals from feed supplements.

Food

The food industry is a major consumer of amino acids, especially glutamic acid, which is used as a flavor enhancer, and aspartame (aspartylphenylalanine 1-methyl ester), which is used as an artificial sweetener. Amino acids are sometimes added to food by manufacturers to alleviate symptoms of mineral deficiencies, such as anemia, by improving mineral absorption and reducing negative side effects from inorganic mineral supplementation.

Chemical Building Blocks

Amino acids are low-cost feedstocks used in chiral pool synthesis as enantiomerically pure building blocks. Amino acids are used in the synthesis of some cosmetics.

15.1.6 Aspirational Uses

Fertilizer

The chelating ability of amino acids is sometimes used in fertilizers to facilitate the delivery of minerals to plants in order to correct mineral deficiencies, such as iron chlorosis. These fertilizers are also used to prevent deficiencies from occurring and to improve the overall health of the plants.

Biodegradable Plastics

Amino acids have been considered as components of biodegradable polymers, which have applications as environmentally friendly packaging and in medicine in drug delivery and the construction of prosthetic implants. An interesting example of such materials is polyaspartate, a water-soluble biodegradable polymer that may have applications in disposable diapers and agriculture. Due to its solubility and ability to chelate metal ions, polyaspartate is also being used as a biodegradable antiscaling agent and a corrosion inhibitor.

15.1.7 Synthesis

Chemical Synthesis

The commercial production of amino acids usually relies on mutant bacteria that overproduce individual amino acids using glucose as a carbon source. Some amino acids are produced by enzymatic conversions of synthetic intermediates. 2-Aminothiazoline-4-carboxylic acid is an intermediate in one industrial synthesis of L-cysteine for example. Aspartic acid is produced by the addition of ammonia to fumarate using a lyase.

Biosynthesis

In plants, nitrogen is first assimilated into organic compounds in the form of glutamate, formed from alpha-ketoglutarate and ammonia in the mitochondrion. For other amino acids, plants use transaminases to move the amino group from glutamate to another alpha-keto acid. For example, aspartate aminotransferase converts glutamate and oxaloacetate to alpha-ketoglutarate and aspartate. Other organisms use transaminases for amino acid synthesis, too.

Nonstandard amino acids are usually formed through modifications to standard amino acids. For example, homocysteine is formed through the transsulfuration pathway or by the demethylation of methionine via the intermediate metabolite S-adenosylmethionine, while hydroxyproline is made by a post translational modification of proline.

Microorganisms and plants synthesize many uncommon amino acids. For example, some microbes make 2-aminoisobutyric acid and lanthionine, which is a sulfide-bridged derivative of alanine. Both of these amino acids are found in peptidic lantibiotics such as alamethicin. However, in plants, 1-aminocyclopropane-1-carboxylic acid is a small disubstituted cyclic amino acid that is an intermediate in the production of the plant hormone ethylene.

Primordial Synthesis

The formation of amino acids and peptides are assumed to precede and perhaps induce the emergence of life on earth. Amino acids can form from simple precursors under various conditions. Surface-based chemical metabolism of amino acids and very small compounds may have led to the build-up of amino acids, coenzymes and phosphate-based small carbon molecules. Amino acids and similar building blocks could have been elaborated into proto-peptides, with peptides being considered key players in the origin of life.

In the famous Urey-Miller experiment, the passage of an electric arc through a mixture of methane, hydrogen, and ammonia produces a large number of amino acids. Since then, scientists have discovered a range of ways and components by which the potentially prebiotic formation and chemical evolution of peptides may have occurred, such as condensing agents, the design of self-replicating peptides and a number of non-enzymatic mechanisms by which amino acids could have emerged and elaborated into peptides. Several hypotheses invoke the Strecker synthesis whereby hydrogen cyanide, simple aldehydes, ammonia, and water produce amino acids.

According to a review, amino acids, and even peptides, "turn up fairly regularly in the various experimental broths that have been allowed to be cooked from simple chemicals. This is because nucleotides are far more difficult to synthesize chemically than amino acids." For a chronological order, it suggests that there must have been a 'protein world' or at least a 'polypeptide world', possibly later followed by the 'RNA world' and the 'DNA world'. Codon–amino acids mappings may be the biological information system at the primordial origin of life on Earth. While amino acids and consequently simple peptides must have formed under different experimentally probed geochemical scenarios, the transition from an abiotic world to the first life forms is to a large extent still unresolved.

15.1.8 Reactions

Amino acids undergo the reactions expected of the constituent functional groups.

Peptide Bond Formation

As both the amine and carboxylic acid groups of amino acids can react to form amide bonds, one amino acid molecule can react with another and become joined through an amide linkage. This polymerization of amino acids is what creates proteins. This condensation reaction yields the newly formed peptide bond and a molecule of water. In cells, this reaction does not occur directly; instead, the amino acid is first activated by attachment to a transfer RNA molecule through an ester bond. This aminoacyl-tRNA is produced in an ATP-dependent reaction carried out by an aminoacyl tRNA synthetase. This aminoacyl-tRNA is then a substrate for the ribosome, which catalyzes the attack of the amino group of the elongating protein chain on the ester bond. As a result of this mechanism, all proteins made by ribosomes are synthesized starting at their *N*-terminus and moving toward their *C*-terminus.

However, not all peptide bonds are formed in this way. In a few cases, peptides are synthesized by specific enzymes. For example, the tripeptide glutathione is an essential part of the defenses of cells against oxidative stress. This peptide is synthesized in two steps from free amino acids. In the first step, gamma-glutamylcysteine synthetase condenses cysteine and glutamate through a peptide bond formed between the side chain carboxyl of the glutamate (the gamma carbon of this side chain) and the amino group of the cysteine. This dipeptide is then condensed with glycine by glutathione synthetase to form glutathione.

In chemistry, peptides are synthesized by a variety of reactions. One of the most-used in solid-phase peptide synthesis uses the aromatic oxime derivatives of amino acids as activated units. These are added in sequence onto the growing peptide chain, which is attached to a solid resin support. Libraries of peptides are used in drug discovery through high-throughput screening.

The combination of functional groups allow amino acids to be effective polydentate ligands for metal–amino acid chelates. The multiple side chains of amino acids can also undergo chemical reactions.

Catabolism

Degradation of an amino acid often involves deamination by moving its amino group to α-ketoglutarate, forming glutamate. This process involves transaminases, often the same as those used in amination during synthesis. In many vertebrates, the amino group is then removed through the urea cycle and is excreted in the form of urea. However, amino acid degradation can produce uric acid or ammonia instead. For example, serine dehydratase converts serine to pyruvate and ammonia. After removal of one or more amino groups, the remainder of the molecule can sometimes be used to synthesize new amino acids, or it can be used for energy by entering glycolysis or the citric acid cycle, as detailed in image at right.

Complexation

Amino acids are bidentate ligands, forming transition metal amino acid complexes.

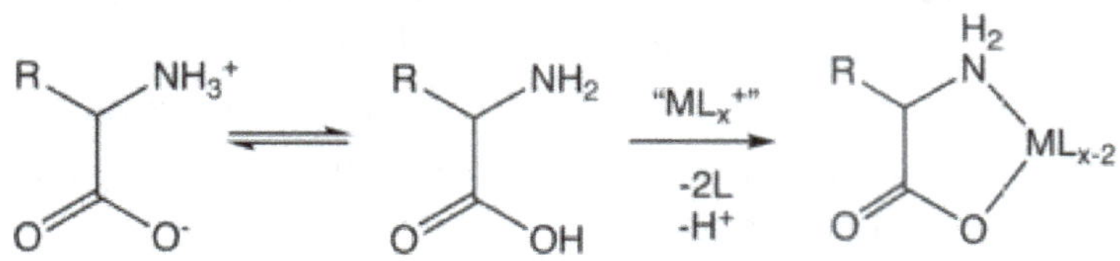

15.1.9 Chemical Analysis

The total nitrogen content of organic matter is mainly formed by the amino groups in proteins. The Total Kjeldahl Nitrogen (TKN) is a measure of nitrogen widely used in the analysis of (waste) water, soil, food, feed and organic matter in general. As the name suggests, the Kjeldahl method is applied. More sensitive methods are available.

15.2 IMPORTANCE OF AMINO ACIDS

Amino acids are vital compounds in both human and animal nutrition, serving as the precursors for protein synthesis. There are 20 standard amino acids, nine of which are classified as essential for humans, meaning they must be obtained from the diet since the body cannot synthesize them. Amino acids are also involved in key metabolic processes such as neurotransmission, hormone synthesis, and immune system function.

In the food and feed industries, amino acids like lysine, methionine, and threonine are added to animal feed to enhance growth and overall health, especially in livestock and poultry. In the pharmaceutical industry, amino acids such as glutamine, glycine, and tryptophan are used as dietary supplements and in the formulation of drugs. Additionally, amino acids are incorporated into cosmetic products for their moisturizing and anti-aging properties.

Given the wide range of applications, the global demand for amino acids continues to increase, driving interest in sustainable, scalable production methods.

Fermentation, a biological process involving the use of microorganisms, has become a popular method for producing amino acids on an industrial scale.

15.3 OVERVIEW OF FERMENTATION

Fermentation is a biochemical process in which microorganisms, such as bacteria, fungi, and yeast, convert organic substrates into useful products through their metabolic activities. In the case of amino acid production, fermentation typically involves using genetically engineered microorganisms to overproduce specific amino acids, which are then recovered and purified for industrial use.

There are two main types of fermentation processes used in the production of amino acids: batch fermentation and continuous fermentation.

- **Batch Fermentation:** In this method, microorganisms are grown in a closed system with a fixed amount of substrate. Once the substrate is consumed, the process is halted, and the amino acids are harvested. Batch fermentation is ideal for producing high-value amino acids but requires frequent cleaning and restarting of the system, which can reduce overall efficiency.
- **Continuous Fermentation:** Continuous fermentation allows for the steady feeding of substrate into the system while the product and byproducts are removed simultaneously. This method is more efficient for large-scale amino acid production and is often used for amino acids that are in high demand, such as glutamic acid and lysine.

Fermentation for amino acid production typically begins with the selection of a suitable microorganism, such as *Corynebacterium glutamicum* or *Escherichia coli*, which are known for their ability to produce amino acids. These microorganisms are often genetically modified to enhance their metabolic pathways, ensuring the overproduction of the desired amino acid.

15.4 TYPES OF AMINO ACIDS PRODUCED BY FERMENTATION

Several amino acids are produced on an industrial scale through fermentation, with each type serving specific functions in various industries.

15.4.1 Glutamic Acid

Glutamic acid is one of the most widely produced amino acids globally, primarily due to its use as a flavor enhancer in the food industry. Its sodium salt form, monosodium glutamate (MSG), is commonly used to enhance the savory taste of foods such as soups, sauces, and snacks. The production of glutamic acid involves the fermentation of glucose or other carbohydrate-rich feedstocks by *Corynebacterium glutamicum*, a bacterium known for its high glutamic acid production capacity. During fermentation, the bacterium metabolizes the glucose, producing glutamic acid as a byproduct, which is then extracted from the fermentation broth through filtration and crystallization processes. Glutamic acid production through fermentation is highly efficient, and the demand for MSG in the food industry ensures that it remains one of the most economically viable amino acids to produce on a large scale.

15.4.2 Lysine

Lysine is an essential amino acid that is widely used as a supplement in animal feed, particularly for poultry and swine. Since animals cannot synthesize lysine on their own, it must be provided through their diet. Lysine enhances growth rates and improves feed efficiency, making it an essential component in the livestock industry. Lysine is typically produced through the fermentation of sugar-rich substrates using bacteria such as *Corynebacterium glutamicum* or *Brevibacterium flavum*. Genetic engineering techniques have been used to enhance the lysine production capacity of these microorganisms by optimizing their metabolic pathways. After fermentation, lysine is recovered through precipitation, filtration, and crystallization. The production of lysine via fermentation is cost-effective, making it one of the most commercially significant amino acids globally.

15.4.3 Threonine

Threonine, like lysine, is an essential amino acid that is used as a dietary supplement in animal feed. It plays a critical role in protein synthesis and helps improve immune function in livestock. Threonine is also used in the pharmaceutical industry for the production of certain drugs and supplements. The fermentation process for threonine production typically involves the use of *Escherichia coli* or *Corynebacterium glutamicum* strains that have been genetically engineered to overproduce threonine. The fermentation process involves the breakdown of glucose or other sugars by the microorganisms, which convert them into threonine. After fermentation, the amino acid is extracted from the fermentation broth and purified. Due to its importance in animal nutrition, threonine production has become a critical component of the global amino acid market.

15.4.4 Tryptophan

Tryptophan is an essential amino acid that serves as a precursor for serotonin, a neurotransmitter involved in mood regulation and sleep. It is used in both animal feed and human dietary supplements. In animals, tryptophan supplementation improves growth rates and reduces aggression, while in humans, it is used in treatments for mood disorders and insomnia. Tryptophan is produced through the fermentation of glucose using genetically modified strains of *Escherichia coli*. The fermentation process is optimized to ensure high yields of tryptophan, which is then recovered and purified for use in various industries.

15.4.5 Methionine

Methionine is an essential amino acid used as a supplement in animal feed, particularly in poultry and fish farming. It plays a crucial role in protein synthesis and helps improve growth rates in animals. While methionine is primarily produced through chemical synthesis, microbial fermentation methods are being explored as a more sustainable alternative.

In fermentation-based production, methionine is produced using genetically engineered microorganisms such as *Escherichia coli* or *Corynebacterium glutamicum*. These microorganisms metabolize sugars to produce methionine, which is then recovered and purified. Although fermentation-based methionine production is still in its early stages, it holds promise as a sustainable alternative to traditional chemical synthesis.

15.5 FACTORS AFFECTING AMINO ACID FERMENTATION

Several factors influence the efficiency and yield of amino acid production through fermentation. These factors include the choice of microorganism, the composition of the fermentation medium, and environmental conditions such as pH, temperature, and oxygen levels.

- **Microorganism Selection:** The choice of microorganism is critical for efficient amino acid production. Bacteria such as *Corynebacterium glutamicum* and *Escherichia coli* are commonly used due to their ability to produce large quantities of amino acids. Genetic modifications can further enhance their production capacity.
- **Fermentation Medium:** The composition of the fermentation medium, including the carbon source (e.g., glucose, sucrose), nitrogen source, and trace elements, significantly affects the growth of microorganisms and the yield of amino acids. Optimizing the medium composition ensures efficient microbial growth and amino acid production.
- **Environmental Conditions:** The pH, temperature, and oxygen levels during fermentation must be carefully controlled to optimize microbial activity and amino acid production. For example, glutamic acid production requires a slightly acidic environment (pH 5.5-6.0), while lysine production typically occurs at neutral pH levels (pH 7.0-7.5).

15.6 INDUSTRIAL APPLICATIONS OF AMINO ACIDS

The industrial applications of amino acids produced through fermentation are vast, with significant use in the food, pharmaceutical, and agricultural sectors.

- **Food and Beverage Industry:** Amino acids such as glutamic acid (in the form of MSG) are widely used as flavor enhancers in processed foods. Additionally, amino acids like glycine and cysteine are used in the production of dietary supplements and functional foods.

- **Animal Feed:** Essential amino acids such as lysine, threonine, and tryptophan are added to animal feed to enhance growth rates and improve overall health in livestock. This practice helps optimize feed efficiency and reduces the environmental impact of animal farming.
- **Pharmaceutical Industry:** Amino acids play an important role in drug formulation and the production of nutritional supplements. For example, glutamine is used to support immune function in patients undergoing chemotherapy, while tryptophan supplements are used to treat mood disorders and sleep disturbances.
- **Cosmetics:** Amino acids such as serine and proline are used in cosmetics and skin care products for their moisturizing and anti-aging properties. These amino acids help improve skin hydration and elasticity, making them valuable ingredients in beauty formulations.

15.7 CONCLUSION

The production of amino acids by fermentation has become a cornerstone of the biochemical industry, offering a sustainable and efficient method for meeting the growing global demand for these essential compounds. Through microbial fermentation, amino acids such as glutamic acid, lysine, threonine, and tryptophan can be produced in large quantities and used in various industries, including food and feed, pharmaceuticals, and cosmetics. Advances in genetic engineering and fermentation technologies will continue to enhance the efficiency and yield of amino acid production, making this process a vital component of modern industrial biotechnology.

CHAPTER 16

MINING MICROBIOLOGY: BIOLEACHING BY MICROORGANISMS

Mining microbiology is a field that explores the use of microorganisms in various aspects of mining, with one of the most important applications being bioleaching. Bioleaching, also known as microbial leaching or biomining, refers to the extraction of metals from their ores through the activity of microorganisms. This process has gained prominence due to its environmental benefits and cost-effectiveness, especially when dealing with low-grade ores that are otherwise uneconomical to process using traditional mining methods.

16.1 BIOLEACHING

Bioleaching is the extraction or liberation of metals from their ores through the use of living organisms. Bioleaching is one of several applications within biohydrometallurgy and several methods are used to treat ores or concentrates containing copper, zinc, lead, arsenic, antimony, nickel, molybdenum, gold, silver, and cobalt. Bioleaching falls into two broad categories.

The first, is the use of microorganisms to oxidize refractory minerals to release valuable metals such and gold and silver. Most commonly the minerals that are the target of oxidization are pyrite and arsenopyrite. The second category is leaching of sulphide minerals to release the associated metal, for example, leaching of pentlandite to release nickel, or the leaching of chalcocite, covellite or chalcopyrite to release copper.

16.1.1 Process

Bioleaching can involve numerous ferrous iron and sulfur oxidizing bacteria, including *Acidithiobacillus ferrooxidans* (formerly known as *Thiobacillus ferrooxidans*) and *Acidithiobacillus thiooxidans* (formerly known as *Thiobacillus thiooxidans*). As a general principle, in one proposed method of bacterial leaching known as Indirect Leaching, Fe^{3+} ions are used to oxidize the ore. This step is entirely independent of microbes. The role of the bacteria is further oxidation of the ore, but also the regeneration of the chemical oxidant Fe^{3+} from Fe^{2+}. For example, bacteria catalyse the breakdown of the mineral pyrite (FeS_2) by oxidising the sulfur and metal (in this case ferrous iron, (Fe^{2+})) using oxygen. This yields soluble products that can be further purified and refined to yield the desired metal.

Pyrite Leaching (FeS2)

In the first step, disulfide is spontaneously oxidized to thiosulfate by ferric ion (Fe^{3+}), which in turn is reduced to give ferrous ion (Fe^{2+}):

$$FeS_2 + 6\,Fe^{3+} + 3\,H_2O \rightarrow 7\,Fe^{2+} + S_2O_3^{2-} + 6\,H^+ \text{ spontaneous}$$

The ferrous ion is then oxidized by bacteria using oxygen:

$$4\,Fe^{2+} + O_2 + 4\,H^+ \rightarrow 4\,Fe^{3+} + 2\,H_2O \text{ (iron oxidizers)}$$

Thiosulfate is also oxidized by bacteria to give sulfate:

$$S_2O_3^{2-} + 2\,O_2 + H_2O \rightarrow 2\,SO_4^{2-} + 2\,H^+ \text{ (sulfur oxidizers)}$$

The ferric ion produced in reaction (2) oxidized more sulfide as in reaction (1), closing the cycle and given the net reaction:

$$2\,FeS_2 + 7\,O_2 + 2\,H_2O \rightarrow 2\,Fe^{2+} + 4\,SO_4^{2-} + 4\,H^+$$

The net products of the reaction are soluble ferrous sulfate and sulfuric acid. The microbial oxidation process occurs at the cell membrane of the bacteria. The electrons pass into the cells and are used in biochemical processes to produce energy for the bacteria while reducing oxygen to water. The critical reaction is the oxidation of sulfide by ferric iron. The main role of the bacterial step is the regeneration of this reactant.

The process for copper is very similar, but the efficiency and kinetics depend on the copper mineralogy. The most efficient minerals are supergene minerals such as chalcocite, Cu_2S and covellite, CuS. The main copper mineral chalcopyrite ($CuFeS_2$) is not leached very efficiently, which is why the dominant copper-producing technology remains flotation, followed by smelting and refining. The leaching of $CuFeS_2$ follows the two stages of being dissolved and then further oxidised, with Cu^{2+} ions being left in solution.

Chalcopyrite Leaching

- $CuFeS_2 + 4\,Fe^{3+} \rightarrow Cu^{2+} + 5\,Fe^{2+} + 2\,S_0$ (spontaneous).
- $4\,Fe^{2+} + O_2 + 4\,H^+ \rightarrow 4\,Fe^{3+} + 2\,H_2O$ (iron oxidizers).
- $2\,S^0 + 3\,O_2 + 2\,H_2O \rightarrow 2\,SO_4^{2-} + 4\,H^+$ (sulfur oxidizers).

Net reaction:

- $CuFeS_2 + 4\,O_2 \rightarrow Cu^{2+} + Fe^{2+} + 2\,SO_4^{2-}$

In general, sulfides are first oxidized to elemental sulfur, whereas disulfides are oxidized to give thiosulfate, and the processes above can be applied to other sulfidic ores. Bioleaching of non-sulfidic ores such as pitchblende also uses ferric iron as an oxidant (e.g., $UO_2 + 2\ Fe^{3+} \Rightarrow UO_2^{2+} + 2\ Fe^{2+}$). In this case, the sole purpose of the bacterial step is the regeneration of Fe^{3+}. Sulfidic iron ores can be added to speed up the process and provide a source of iron. Bioleaching of non-sulfidic ores by layering of waste sulfides and elemental sulfur, colonized by *Acidithiobacillus* spp., has been accomplished, which provides a strategy for accelerated leaching of materials that do not contain sulfide minerals.

16.1.2 Further Processing

The dissolved copper (Cu^{2+}) ions are removed from the solution by ligand exchange solvent extraction, which leaves other ions in the solution. The copper is removed by bonding to a ligand, which is a large molecule consisting of a number of smaller groups, each possessing a lone electron pair. The ligand-copper complex is extracted from the solution using an organic solvent such as kerosene:

$$Cu^{2+}{}_{(aq)} + 2LH\,(organic) \rightarrow CuL_2\,(organic) + 2H^{+}{}_{(aq)}$$

The ligand donates electrons to the copper, producing a complex - a central metal atom (copper) bonded to the ligand. Because this complex has no charge, it is no longer attracted to polar water molecules and dissolves in the kerosene, which is then easily separated from the solution. Because the initial reaction is reversible, it is determined by pH. Adding concentrated acid reverses the equation, and the copper ions go back into an aqueous solution.

Then the copper is passed through an electro-winning process to increase its purity: An electric current is passed through the resulting solution of copper ions. Because copper ions have a 2+ charge, they are attracted to the negative cathodes and collect there.

The copper can also be concentrated and separated by displacing the copper with Fe from scrap iron:

$$Cu^{2+}{}_{(aq)} + Fe_{(s)} \rightarrow Cu_{(s)} + Fe^{2+}{}_{(aq)}$$

The electrons lost by the iron are taken up by the copper. Copper is the oxidising agent (it accepts electrons), and iron is the reducing agent (it loses electrons).

Traces of precious metals such as gold may be left in the original solution. Treating the mixture with sodium cyanide in the presence of free oxygen dissolves the gold. The gold is removed from the solution by adsorbing (taking it up on the surface) to charcoal.

16.1.3 With Fungi

Several species of fungi can be used for bioleaching. Fungi can be grown on many different substrates, such as electronic scrap, catalytic converters, and fly ash from municipal waste incineration. Experiments have shown that two fungal strains

(*Aspergillus niger, Penicillium simplicissimum*) were able to mobilize Cu and Sn by 65%, and Al, Ni, Pb, and Zn by more than 95%. *Aspergillus niger* can produce some organic acids such as citric acid. This form of leaching does not rely on microbial oxidation of metal but rather uses microbial metabolism as source of acids that directly dissolve the metal.

16.1.4 Feasibility

Economic Feasibility

Bioleaching is in general simpler and, therefore, cheaper to operate and maintain than traditional processes, since fewer specialists are needed to operate complex chemical plants. And low concentrations are not a problem for bacteria because they simply ignore the waste that surrounds the metals, attaining extraction yields of over 90% in some cases. These microorganisms actually gain energy by breaking down minerals into their constituent elements. The company simply collects the ions out of the solution after the bacteria have finished.

Bioleaching can be used to extract metals from low concentration ores such as gold that are too poor for other technologies. It can be used to partially replace the extensive crushing and grinding that translates to prohibitive cost and energy consumption in a conventional process. Because the lower cost of bacterial leaching outweighs the time it takes to extract the metal.

High concentration ores, such as copper, are more economical to smelt rather bioleach due to the slow speed of the bacterial leaching process compared to smelting. The slow speed of bioleaching introduces a significant delay in cash flow for new mines. Nonetheless, at the largest copper mine of the world, Escondida in Chile the process seems to be favorable.

Economically it is also very expensive and many companies once started can not keep up with the demand and end up in debt.

In Space

In 2020 scientists showed, with an experiment with different gravity environments on the ISS, that microorganisms could be employed to mine useful elements from basaltic rocks via bioleaching in space.

16.1.5 Environmental Impact

The process is more environmentally friendly than traditional extraction methods. For the company this can translate into profit, since the necessary limiting of sulfur dioxide emissions during smelting is expensive. Less landscape damage occurs, since the bacteria involved grow naturally, and the mine and surrounding area can be left relatively untouched. As the bacteria breed in the conditions of the mine, they are easily cultivated and recycled.

Toxic chemicals are sometimes produced in the process. Sulfuric acid and H^+ ions that have been formed can leak into the ground and surface water turning it acidic, causing environmental damage. Heavy ions such as iron, zinc, and arsenic leak during acid mine drainage. When the pH of this solution rises, as a result of dilution by fresh water, these ions precipitate, forming "Yellow Boy" pollution. For these reasons, a setup of bioleaching must be carefully planned, since the process can lead to a biosafety failure. Unlike other methods, once started, bioheap leaching cannot be quickly stopped, because leaching would still continue with rainwater and natural bacteria. Projects like Finnish Talvivaara proved to be environmentally and economically disastrous.

16.2 MICROORGANISMS INVOLVED IN BIOLEACHING

Several microorganisms are involved in the bioleaching process, and they play a crucial role in the efficiency and speed of metal extraction. The key microorganisms used in bioleaching belong primarily to the groups of bacteria and archaea. These microorganisms are typically acidophiles, thriving in environments with low pH and high metal concentrations.

- **Acidithiobacillus Ferrooxidans:** This bacterium is the most widely studied and used microorganism in bioleaching. It is a chemolithotrophic, acidophilic bacterium that derives its energy by oxidizing ferrous iron (Fe^{2+}) to ferric iron (Fe^{3+}) and reduced sulfur compounds like sulfides into sulfate. The ferric iron produced by this bacterium acts as an oxidizing agent that can dissolve metal sulfides, liberating valuable metals like copper and uranium.
- **Leptospirillum Ferrooxidans:** Similar to *Acidithiobacillus ferrooxidans*, *Leptospirillum ferrooxidans* oxidizes ferrous iron but lacks the ability to oxidize sulfur compounds. It is often found in environments where ferrous iron is abundant, such as mine drainage and tailings piles. This microorganism plays an important role in the bioleaching of iron and copper ores.
- **Sulfolobus Metallicus:** This extremophilic archaeon thrives in thermophilic conditions and is known for its ability to oxidize metal sulfides at high temperatures. *Sulfolobus metallicus* is particularly effective in bioleaching operations involving higher temperature environments, which can accelerate the breakdown of sulfide ores.
- **Thiobacillus Thiooxidans:** This bacterium specializes in the oxidation of reduced sulfur compounds, converting them into sulfuric acid. The sulfuric acid produced helps solubilize metal sulfides and contributes to the bioleaching process by creating an acidic environment that enhances metal dissolution.

These microorganisms are naturally found in environments where metals and sulfur are present, such as mine tailings, volcanic regions, and geothermal areas. They have been harnessed and optimized in controlled bioleaching environments to increase the efficiency of metal recovery.

16.3 THE BIOLEACHING PROCESS

Bioleaching can be carried out in different settings depending on the type of ore being processed and the desired scale of metal extraction. The most common bioleaching methods include heap leaching, tank leaching, and in situ leaching.

16.3.1 Heap Leaching

Heap leaching is an industrial mining process used to extract precious metals, copper, uranium, and other compounds from ore using a series of chemical reactions that absorb specific minerals and re-separate them after their division from other earth materials. Similar to in situ mining, heap leach mining differs in that it places ore on a liner, then adds the chemicals via drip systems to the ore, whereas *in situ* mining lacks these liners and pulls pregnant solution up to obtain the minerals. Heap leaching is widely used in modern large-scale mining operations as it produces the desired concentrates at a lower cost compared to conventional processing methods such as flotation, agitation, and vat leaching.

Additionally, dump leaching is an essential part of most copper mining operations and determines the quality grade of the produced material along with other factors Due to the profitability that the dump leaching has on the mining process, i.e. it can contribute substantially to the economic viability of the mining process, it is advantageous to include the results of the leaching operation in the economic overall project evaluation.

The process has ancient origins; one of the classical methods for the manufacture of copperas (iron sulfate) was to heap up iron pyrite and collect the leachate from the heap, which was then boiled with iron to produce iron(II) sulfate.

Process

The mined ore is usually crushed into small chunks and heaped on an impermeable plastic or clay lined leach pad where it can be irrigated with a leach solution to dissolve the valuable metals. While sprinklers are occasionally used for irrigation, more often operations use drip irrigation to minimize evaporation, provide more uniform distribution of the leach solution, and avoid damaging the exposed mineral. The solution then percolates through the heap and leaches both the target and other minerals. This process, called the "leach cycle," generally takes from one or two months for simple oxide ores (e.g. most gold ores) to two years for nickel laterite ores. The leach solution containing the dissolved minerals is then collected, treated in a process plant to recover the target mineral and in some cases precipitate other minerals, and recycled to the heap after reagent levels are adjusted. Ultimate recovery of the target mineral can range from 30% of contained run-of-mine dump leaching sulfide copper ores to over 90% for the ores that are easiest to leach, some oxide gold ores.

The essential questions to address during the process of the heap leaching are:

- Can the investment of crushing the ore be justified by the potential increase in recovery and rate of recovery?
- How should the concentration of acid be altered over time in order to produce a solution that can be economically treated?
- How does the form of a heap affect the recovery and solution grade?
- Under any given set of circumstances, what type of recovery can be expected before the leach solution quality drops below a critical limit?
- What recovery (quantifiable measure) can be expected?

In recent years, the addition of an agglomeration drum has improved on the heap leaching process by allowing for a more efficient leach. The rotary drum agglomerator works by taking the crushed ore fines and agglomerating them into more uniform particles. This makes it much easier for the leaching solution to percolate through the pile, making its way through the channels between particles.

The addition of an agglomeration drum also has the added benefit of being able to pre-mix the leaching solution with the ore fines to achieve a more concentrated, homogeneous mixture and allow the leach to begin prior to the heap.

Although heap leach design has made significant progress over the last few years through the use of new materials and improved analytical tools, industrial experience shows that there are significant benefits from extending the design process beyond the liner and into the rock pile itself. Characterization of the physical and hydraulic (hydrodynamic) properties of ore-for-leach focuses on the direct measurement of the key properties of the ore, namely:

- The relationship between heap height and ore bulk density (density profile).
- The relationship between bulk density and percolation capacity (conductivity profile).
- The relationship between the bulk density, porosity and its components (micro and macro).
- The relationship between the moisture content and percolation capacity (conductivity curve).

- The relationship between the aforementioned parameters and the ore preparation practices (mining, crushing, agglomeration, curing, and method of placement).

Theoretical and numerical analysis, and operational data show that these fundamental mechanisms are controlled by scale, dimensionality, and heterogeneity, all of which adversely affect the scalability of metallurgical and hydrodynamic properties from the lab to the field. The dismissal of these mechanisms can result in a number of practical and financial problems that will resonate throughout the life of the heap impacting the financial return of the operation. Through procedures that go beyond the commonly employed metallurgical testing and the integration of data gleaned through real time 3D monitoring, a more complete representative characterization of the physicochemical properties of the heap environment is obtained.

This improved understanding results in a significantly higher degree of accuracy in terms of creating a truly representative sample of the environment within the heap.

By adhering to the characterization identified above, a more comprehensive view of heap leach environments can be realized, allowing the industry to move away from the *de facto* black-box approach to a physicochemically inclusive industrial reactor model.

Precious Metals

The crushed ore is irrigated with a dilute alkaline cyanide solution. The solution containing the dissolved precious metals in a pregnant solution continues percolating through the crushed ore until it reaches the liner at the bottom of the heap where it drains into a storage (pregnant solution) pond.

After separating the precious metals from the pregnant solution, the dilute cyanide solution (now called "barren solution") is normally re-used in the heap-leach-process or occasionally sent to an industrial water treatment facility where the residual cyanide is treated and residual metals are removed. In very high rainfall areas, such as the tropics, in some cases there is surplus water that is then discharged to the environment, after treatment, posing possible water pollution if treatment is not properly carried out.

The production of one gold ring through this method, can generate 20 tons of waste material.

During the extraction phase, the gold ions form complex ions with the cyanide:

$$Au + (s) + 2CN^{-}(aq) \rightarrow Au(CN)2^{-}(aq)$$

Recuperation of the gold is readily achieved with a redox-reaction:

$$2Au(CN)2^{-}(aq)+Zn(s)\rightarrow Zn(CN)4^{2}-(aq)+2Au(s)$$

The most common methods to remove the gold from solution are either using activated carbon to selectively absorb it, or the Merrill-Crowe process where zinc powder is added to cause a precipitation of gold and zinc. The fine product can be either doré (gold-silver bars) or zinc-gold sludge that is then refined elsewhere.

Copper Ores

The method is similar to the cyanide method above, except sulfuric acid is used to dissolve copper from its ores. The acid is recycled from the solvent extraction circuit and reused on the leach pad. A byproduct is iron(II) sulfate, jarosite, which is produced as a byproduct of leaching pyrite, and sometimes even the same sulfuric acid that is needed for the process. Both oxide and sulfide ores can be leached, though the leach cycles are much different and sulfide leaching requires a bacterial, or bio-leach, component.

In 2011 leaching, both heap leaching and in-situ leaching, produced 3.4 million metric tons of copper, 22 percent of world production. The largest copper heap leach operations are in Chile, Peru, and the southwestern United States.

Although heap leaching is a low cost-process, it normally has recovery rates of 60-70%. It is normally most profitable with low-grade ores. Higher-grade ores are usually put through more complex milling processes where higher recoveries justify the extra cost. The process chosen depends on the properties of the ore.

The final product is cathode copper.

Nickel Ores

This method is an acid heap leaching method like that of the copper method in that it utilises sulfuric acid instead of cyanide solution to dissolve the target minerals from crushed ore. The amount of sulfuric acid required is much higher than for copper ores, as high as 1,000 kg of acid per tonne of ore, but 500 kg is more common. The method was originally patented by Australian miner BHP and is being commercialized by Cerro Matoso in Colombia, a wholly owned subsidiary of BHP; Vale in Brazil; and European Nickel for the rock laterite deposits of Turkey, Talvivaara mine in Finland, the Balkans, and the Philippines. There currently are no operating commercial scale nickel laterite heap leach operations, but there is a sulphide HL operating in Finland.

Nickel recovery from the leach solutions is much more complex than for copper and requires various stages of iron and magnesium removal, and the process produces both leached ore residue ("ripios") and chemical precipitates from the recovery plant (principally iron oxide residues, magnesium sulfate and calcium sulfate) in roughly equal proportions. Thus, a unique feature of nickel heap leaching is the need for a tailings disposal area. The final product can be nickel hydroxide precipitates (NHP) or mixed metal hydroxide precipitates (MHP), which are then subject to conventional smelting to produce metallic nickel.

Uranium Ores

Similar to copper oxide heap leaching, also using dilute sulfuric acid. Rio Tinto is commercializing this technology in Namibia and Australia; the French nuclear fuel company Orano, in Niger with two mines and Namibia; and several other companies are studying its feasibility. The final product is yellowcake and requires significant further processing to produce fuel-grade feed.

Apparatus

While most mining companies have shifted from a previously accepted sprinkler method to the percolation of slowly dripping choice chemicals including cyanide or sulfuric acid closer to the actual ore bed, heap leach pads have not changed too much throughout the years.

There are still four main categories of pads: conventional, dump leach, valley fills, and on/off pads. Typically, each pad only has a single, geomembrane liner for each pad, with a minimum thickness of 1.5mm, usually thicker.

The conventional pads simplest in design are used for mostly flat or gentle areas and hold thinner layers of crushed ore. Dump leach pads hold more ore and can usually handle a less flat terrain.

Valley fills are pads situated at valley bottoms or levels that can hold everything falling into it. On/off pads involve putting significantly larger loads on the pads and removing and reloading it after every cycle.

Many of these mines which previously had digging depths of about 15 meters are digging deeper than ever before to mine materials, approximately 50 meters, sometimes more, which means that, in order to accommodate all of the ground being displaced, pads will have to hold higher weights from more crushed ore being contained in a smaller area.

With that increase in build up comes in potential for decrease in yield or ore quality, as well as potential either weak spots in the lining or areas of increased pressure buildup. This build up still has the potential to lead to punctures in the liner.

As of 2004 cushion fabrics, which could reduce potential punctures and their leaking, were still being debated due to their tendency to increase risks if too much weight on too large a surface was placed on the cushioning.

In addition, some liners, depending on their composition, may react with salts in the soil as well as acid from the chemical leaching to affect the successfulness of the liner. This can be amplified over time.

Environmental Concerns

Effectiveness

Heap leach mining works well for large volumes of low grade ores, as reduced metallurgical treatment (comminution) of the ore is required in order to extract an equivalent amount of minerals when compared to milling. The significantly reduced processing costs are offset by the reduced yield of usually approximately 60-70%. The amount of overall environmental impact caused by heap leaching is often lower than more traditional techniques.

It also requires less energy consumption to use this method, which many consider to be an environmental alternative.

Government Regulation

In the United States, the General Mining Law of 1872 gave rights to explore and mine on public domain land; the original law did not require post-mining reclamation (Woody et al. 2011).

Mined land reclamation requirements on federal land depended on state requirements until the passage of the Federal Land Policy and Management Act in 1976. Currently, mining on federal land must have a government-approved mining and reclamation plan before mining can start. Reclamation bonds are required. Mining on either federal, state, or private land is subject to the requirements of the Clean Air Act and the Clean Water Act. One solution proposed to reclamation problems is the privatization of the land to be mined (Woody et al. 2011).

Cultural and Social Concerns

With the rise of the environmentalist movement has also come an increased appreciation for social justice, and mining has showed similar trends lately. Societies located near potential mining sites are at increased risk to be subjected to injustices as their environment is affected by the changes made to mined lands—either public or private—that could eventually lead to problems in social structure, identity, and physical health (Franks 2009). Many have argued that by cycling mine power through local citizens, this disagreement can be alleviated, since both interest groups would have shared and equal voice and understanding in future goals.

However, it is often difficult to match corporate mining interests with local social interests, and money is often a deciding factor in the successes of any disagreements. If communities are able to feel like they have a valid understanding and power in issues concerning their local environment and society, they are more likely to tolerate and encourage the positive benefits that come with mining, as well as more effectively promote alternative methods to heap leach mining using their intimate knowledge of the local geography.

16.3.2 Tank Leaching

In metallurgical processes tank leaching is a hydrometallurgical method of extracting valuable material (usually metals) from ore.

Tank vs. Vat Leaching

Factors

Tank leaching is usually differentiated from vat leaching on the following factors:

- In tank leaching the material is ground sufficiently fine to form a slurry or pulp, which can flow under gravity or when pumped. In vat leaching typically a coarser material is placed in the vat for leaching, this reduces the cost of size reduction.
- Tanks are typically equipped with agitators, baffles, gas introduction equipment designed to maintain the solids in suspension in the slurry, and achieve leaching. Vats usually do not contain much internal equipment, except for agitators.
- Tank leaching is typically continuous, while vat leaching is operated in a batch fashion, this is not always the case, and commercial processes using continuous vat leaching have been tested.
- Typically the retention time required for vat leaching is more than that for tank leaching to achieve the same percentage of recovery of the valuable material being leached.
- In a tank leach the slurry is moved, while in a vat leach the solids remain in the vat, and solution is moved.

Processes

Tank and vat leaching involves placing ore, usually after size reduction and classification, into large tanks or vats at ambient operating conditions containing a leaching solution and allowing the valuable material to leach from the ore into solution.

In tank leaching the ground, classified solids are already mixed with water to form a slurry or pulp, and this is pumped into the tanks. Leaching reagents are added to the tanks to achieve the leaching reaction. In a continuous system the slurry will then either overflow from one tank to the next, or be pumped to the next tank. Ultimately the "pregnant" solution is separated from the slurry using some form of liquid/solid separation process, and the solution passes on to the next phase of recovery.

In vat leaching the solids are loaded into the vat, once full the vat is flooded with a leaching solution. The solution drains from the tank, and is either recycled back into the vat or is pumped to the next step of the recovery process. Vat leach units are rectangular containers (drums, barrels, tanks or vats), usually very big and made of wood or concrete, lined with material resistant to the leaching media. The treated ore is usually coarse.

The vats are usually run sequentially to maximize the contact time between the ore and the reagent. In such a series the leachate collected from one container is added to another vat with fresher ore

As mentioned previously tanks are equipped with agitators to keep the solids in suspension in the vats and improve the solid to liquid to gas contact. Agitation is further assisted by the use of tank baffles to increase the efficiency of agitation and prevent centrifuging of slurries in circular tanks...

Extraction Efficiency Factors

Aside from chemical requirements several key factors influence extraction efficiency:

- **Retention Time:** Refers to the Time Spent in the Leaching System by the Solids: This is calculated as the total volumetric capacity of the leach tank/s divided by the volumetric throughput of the solid/liquid slurry. Retention time is commonly measured in hours for precious metals recovery. A sequence of leach tanks is referred to as a leach "train", and retention time is measured considering the total volume of the leach train. The desired retention time is determined during the testing phase, and the system is then designed to achieve this.
- **Size:** The ore must be ground to a size that exposes the desired mineral to the leaching agent (referred to as "liberation"), and in tank leaching this must be a size that can be suspended by the agitator. In vat leaching this is the size that is the most economically viable, where the recovery achieved as ore is ground finer is balanced against the increased cost of processing the material.
- **Slurry Density:** The slurry density (percent solids) determines retention time. The settling rate and viscosity of the slurry are functions of the slurry density. The viscosity, in turn, controls the gas mass transfer and the leaching rate.
- **Numbers of Tanks:** Agitated tank leach circuits are typically designed with no less than four tanks and preferably more to prevent short-circuiting of the slurry through the tanks.
- **Dissolved Gas:** Gas is often injected below the agitator or into the vat to obtain the desired dissolved gas levels – typically oxygen, in some base metal plants sulphur dioxide may be required.

- **Reagents:** Adding and maintaining the appropriate amount of reagents throughout the leach circuit is critical to a successful operation. Adding insufficient quantities of reagents reduces the metal recovery but adding excess reagents increases the operating costs without recovering enough additional metal to cover the cost •of the reagents.

16.3.3 In Situ Leaching

In-situ leaching (ISL), also called in-situ recovery (ISR) or solution mining, is a mining process used to recover minerals such as copper and uranium through boreholes drilled into a deposit, *in situ*. In-situ leach works by artificially dissolving minerals occurring naturally in the solid state. The process initially involves the drilling of boreholes into the ore deposit. Explosive or hydraulic fracturing can be used to create open pathways in the deposit for the solution to penetrate. Leaching solution is pumped into the deposit where it comes in contact with the ore. The solution bearing the dissolved ore content is then pumped to the surface and processed. This process allows the extraction of metals and salts from an ore body without the need for conventional mining involving drill-and-blast, open-cut or underground mining.

Process

In-situ leach mining involves pumping of a lixiviant into the ore body via a borehole, which circulates through the porous rock dissolving the ore and is extracted via a second borehole. The lixiviant varies according to the ore deposit: for salt deposits the leachate can be fresh water into which salts can readily dissolve. For copper, acids are generally needed to enhance solubility of the ore minerals within the solution. For uranium ores, the lixiviant may be acid or sodium bicarbonate.

Minerals

Potash and Soluble Salts

In-situ leach is widely used to extract deposits of water-soluble salts such as potash (sylvite and carnallite), rock salt (halite), sodium chloride, and sodium sulfate. It has been used in the US state of Colorado to extract nahcolite (sodium bicarbonate). In-situ leaching is often used for deposits that are too deep, or beds that are too thin, for conventional Underground Mining.

Uranium

In-situ leach for uranium has expanded rapidly since the 1990s, and is now the predominant method for mining uranium, accounting for 45 percent of the uranium mined worldwide in 2012. Unlike open-pit and underground mining, in-situ leaching does not rely on burial depth as a criterion but is based on the properties of the uranium deposit. In-situ leaching techniques are systematically categorized based on the primary components of the leaching solution, encompassing acid leaching, alkaline leaching, neutral leaching and bioleaching.

Acid leaching is applicable to low-carbonate uranium deposits, with U(VI) dissolving in acid solution while U(IV) dissolving in acid solution with oxidizing agent; Alkaline leaching is effective for high-carbonate uranium deposits but unsuitable for high-pyrite deposits, with U(VI) dissolving in alkaline solution while U(IV) dissolving in alkaline solution with oxidizing agent; Neutral leaching, including CO2-O2 leaching and weak acid leaching, is widely applicable; and Bioleaching is also widely applicable, especially ideal for pyrite-rich uranium deposits.

ISL of uranium ores started in the United States and the Soviet Union in the early 1960s. The first uranium ISL in the US was in the Shirley Basin in the state of Wyoming, which operated from 1961-1970 using sulfuric acid. Since 1970, all commercial-scale ISL mines in the US have used carbonate solutions. ISL mining in Australia uses acid solutions. In-situ recovery involves the extraction of uranium-bearing water (grading as low as 0.05 %, or 500 ppm, U_3O_8). The extracted uranium solution is then filtered through resin beads. Through an ion exchange process, the resin beads attract uranium from the solution. Uranium loaded resins are then transported to a processing plant, where U_3O_8 is separated from the resin beads and yellowcake is produced. The resin beads can then be returned to the ion exchange facility where they are reused.

At the end of 2008 there were four in-situ leaching uranium mines operating in the United States, operated by Cameco, Mestena and Uranium Resources, Inc., all using sodium bicarbonate. ISL produces 90 % of the uranium mined in the US. In 2010, Uranium Energy Corporation began in-situ leach operations at their Palangana project in Duval County, Texas.

In July 2012 Cameco delayed development of their Kintyre project, due to challenging project economics based on $45.00 U_3O_8. One ISR reclamation project was also in operation as of 2009. Significant ISL mines are operating in Kazakhstan and Australia.

Examples of in-situ uranium mines include:

- **The Beverley Uranium Mine,** South Australia, is an operating ISL uranium mine and Australia's first such mine.
- **The Honeymoon Uranium Mine,** South Australia, opened in 2011 and is Australia's second ISL uranium mine.
- **Crow Butte (Operating),** Smith Ranch-Highland (operating), Christensen Ranch (reclamation), Irigaray (reclamation), Churchrock (proposed), Crownpoint (proposed), Alta Mesa (operating), Hobson (standby), La Palangana (operating), Kingsville Dome (operating),

Rosita (standby) and Vasquez (restoration) are ISL uranium operations in the United States.

- **In 2010, Uranium Energy Corp.** began an ISL mining operation in the Palangana deposit in Duval County, Texas. The ion exchange facility at Palangana trucks uranium-loaded resin beads to the company's Hobson processing plant, where yellowcake is produced. Uranium Energy Corp. has three additional South Texas deposits permitted or in development.

Rhenium

There are technologies for the associated extraction of rhenium from productive solutions of underground leaching of uranium ores.

Copper

In-situ leaching of copper was done by the Chinese by 907 AD, and perhaps as early as 177 BC. Copper is usually leached using acid (sulfuric acid or hydrochloric acid), then recovered from solution by solvent extraction electrowinning (SX-EW) or by chemical precipitation.

Ores most amenable to leaching include the copper carbonates malachite and azurite, the oxide tenorite, and the silicate chrysocolla. Other copper minerals, such as the oxide cuprite and the sulfide chalcocite may require addition of oxidizing agents such as ferric sulfate and oxygen to the leachate before the minerals are dissolved.

The ores with the highest sulfide contents, such as bornite and chalcopyrite will require more oxidants and will dissolve more slowly. Sometimes oxidation is sped up by the bacteria *Thiobacillus ferrooxidans*, which feeds on sulfide compounds.

Copper ISL is often done by *stope leaching*, in which broken low-grade ore is leached in a current or former conventional underground mine. The leaching may take place in backfilled stopes or caved areas. In 1994, stope leaching of copper was reported at 16 mines in the US.

At the San Manuel Mine in the US state of Arizona, ISL was initially used by collecting the resultant solution underground but in 1995 this was converted to a well-to-well recovery method which was the first large scale implementation of that method. This well-to-well method has been proposed for other copper deposits in Arizona.

Gold

In-situ leaching has not been used on a commercial scale for gold mining. A three-year pilot program was undertaken in the 1970s to in-situ leach gold ore at the Ajax mine in the Cripple Creek district in the US, using a chloride and iodide solution. After obtaining poor results, perhaps because of the complex telluride ore, the test was halted.

Environmental Concerns

According to the World Nuclear Organization:

- In the USA legislation requires that the water quality in the effected aquifer be restored so as to enable its pre-mining use. Usually this is potable water or stock water (usually less than 500 ppm total dissolved solids), and while not all chemical characteristics can be returned to those pre-mining, the water must be usable for the same purposes as before. Often it needs to be treated by reverse osmosis, giving rise to a problem in disposing of the concentrated brine stream from this.
- The usual radiation safeguards are applied at an ISL Uranium mining operation, despite the fact that most of the orebody's radioactivity remains well underground and there is hence minimal increase in radon release and no ore dust. Employees are monitored for alpha radiation contamination and personal dosimeters are worn to measure exposure to gamma radiation. Routine monitoring of air, dust and surface contamination are undertaken.

The advantages of this technology are:

- Reduced hazards for the employees from accidents, dust, and radiation, and
- Low cost, no need for large uranium mill tailings deposits.

After termination of an in-situ leaching operation, the waste slurries produced must be safely disposed, and the aquifer, contaminated from the leaching activities, must be restored. Groundwater restoration is a very tedious process that is not yet fully understood. The best results have been obtained with the following treatment scheme, consisting of a series of different steps:

- **Phase 1:** Pumping of contaminated water: the injection of the leaching solution is stopped and the contaminated liquid is pumped from the leaching zone. Subsequently, clean groundwater flows in from outside of the leaching zone.
- **Phase 2:** As 1, but with treatment of the pumped liquid (by reverse osmosis) and re-injection into the former leaching zone. This scheme results in circulation of the liquid.
- **Phase 3:** As 2, with the addition of a reducing chemical (for example hydrogen sulfide (H_2S) or sodium sulfide (Na_2S). This causes the chemical precipitation and thus immobilization of major contaminants.
- **Phase 4:** Circulation of the liquid by pumping and re- injection, to obtain uniform conditions in the whole former leaching zone.

But, even with this treatment scheme, various problems remain unresolved:

- Contaminants that are mobile under chemically reducing conditions, such as radium, cannot be controlled.
- If chemically reducing conditions are later disturbed for any reasons, the precipitated contaminants are re-mobilized.
- The restoration process takes very long periods of time, not all parameters can be lowered appropriately.

Most restoration experiments reported refer to the alkaline leaching scheme, since this scheme is the only one used in Western world commercial in-situ operations. Therefore, nearly no experience exists with groundwater restoration after acid in- situ leaching, the scheme that was applied in most instances in Eastern Europe. The only Western in-situ leaching site restored after sulfuric acid leaching so far, is the small pilot scale facility Nine Mile Lake near Casper, Wyoming (USA). The results can therefore not simply be transferred to production scale facilities. The restoration scheme applied included the first two steps mentioned above. It turned out that a water volume of more than 20 times the pore volume of the leaching zone had to be pumped, and still several parameters did not reach background levels. Moreover, the restoration required about the same time as used for the leaching period.

In USA, the Pawnee, Lamprecht, and Zamzow ISL Sites in Texas were restored using steps 1 and 2 of the above listed treatment scheme. Relaxed groundwater restoration standards have been granted at these and other sites, since the restoration criteria could not be met.

A study published by the U.S. Geological Survey in 2009 found that "To date, no remediation of an ISR operation in the United States has successfully returned the aquifer to baseline conditions." Baseline conditions include commercial quantities of radioactive U_3O_8. Efficient in-situ recovery reduces U_3O_8 values of the aquifer. Speaking at an EPA Region 8 workshop, on September 29, 2010, Ardyth Simmons, PhD, Los Alamos National Laboratory (Los Alamos, NM) on the subject "Establishing Baseline and Comparison to Restoration Values at Uranium In-Situ Recovery Sites" stated "These results indicated that it may be unrealistic for ISR operations to restore aquifers to the mean, because in some cases, this means that there would have to be less uranium present than there was pre-mining. Pursuing more conservative concentrations results in a considerable amount of water usage, and many of these aquifers were not suitable for drinking water before mining initiated."

The EPA is considering the need to update the environmental protection standards for uranium mining because current regulations, promulgated in response to the

Uranium Mill Tailings Radiation Control Act of 1978, do not address the relatively recent process of in-situ leaching (ISL) of uranium from underground ore bodies. In a February, 2012 letter the EPA states, "Because the ISL process affects groundwater quality, the EPA's Office of Radiation and Indoor Air requested advice from the Science Advisory Board (SAB) on issues related to design and implementation of groundwater monitoring at ISL mining sites."

The SAB makes recommendations concerning monitoring to characterize baseline groundwater quality prior to the start of mining operations, monitoring to detect any leachate excursions during mining, and monitoring to determine when groundwater quality has stabilized after mining operations have been completed. The SAB also reviews the advantages and disadvantages of alternative statistical techniques to determine whether post-operational groundwater quality has returned to near pre-mining conditions and whether mine operation can be predicted not to adversely impact groundwater quality after site closure acceptance.

16.4 METALS EXTRACTED THROUGH BIOLEACHING

Bioleaching is primarily used to extract metals from sulfide ores, and several metals can be efficiently recovered through this process. Some of the most commonly bioleached metals include:

16.4.1 Copper

Copper is the most extensively bioleached metal, primarily due to the abundance of low-grade copper sulfide ores like chalcopyrite. Bioleaching has become a significant part of copper production, especially in regions where traditional mining methods are inefficient or uneconomical. *Acidithiobacillus ferrooxidans* and *Leptospirillum ferrooxidans* play key roles in copper bioleaching, converting the insoluble copper sulfides into soluble copper ions that can be recovered through solvent extraction or electrowinning.

16.4.2 Gold

Bioleaching is used to recover gold from refractory ores, which are difficult to process using conventional cyanide leaching. Microorganisms like *Acidithiobacillus ferrooxidans* help break down the sulfide matrix surrounding the gold particles, making it easier to extract the gold through cyanidation. While gold bioleaching is not as widespread as copper bioleaching, it is gaining attention as a more environmentally friendly alternative to traditional gold extraction methods.

16.4.3 Uranium

Uranium bioleaching is a well-established process, particularly in regions with low-grade uranium ores. Microorganisms like *Acidithiobacillus ferrooxidans* oxidize uranium sulfides, converting the uranium into a soluble form that can be extracted from the leach solution. In situ bioleaching is commonly used for uranium recovery, as it minimizes surface disturbance and reduces the environmental impact of mining.

16.4.4 Zinc

Zinc can also be extracted through bioleaching, particularly from zinc sulfide ores like sphalerite. The process involves the oxidation of zinc sulfides by microorganisms, which solubilizes the zinc ions and allows for their recovery. Bioleaching of zinc is less common than copper or uranium, but it is used in some regions where low-grade zinc ores are abundant.

16.5 ENVIRONMENTAL AND ECONOMIC BENEFITS OF BIOLEACHING

Bioleaching offers several environmental and economic advantages over traditional mining and metallurgical processes.

16.5.1 Environmental Benefits

One of the most significant environmental benefits of bioleaching is its reduced energy consumption compared to traditional methods like smelting. Bioleaching occurs at ambient temperatures, eliminating the need for high-temperature furnaces and the associated greenhouse gas emissions. Additionally, bioleaching minimizes the use of toxic chemicals, such as cyanide and sulfuric acid, which are commonly used in chemical leaching processes. This reduces the risk of chemical spills and environmental contamination. Bioleaching also generates fewer solid wastes, as it can be used to extract metals from mine tailings and waste rock, which would otherwise be discarded. This contributes to the overall sustainability of mining operations by reducing the volume of waste materials.

16.5.2 Economic Benefits

Economically, bioleaching allows for the recovery of metals from low-grade ores that would otherwise be unprofitable to process using conventional methods. This makes bioleaching an attractive option for mining companies looking to maximize resource extraction while minimizing costs. Bioleaching can also be applied to abandoned or closed mine sites, enabling the recovery of valuable metals from previously uneconomical deposits. In addition, bioleaching is a scalable and flexible process, making it suitable for both large and small mining operations. The relatively low infrastructure requirements and operational costs of bioleaching make it a viable alternative to traditional mining methods, particularly in regions with limited access to energy or processing facilities.

SECTION 5:

USE OF WATER CELLS FOR FOOD RELATED PURPOSES

CHAPTER 17
SINGLE CELL PROTEIN PRODUCTION

Single-cell proteins (SCP) or microbial proteins refer to edible unicellular microorganisms. The biomass or protein extract from pure or mixed cultures of algae, yeasts, fungi or bacteria may be used as an ingredient or a substitute for protein-rich foods, and is suitable for human consumption or as animal feeds. Industrial agriculture is marked by a high water footprint, high land use, biodiversity destruction, general environmental degradation and contributes to climate change by emission of a third of all greenhouse gases; production of SCP does not necessarily exhibit any of these serious drawbacks. As of today, SCP is commonly grown on agricultural waste products, and as such inherits the ecological footprint and water footprint of industrial agriculture. However, SCP may also be produced entirely independent of agricultural waste products through autotrophic growth. Thanks to the high diversity of microbial metabolism, autotrophic SCP provides several different modes of growth, versatile options of nutrients recycling, and a substantially increased efficiency compared to crops. A 2021 publication showed that photovoltaic-driven microbial protein production could use 10 times less land for an equivalent amount of protein compared to soybean cultivation.

With the world population reaching 9 billion by 2050, there is strong evidence that agriculture will not be able to meet demand and that there is serious risk of food shortage. Autotrophic SCP represents options of fail-safe mass food-production which can produce food reliably even under harsh climate conditions.

17.1 HISTORY

In 1781, processes for preparing highly concentrated forms of yeast were established. Research on Single Cell Protein Technology started a century ago when Max Delbrück and his colleagues found out the high value of surplus brewer's yeast as a feeding supplement for animals. During World War I and World War II, yeast-SCP was employed on a large scale in Germany to counteract food shortages during the war. Inventions for SCP production often represented milestones for biotechnology in general: for example, in 1919, Sak in Denmark and Hayduck in Germany invented a method named, "Zulaufverfahren", (fed-batch) in which sugar solution was fed continuously to an aerated suspension of yeast instead of adding yeast to diluted sugar solution once (batch). In post war period, the Food and Agriculture Organization of the United Nations (FAO) emphasized on hunger and malnutrition problems of the world in 1960 and introduced the concept of protein gap, showing that 25% of the world population had a deficiency of protein intake in their diet. It was also feared that agricultural production would fail to meet the increasing demands of food by humanity. By the mid 60's, almost quarter of a million tons of food yeast were being produced in different parts of the world and Soviet Union alone produced some 900,000 tons by 1970 of food and fodder yeast.

In the 1960s, researchers at British Petroleum developed what they called "proteins-from-oil process": a technology for producing single-cell protein by yeast fed by waxy n-paraffins, a byproduct of oil refineries. Initial research work was done by Alfred Champagnat at BP's Lavera Oil Refinery in France; a small pilot plant there started operations in March 1963, and the same construction of the second pilot plant, at Grangemouth Oil Refinery in Britain, was authorized.

The term SCP was coined in 1966 by Carroll L. Wilson of MIT. The "food from oil" idea became quite popular by the 1970s, with Champagnat being awarded the UNESCO Science Prize in 1976, and paraffin-fed yeast facilities being built in a number of countries. The primary use of the product was as poultry and cattle feed.

The Soviets were particularly enthusiastic, opening large "BVK" (*belkovo-vitaminny kontsentrat*, i.e., "protein-vitamin concentrate") plants next to their oil refineries in Kstovo (1973) and Kirishi (1974).

The Soviet Ministry of Microbiological Industry had eight plants of this kind by 1989. However, due to concerns of toxicity of alkanes in SCP and pressured by the environmentalist movements, the government decided to close them down, or convert to some other microbiological processes.

Quorn is a range of vegetarian and vegan meat-substitutes made from *Fusarium venenatum* mycoprotein, sold in Europe and North America. Another type of single cell protein-based meat analogue (which does not use fungi however but rather bacteria) is Calysta. Other producers are Unibio (Denmark) Circe Biotechnologie (Austria) and String Bio (India). SCP has been argued to be a source of alternative or resilient food.

17.2 PRODUCTION PROCESS

Single-cell proteins develop when microbes ferment waste materials (including wood, straw, cannery, and food-processing wastes, residues from alcohol production, hydrocarbons, or human and animal excreta). With 'electric food' processes the inputs are electricity, CO_2 and trace minerals and chemicals such as fertiliser. It is also possible to derive SCP from natural gas to use as a resilient food. Similarly SCP can be derived from waste plastic by upcycling.

The problem with extracting single-cell proteins from the wastes is the dilution and cost. They are found in very low concentrations, usually less than 5%. Engineers have developed ways to increase the concentrations including centrifugation, flotation, precipitation, coagulation, and filtration, or the use of semi-permeable membranes.

The single-cell protein must be dehydrated to approximately 10% moisture content and/or acidified to aid in storage and prevent spoilage. The methods to increase the concentrations to adequate levels and the de-watering process require equipment that is expensive and not always suitable for small-scale operations. It is economically prudent to feed the product locally and soon after it is produced.

17.3 PROPERTIES

Large-scale production of microbial biomass has many advantages over the traditional methods for producing proteins for food or feed.

- Microorganisms have a much higher growth rate (algae: 2–6 hours, yeast: 1–3 hours, bacteria: 0.5–2 hours). This also allows selection for strains with high yield and good nutritional composition more quickly and easily compared to breeding.
- Whereas large parts of crops, such as stems, leaves and roots, are not edible, single-cell microorganisms can be used entirely. Whereas parts of the edible fraction of crops are indigestible, many microorganisms are digestible at a much higher fraction.
- Microorganisms usually have a much higher protein content of 30–70% in the dry mass than vegetables or grains. The amino acid profiles of many SCP microorganisms often have excellent nutritional quality, comparable to hen's eggs.
- Some microorganisms can build vitamins and nutrients which eukaryotic organisms such as plants cannot produce or not produce in significant amounts, including vitamin B12.
- Microorganisms can utilize a broad spectrum of raw materials as carbon sources including alkanes, methanol, methane, ethanol and sugars. What was considered "waste product" often can be reclaimed as nutrients and support growth of edible microorganisms.
- Like plants, autotrophic microorganisms are capable of growing on CO_2. Some of them, such as bacteria with the Wood–Ljungdahl pathway or the reductive TCA can fix CO_2 with efficiencies ranging from 2-3 times to 10 times more efficiently than plants, when also considering the effects of photoinhibition.

- Some bacteria, such as several homoacetogenic clostridia, are capable of performing syngas fermentation. This means they can metabolize synthesis gas, a gas mixture of CO, H_2 and CO_2 that can be made by gasification of residual intractable biowastes such as lignocellulose.
- Some bacteria are diazotrophic, i.e. they can fix N_2 from the air and are thus independent of chemical N-fertilizer, whose production, utilization and degradation causes tremendous harm to the environment, deteriorates public health, and fosters climate change.
- Many bacteria can utilize H_2 for energy supply, using enzymes called hydrogenases. Whereas hydrogenases are normally highly O_2-sensitive, some bacteria are capable of performing O_2-dependent respiration of H_2. This feature allows autotrophic bacteria to grow on CO_2 without light at a fast growth rate. Since H_2 can be made efficiently by water electrolysis, in a manner of speaking, those bacteria can be "powered by electricity".
- Microbial biomass production is independent of seasonal and climatic variations, and can easily be shielded from extreme weather events that are expected to cause crop failures with the ongoing climate-change. Light-independent microorganisms such as yeasts can continue to grow at night.
- Cultivation of microorganisms generally has a much lower water footprint than agricultural food production. Whereas the global average blue-green water footprint (irrigation, surface, ground and rain water) of crops reaches about 1800 liters per kg crop due to evaporation, transpiration, drainage and runoff, closed bioreactors producing SCP exhibits none of these causes.
- Cultivation of microorganisms does not require fertile soil and therefore does not compete with agriculture. Thanks to the low water requirements, SCP cultivation can even be done in dry climates with infertile soil and may provide a means of fail-safe food supply in arid countries.
- Photosynthetic microorganisms can reach a higher solar-energy-conversion efficiency than plants, because in photobioreactors supply of water, CO_2 and a balanced light distribution can be tightly controlled.
- Unlike agricultural products which are processed towards a desired quality, it is easier with microorganisms to direct production towards a desired quality. Instead of extracting amino acids from soy beans and throwing away half of the plant body in the process, microorganisms

can be genetically modified to overproduce or even secrete a particular amino acid. However, in order to keep a good consumer acceptance, it is usually easier to obtain similar results by screening for microorganisms which already have the desired trait or train them via selective adaptation.

Although SCP shows very attractive features as a nutrient for humans, however there are some problems that deter its adoption on global basis:

- Fast growing microorganisms such as bacteria and yeast have a high concentration of nucleic acid, notably RNA. Levels must be limited in the diets of monogastric animals to <50 g per day. Ingestion of purine compounds arising from RNA breakdown leads to increased plasma levels of uric acid, which can cause gout and kidney stones. Uric acid can be converted to allantoin, which is excreted in urine. Nucleic acid removal is not necessary from animal feeds but is from human foods. A temperature hold at 64 °C inactivates fungal proteases . However, this problem can be remediated. One common method consists in a heat treatment which kills the cells, inactivates proteases and allows endogenous RNases to hydrolyse RNA with release of nucleotides from cell to culture broth.
- Similar to plant cells, the cell wall of some microorganisms such as algae and yeast contains indigestible components, such as cellulose. The cells of some kind of SCP should be broken up in order to liberate the cell interior and allow complete digestion.
- Some kind of SCP exhibits unpleasant color and flavors.
- Depending on the kind of SCP and the cultivation conditions, care must be taken to prevent and control contamination by other microorganisms because contaminants may produce toxins such as mycotoxins or cyanotoxins. An interesting approach to address this problem was proposed with the fungus *Scytalidium acidophilum* which grows at a pH as low as 1, outside the tolerance of most microorganisms. This allows it to grow on acid-hydrolysed paper waste at low-cost.
- Some yeast and fungal proteins are deficient in methionine.

CHAPTER 18
YEAST PRODUCTION

Yeasts are eukaryotic, single-celled microorganisms classified as members of the fungus kingdom. The first yeast originated hundreds of millions of years ago, and at least 1,500 species are currently recognized. They are estimated to constitute 1% of all described fungal species.

Some yeast species have the ability to develop multicellular characteristics by forming strings of connected budding cells known as pseudohyphae or false hyphae, or quickly evolve into a multicellular cluster with specialised cell organelles function. Yeast sizes vary greatly, depending on species and environment, typically measuring 3–4 µm in diameter, although some yeasts can grow to 40 µm in size. Most yeasts reproduce asexually by mitosis, and many do so by the asymmetric division process known as budding. With their single-celled growth habit, yeasts can be contrasted with molds, which grow hyphae. Fungal species that can take both forms (depending on temperature or other conditions) are called dimorphic fungi.

The yeast species *Saccharomyces cerevisiae* converts carbohydrates to carbon dioxide and alcohols through the process of fermentation. The products of this reaction have been used in baking and the production of alcoholic beverages for thousands of years. *S. cerevisiae* is also an important model organism in modern cell biology research, and is one of the most thoroughly studied eukaryotic microorganisms. Researchers have cultured it in order to understand the biology of the eukaryotic cell and ultimately human biology in great detail. Other species of yeasts, such as *Candida albicans*, are opportunistic pathogens and can cause infections in humans. Yeasts have recently been used to generate electricity in microbial fuel cells and to produce ethanol for the biofuel industry.

Yeasts do not form a single taxonomic or phylogenetic grouping. The term "yeast" is often taken as a synonym for *Saccharomyces cerevisiae*, but the phylogenetic diversity of yeasts is shown by their placement in two separate phyla: the Ascomycota and the Basidiomycota. The budding yeasts or "true yeasts" are classified in the order Saccharomycetales, within the phylum Ascomycota.

18.1 HISTORY

The word "yeast" comes from Old English *gist*, *gyst*, and from the Indo-European root *yes-*, meaning "boil", "foam", or "bubble". Yeast microbes are probably one of the earliest domesticated organisms. Archaeologists digging in Egyptian ruins found early grinding stones and baking chambers for yeast-raised bread, as well as drawings of 4,000-year-old bakeries and breweries. Vessels studied from several archaeological sites in Israel (dating to around 5,000, 3,000 and 2,500 years ago), which were believed to have contained alcoholic beverages (beer and mead), were found to contain yeast colonies that had survived over the millennia, providing the first direct biological evidence of yeast use in early cultures. In 1680, Dutch naturalist Anton van Leeuwenhoek first microscopically observed yeast, but at the time did not consider them to be living organisms, but rather globular structures as researchers were doubtful whether yeasts were algae or fungi. Theodor Schwann recognized them as fungi in 1837.

In 1857, French microbiologist Louis Pasteur showed that by bubbling oxygen into the yeast broth, cell growth could be increased, but fermentation was inhibited – an observation later called the "Pasteur effect". In the paper "*Mémoire sur la fermentation alcoolique,*" Pasteur proved that alcoholic fermentation was conducted by living yeasts and not by a chemical catalyst.

By the late 18th century two yeast strains used in brewing had been identified: *Saccharomyces cerevisiae* (top-fermenting yeast) and *S. pastorianus* (bottom-fermenting yeast). *S. cerevisiae* has been sold commercially by the Dutch for bread-making since 1780; while, around 1800, the Germans started producing *S. cerevisiae* in the form of cream. In 1825, a method was developed to remove the liquid so the yeast could be prepared as solid blocks. The industrial production of yeast blocks was enhanced by the introduction of the filter press in 1867. In 1872, Baron Max de Springer developed a manufacturing process to create granulated yeast from beetroot molasses, a technique that was used until the first World War. In the United States, naturally occurring airborne yeasts were used almost exclusively until commercial yeast was marketed at the Centennial Exposition in 1876 in Philadelphia, where Charles L. Fleischmann exhibited the product and a process to use it, as well as serving the resultant baked bread.

The mechanical refrigerator (first patented in the 1850s in Europe) liberated brewers and winemakers from seasonal constraints for the first time and allowed them to exit cellars and other earthen environments. For John Molson, who made his livelihood in Montreal prior to the development of the fridge, the brewing season lasted from September through to May. The same seasonal restrictions formerly governed the distiller's art.

18.2 NUTRITION AND GROWTH

Yeasts are chemoorganotrophs, as they use organic compounds as a source of energy and do not require sunlight to grow. Carbon is obtained mostly from hexose sugars, such as glucose and fructose, or disaccharides such as sucrose and maltose.

Some species can metabolize pentose sugars such as ribose, alcohols, and organic acids. Yeast species either require oxygen for aerobic cellular respiration (obligate aerobes) or are anaerobic, but also have aerobic methods of energy production (facultative anaerobes). Unlike bacteria, no known yeast species grow only anaerobically (obligate anaerobes). Most yeasts grow best in a neutral or slightly acidic pH environment.

Yeasts vary in regard to the temperature range in which they grow best. For example, *Leucosporidium frigidum* grows at -2 to 20 °C (28 to 68 °F), *Saccharomyces telluris* at 5 to 35 °C (41 to 95 °F), and *Candida slooffi* at 28 to 45 °C (82 to 113 °F). The cells can survive freezing under certain conditions, with viability decreasing over time.

In general, yeasts are grown in the laboratory on solid growth media or in liquid broths. Common media used for the cultivation of yeasts include potato dextrose agar or potato dextrose broth, Wallerstein Laboratories nutrient agar, yeast peptone dextrose agar, and yeast mould agar or broth. Home brewers who cultivate yeast frequently use dried malt extract and agar as a solid growth medium. The fungicide cycloheximide is sometimes added to yeast growth media to inhibit the growth of *Saccharomyces* yeasts and select for wild/indigenous yeast species. This will change the yeast process.

The appearance of a white, thready yeast, commonly known as kahm yeast, is often a byproduct of the lactofermentation (or pickling) of certain vegetables. It is usually the result of exposure to air. Although harmless, it can give pickled vegetables a bad flavor and must be removed regularly during fermentation.

18.3 ECOLOGY

Yeasts are very common in the environment, and are often isolated from sugar-rich materials. Examples include naturally occurring yeasts on the skins of fruits and berries (such as grapes, apples, or peaches), and exudates from plants (such as plant saps or cacti). Some yeasts are found in association with soil and insects. Yeasts from the soil and from the skins of fruits and berries have been shown to dominate fungal succession during fruit decay. The ecological function and biodiversity of yeasts are relatively unknown compared to those of other microorganisms. Yeasts, including *Candida albicans*, *Rhodotorula rubra*, *Torulopsis* and *Trichosporon cutaneum*, have been found living in between people's toes as part of their skin flora. Yeasts are also present in the gut flora of mammals and some insects and even deep-sea environments host an array of yeasts.

An Indian study of seven bee species and nine plant species found 45 species from 16 genera colonize the nectaries of flowers and honey stomachs of bees. Most were members of the genus *Candida*; the most common species in honey stomachs was *Dekkera intermedia* and in flower nectaries, *Candida blankii*. Yeast colonising nectaries of the stinking hellebore have been found to raise the temperature of the flower, which may aid in attracting pollinators by increasing the evaporation of volatile organic compounds. A black yeast has been recorded as a partner in a complex relationship between ants, their mutualistic fungus, a fungal parasite of the fungus and a bacterium that kills the parasite. The yeast has a negative effect on the bacteria that normally produce antibiotics to kill the parasite, so may affect the ants' health by allowing the parasite to spread.

Certain strains of some species of yeasts produce proteins called yeast killer toxins that allow them to eliminate competing strains. This can cause problems for winemaking but could potentially also be used to advantage by using killer toxin-producing strains to make the wine. Yeast killer toxins may also have medical applications in treating yeast infections.

Marine yeasts, defined as the yeasts that are isolated from marine environments, are able to grow better on a medium prepared using seawater rather than freshwater. The first marine yeasts were isolated by Bernhard Fischer in 1894 from the Atlantic Ocean, and those were identified as *Torula* sp. and *Mycoderma* sp. Following this discovery, various other marine yeasts have been isolated from around the world from different sources, including seawater, seaweeds, marine fish and mammals. Among these isolates, some marine yeasts originated from terrestrial habitats (grouped as facultative marine yeast), which were brought to and survived in marine environments. The other marine yeasts were grouped as obligate or indigenous marine yeasts, which are confined to marine habitats. However, no sufficient evidence has been found to explain the indispensability of seawater for obligate marine yeasts. It has been reported that marine yeasts are able to produce many bioactive substances, such as amino acids, glucans, glutathione, toxins, enzymes, phytase, and vitamins with potential applications in the food, pharmaceutical, cosmetic, and chemical industries as well as for marine culture and environmental protection. Marine yeast was successfully used to produce bioethanol using seawater-based media which will potentially reduce the water footprint of bioethanol.

18.4 REPRODUCTION

Yeasts, like all fungi, may have asexual and sexual reproductive cycles. The most common mode of vegetative growth in yeast is asexual reproduction by budding, where a small bud (also known as a bleb or daughter cell) is formed on the parent cell. The nucleus of the parent cell splits into a daughter nucleus and migrates into the daughter cell. The bud then continues to grow until it separates from the parent cell, forming a new cell. The daughter cell produced during the budding process is generally smaller than the mother cell. Some yeasts, including *Schizosaccharomyces pombe*, reproduce by fission instead of budding, and thereby creating two identically sized daughter cells.

In general, under high-stress conditions such as nutrient starvation, haploid cells will die; under the same conditions, however, diploid cells can undergo sporulation, entering sexual reproduction (meiosis) and producing a variety of haploid spores, which can go on to mate (conjugate), reforming the diploid.

The haploid fission yeast *Schizosaccharomyces pombe* is a facultative sexual microorganism that can undergo mating when nutrients are limited. Exposure of *S. pombe* to hydrogen peroxide, an agent that causes oxidative stress leading to oxidative DNA damage, strongly induces mating and the formation of meiotic spores. The budding yeast *Saccharomyces cerevisiae* reproduces by mitosis as diploid cells when nutrients are abundant, but when starved, this yeast undergoes meiosis to form haploid spores. Haploid cells may then reproduce asexually by mitosis. Katz Ezov et al. presented evidence that in natural *S. cerevisiae* populations clonal reproduction and selfing (in the form of intratetrad mating) predominate. In nature, the mating of haploid cells to form diploid cells is most often between members of the same clonal population and out-crossing is uncommon. Analysis of the ancestry of natural *S. cerevisiae* strains led to the conclusion that out-crossing occurs only about once every 50,000 cell divisions. These observations suggest that the possible long-term benefits of outcrossing (e.g. generation of diversity) are likely to be insufficient for generally maintaining sex from one generation to the next. Rather, a short-term benefit, such as recombinational repair during meiosis, may be the key to the maintenance of sex in *S. cerevisiae*. Some pucciniomycete yeasts, in particular species of *Sporidiobolus* and *Sporobolomyces*, produce aerially dispersed, asexual ballistoconidia.

18.5 USES

The useful physiological properties of yeast have led to their use in the field of biotechnology. Fermentation of sugars by yeast is the oldest and largest application of this technology. Many types of yeasts are used for making many foods: baker's yeast in bread production, brewer's yeast in beer fermentation, and yeast in wine fermentation and for xylitol production. So-called red rice yeast is actually a mold, *Monascus purpureus*. Yeasts include some of the most widely used model organisms for genetics and cell biology.

18.5.1 Alcoholic Beverages

Alcoholic beverages are defined as beverages that contain ethanol (C_2H_5OH). This ethanol is almost always produced by fermentation – the metabolism of carbohydrates by certain species of yeasts under anaerobic or low-oxygen conditions. Beverages such as mead, wine, beer, or distilled spirits all use yeast at some stage of their production. A distilled beverage is a beverage containing ethanol that has been purified by distillation. Carbohydrate-containing plant material is fermented by yeast, producing a dilute solution of ethanol in the process. Spirits such as whiskey and rum are prepared by distilling these dilute solutions of ethanol. Components other than ethanol are collected in the condensate, including water, esters, and other alcohols, which (in addition to that provided by the oak in which it may be aged) account for the flavour of the beverage.

Beer

Brewing yeasts may be classed as "top-cropping" (or "top-fermenting") and "bottom-cropping" (or "bottom-fermenting"). Top-cropping yeasts are so called because they form a foam at the top of the wort during fermentation. An example of a top-cropping yeast is *Saccharomyces cerevisiae*, sometimes called an "ale yeast". Bottom-cropping yeasts are typically used to produce lager-type beers, though they can also produce ale-type beers. These yeasts ferment well at low temperatures. An example of bottom-cropping yeast is *Saccharomyces pastorianus*, formerly known as *S. carlsbergensis*.

Decades ago, taxonomists reclassified *S. carlsbergensis* (uvarum) as a member of *S. cerevisiae*, noting that the only distinct difference between the two is metabolic. Lager strains of *S. cerevisiae* secrete an enzyme called melibiase, allowing them to hydrolyse melibiose, a disaccharide, into more fermentable monosaccharides. Top- and bottom-cropping and cold- and warm-fermenting distinctions are largely generalizations used by laypersons to communicate to the general public.

The most common top-cropping brewer's yeast, *S. cerevisiae*, is the same species as the common baking yeast. Brewer's yeast is also very rich in essential minerals and the B vitamins (except B_{12}), a feature exploited in food products made from leftover (by-product) yeast from brewing. However, baking and brewing yeasts typically belong to different strains, cultivated to favour different characteristics: baking yeast strains are more aggressive, to carbonate dough in the shortest amount of time possible; brewing yeast strains act more slowly but tend to produce fewer off-flavours and tolerate higher alcohol concentrations (with some strains, up to 22%).

Dekkera/Brettanomyces is a genus of yeast known for its important role in the production of 'lambic' and specialty sour ales, along with the secondary conditioning of a particular Belgian Trappist beer. The taxonomy of the genus *Brettanomyces* has been debated since its early discovery and has seen many reclassifications over the years. Early classification was based on a few species that reproduced asexually (anamorph form) through multipolar budding. Shortly after, the formation of ascospores was observed and the genus *Dekkera*, which reproduces sexually (teleomorph form), was introduced as part of the taxonomy. The current taxonomy includes five species within the genera of *Dekkera/Brettanomyces*. Those are the anamorphs *Brettanomyces bruxellensis*, *Brettanomyces anomalus*, *Brettanomyces custersianus*, *Brettanomyces naardenensis*, and *Brettanomyces nanus*, with teleomorphs existing for the first two species, *Dekkera bruxellensis* and *Dekkera anomala*. The distinction between *Dekkera* and *Brettanomyces* is arguable, with Oelofse et al. (2008) citing Loureiro and Malfeito-Ferreira from 2006 when they affirmed that current molecular DNA detection techniques have uncovered no variance between the anamorph and teleomorph states. Over the past decade, *Brettanomyces* spp. have seen an increasing use in the craft-brewing sector of the industry, with a handful of breweries having produced beers that were primarily fermented with pure cultures of *Brettanomyces* spp.

This has occurred out of experimentation, as very little information exists regarding pure culture fermentative capabilities and the aromatic compounds produced by various strains. *Dekkera/Brettanomyces* spp. have been the subjects of numerous studies conducted over the past century, although a majority of the recent research has focused on enhancing the knowledge of the wine industry. Recent research on eight *Brettanomyces* strains available in the brewing industry focused on strain-specific fermentations and identified the major compounds produced during pure culture anaerobic fermentation in wort.

Wine

Yeast is used in winemaking, where it converts the sugars present (glucose and fructose) in grape juice (must) into ethanol. Yeast is normally already present on grape skins. Fermentation can be done with this endogenous "wild yeast", but this procedure gives unpredictable results, which depend upon the exact types of yeast species present. For this reason, a pure yeast culture is usually added to the must; this yeast quickly dominates the fermentation. The wild yeasts are repressed, which ensures a reliable and predictable fermentation.

Most added wine yeasts are strains of *S. cerevisiae*, though not all strains of the species are suitable. Different *S. cerevisiae* yeast strains have differing physiological and fermentative properties, therefore the actual strain of yeast selected can have a direct impact on the finished wine. Significant research has been undertaken into the development of novel wine yeast strains that produce atypical flavour profiles or increased complexity in wines.

The growth of some yeasts, such as *Zygosaccharomyces* and *Brettanomyces*, in wine can result in wine faults and subsequent spoilage. *Brettanomyces* produces an array of metabolites when growing in wine, some of which are volatile phenolic compounds. Together, these compounds are often referred to as "*Brettanomyces* character", and are often described as "antiseptic" or "barnyard" type aromas. *Brettanomyces* is a significant contributor to wine faults within the wine industry.

Researchers from the University of British Columbia, Canada, have found a new strain of yeast that has reduced amines. The amines in red wine and Chardonnay produce off-flavors and cause headaches and hypertension in some people. About 30% of people are sensitive to biogenic amines, such as histamines.

18.5.2 Baking

Yeast, most commonly *S. cerevisiae*, is used in baking as a leavening agent, converting the fermentable sugars present in dough into carbon dioxide. This causes the dough to expand or rise as gas forms pockets or bubbles. When the dough is baked, the yeast dies and the air pockets "set", giving the baked product a soft and spongy texture. The use of potatoes, water from potato boiling, eggs, or sugar in a bread dough accelerates the growth of yeast. Most yeasts used in baking are of the same species common in alcoholic fermentation. In addition, *Saccharomyces exiguus* (also known as *S. minor*), a wild yeast found on plants, fruits, and grains, is occasionally used for baking. In breadmaking, the yeast initially respires aerobically, producing carbon dioxide and water. When the oxygen is depleted, fermentation begins, producing ethanol as a waste product; however, this evaporates during baking.

It is not known when yeast was first used to bake bread. The first records that show this use came from Ancient Egypt. Researchers speculate a mixture of flour meal and water was left longer than usual on a warm day and the yeasts that occur in natural contaminants of the flour caused it to ferment before baking. The resulting bread would have been lighter and tastier than the normal flat, hard cake.

Today, there are several retailers of baker's yeast; one of the earlier developments in North America is Fleischmann's Yeast, in 1868. During World War II, Fleischmann's developed a granulated active dry yeast which did not require refrigeration, had a longer shelf life than fresh yeast, and rose twice as fast. Baker's yeast is also sold as a fresh yeast compressed into a square "cake". This form perishes quickly, so must be used soon after production. A weak solution of water and sugar can be used to determine whether yeast is expired. In the solution, active yeast will foam and bubble as it ferments the sugar into ethanol and carbon dioxide. Some recipes refer to this as proofing the yeast, as it "proves" (tests) the viability of the yeast before the other ingredients are added. When a sourdough starter is used, flour and water are added instead of sugar; this is referred to as proofing the sponge.

When yeast is used for making bread, it is mixed with flour, salt, and warm water or milk. The dough is kneaded until it is smooth, and then left to rise, sometimes until it has doubled in size. The dough is then shaped into loaves. Some bread doughs are knocked back after one rising and left to rise again (this is called dough proofing) and then baked. A longer rising time gives a better flavor, but the yeast can fail to raise the bread in the final stages if it is left for too long initially.

18.5.3 Bioremediation

Some yeasts can find potential application in the field of bioremediation. One such yeast, *Yarrowia lipolytica,* is known to degrade palm oil mill effluent, TNT (an explosive material), and other hydrocarbons, such as alkanes, fatty acids, fats and oils. It can also tolerate high concentrations of salt and heavy metals, and is being investigated for its potential as a heavy metal biosorbent. *Saccharomyces cerevisiae* has potential to bioremediate toxic pollutants like arsenic from industrial effluent. Bronze statues are known to be degraded by certain species of yeast. Different yeasts from Brazilian gold mines bioaccumulate free and complexed silver ions.

18.5.4 Industrial Ethanol Production

The ability of yeast to convert sugar into ethanol has been harnessed by the biotechnology industry to produce ethanol fuel. The process starts by milling a feedstock, such as sugar cane, field corn, or other cereal grains, and then adding dilute sulfuric acid, or fungal alpha amylase enzymes, to break down the starches into complex sugars. A glucoamylase is then added to break the complex sugars down into simple sugars. After this, yeasts are added to convert the simple sugars to ethanol, which is then distilled off to obtain ethanol up to 96% in purity. *Saccharomyces* yeasts have been genetically engineered to ferment xylose, one of the major fermentable sugars present in cellulosic biomasses, such as agriculture residues, paper wastes, and wood chips. Such a development means ethanol can be efficiently produced from more inexpensive feedstocks, making cellulosic ethanol fuel a more competitively priced alternative to gasoline fuels.

18.5.5 Nonalcoholic Beverages

A number of sweet carbonated beverages can be produced by the same methods as beer, except the fermentation is stopped sooner, producing carbon dioxide, but only trace amounts of alcohol, leaving a significant amount of residual sugar in the drink.

- Root beer, originally made by Native Americans, commercialized in the United States by Charles Elmer Hires and especially popular during Prohibition
- Kvass, a fermented drink made from rye, popular in Eastern Europe. It has a recognizable, but low alcoholic content.
- Kombucha, a fermented sweetened tea. Yeast in symbiosis with acetic acid bacteria is used in its preparation. Species of yeasts found in the tea can vary, and may include: *Brettanomyces bruxellensis*, *Candida stellata*, *Schizosaccharomyces pombe*, *Torulaspora delbrueckii* and *Zygosaccharomyces bailii*. Also popular in Eastern Europe and some former Soviet republics under the name *chajnyj grib* which means "tea mushroom".

- Kefir and kumis are made by fermenting milk with yeast and bacteria.
- Mauby made by fermenting sugar with the wild yeasts naturally present on the bark of the *Colubrina elliptica* tree, popular in the Caribbean

18.5.6 Foods and Nutritional Supplements

Yeast is used as an ingredient in foods for its umami flavor, in much of the same way that monosodium glutamate (MSG) is used and, like MSG, yeast often contains free glutamic acid. Examples include:

- **Yeast Extract,** made from the intracellular contents of yeast and used as food additives or flavours. The general method for making yeast extract for food products such as Vegemite and Marmite on a commercial scale is *heat autolysis*, i.e. to add salt to a suspension of yeast, making the solution hypertonic, which leads to the cells' shrivelling up. This triggers autolysis, wherein the yeast's digestive enzymes break their own proteins down into simpler compounds, a process of self-destruction. The dying yeast cells are then heated to complete their breakdown, after which the husks (yeast with thick cell walls that would give poor texture) are removed. Yeast autolysates are used in Vegemite and Promite (Australia); Marmite (the United Kingdom); the unrelated Marmite (New Zealand); Vitam-R (Germany); and Cenovis (Switzerland).
- **Nutritional Yeast,** which is whole dried, deactivated yeast cells, usually *S. cerevisiae*. Usually in the form of yellow flake or powder, its nutty and umami flavor makes it a vegan substitute for cheese powder. Another popular use is as a topping for popcorn. It can also be used in mashed and fried potatoes, as well as in scrambled eggs. It comes in the form of flakes, or as a yellow powder similar in texture to cornmeal. In Australia, it is sometimes sold as "savoury yeast flakes".

Both types of yeast foods above are rich in B-complex vitamins (besides vitamin B_{12} unless fortified), making them an attractive nutritional supplement to vegans. The same vitamins are also found in some yeast-fermented products mentioned above, such as kvass.

Nutritional yeast in particular is naturally low in fat and sodium and a source of protein and vitamins as well as other minerals and cofactors required for growth.

Many brands of nutritional yeast and yeast extract spreads, though not all, are fortified with vitamin B_{12}, which is produced separately by bacteria.

In 1920, the Fleischmann Yeast Company began to promote yeast cakes in a "Yeast for Health" campaign. They initially emphasized yeast as a source of vitamins, good for skin and digestion. Their later advertising claimed a much broader range of health benefits, and was censured as misleading by the Federal Trade Commission. The fad for yeast cakes lasted until the late 1930s.

18.5.7 Probiotics

Some probiotic supplements use the yeast *S. boulardii* to maintain and restore the natural flora in the gastrointestinal tract. *S. boulardii* has been shown to reduce the symptoms of acute diarrhea, reduce the chance of infection by *Clostridium difficile* (often identified simply as C. difficile or C. diff), reduce bowel movements in diarrhea-predominant IBS patients, and reduce the incidence of antibiotic-, traveler's-, and HIV/AIDS-associated diarrheas.

18.5.8 Aquarium Hobby

Yeast is often used by aquarium hobbyists to generate carbon dioxide (CO_2) to nourish plants in planted aquaria. CO_2 levels from yeast are more difficult to regulate than those from pressurized CO_2 systems. However, the low cost of yeast makes it a widely used alternative.

18.5.9 Scientific Research

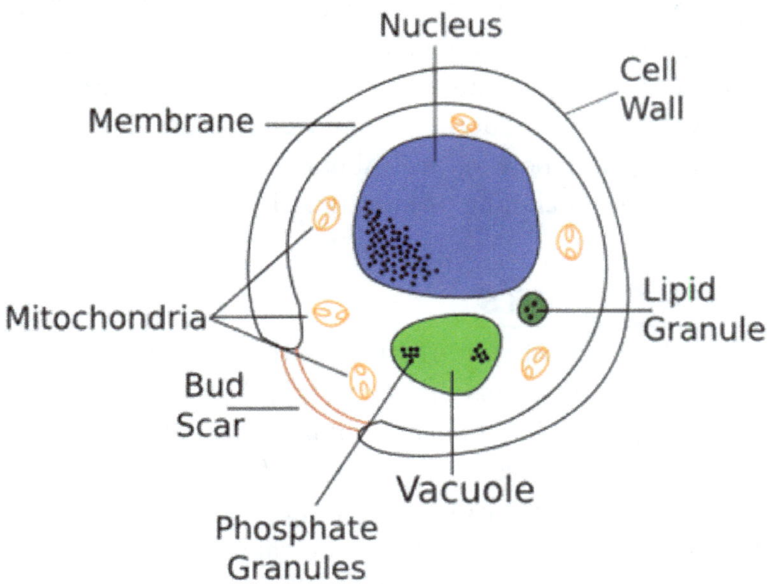

Fig 1: Diagram Showing a Yeast Cell

Several yeasts, in particular *S. cerevisiae* and *S. pombe*, have been widely used in genetics and cell biology, largely because they are simple eukaryotic cells, serving as a model for all eukaryotes, including humans, for the study of fundamental cellular processes such as the cell cycle, DNA replication, recombination, cell division, and metabolism. Also, yeasts are easily manipulated and cultured in the laboratory, which has allowed for the development of powerful standard techniques, such as yeast two-hybrid, synthetic genetic array analysis, and tetrad analysis. Many proteins important in human biology were first discovered by studying their homologues in yeast; these proteins include cell cycle proteins, signaling proteins, and protein-processing enzymes.

On 24 April 1996, *S. cerevisiae* was announced to be the first eukaryote to have its genome, consisting of 12 million base pairs, fully sequenced as part of the Genome Project. At the time, it was the most complex organism to have its full genome sequenced, and the work of seven years and the involvement of more than 100 laboratories to accomplish. The second yeast species to have its genome sequenced was *Schizosaccharomyces pombe*, which was completed in 2002. It was the sixth eukaryotic genome sequenced and consists of 13.8 million base pairs. As of 2014, over 50 yeast species have had their genomes sequenced and published. Genomic and functional gene annotation of the two major yeast models can be accessed via their respective model organism databases: SGD and PomBase.

18.5.10 Genetically Engineered Biofactories

Various yeast species have been genetically engineered to efficiently produce various drugs, a technique called metabolic engineering. *S. cerevisiae* is easy to genetically engineer; its physiology, metabolism and genetics are well known, and it is amenable for use in harsh industrial conditions. A wide variety of chemical in different classes can be produced by engineered yeast, including phenolics, isoprenoids, alkaloids, and polyketides. About 20% of biopharmaceuticals are produced in *S. cerevisiae*, including insulin, vaccines for hepatitis, and human serum albumin.

18.6 PATHOGENIC YEASTS

Some species of yeast are opportunistic pathogens that can cause infection in people with compromised immune systems. *Cryptococcus neoformans* and *Cryptococcus gattii* are significant pathogens of immunocompromised people. They are the species primarily responsible for cryptococcosis, a fungal infection that occurs in about one million HIV/AIDS patients, causing over 600,000 deaths annually. The cells of these yeast are surrounded by a rigid polysaccharide capsule, which helps to prevent them from being recognised and engulfed by white blood cells in the human body.

Yeasts of the genus *Candida*, another group of opportunistic pathogens, cause oral and vaginal infections in humans, known as candidiasis. *Candida* is commonly found as a commensal yeast in the mucous membranes of humans and other warm-blooded animals. However, sometimes these same strains can become pathogenic. The yeast cells sprout a hyphal outgrowth, which locally penetrates the mucosal membrane, causing irritation and shedding of the tissues. A book from the 1980s listed the pathogenic yeasts of candidiasis in probable descending order of virulence for humans as: *C. albicans*, *C. tropicalis*, *C. stellatoidea*, *C. glabrata*, *C. krusei*, *C. parapsilosis*, *C. guilliermondii*, *C. viswanathii*, *C. lusitaniae*, and *Rhodotorula mucilaginosa*. *Candida glabrata* is the second most common *Candida* pathogen after *C. albicans*, causing infections of the urogenital tract, and of the bloodstream (candidemia). *C. auris* has been more recently identified.

18.7 FOOD SPOILAGE

Yeasts are able to grow in foods with a low pH (5.0 or lower) and in the presence of sugars, organic acids, and other easily metabolized carbon sources. During their growth, yeasts metabolize some food components and produce metabolic end products. This causes the physical, chemical, and sensible properties of a food to change, and the food is spoiled. The growth of yeast within food products is often seen on their surfaces, as in cheeses or meats, or by the fermentation of sugars in beverages, such as juices, and semiliquid products, such as syrups and jams. The yeast of the genus *Zygosaccharomyces* have had a long history as spoilage yeasts within the food industry. This is mainly because these species can grow in the presence of high sucrose, ethanol, acetic acid, sorbic acid, benzoic acid, and sulfur dioxide concentrations, representing some of the commonly used food preservation methods. Methylene blue is used to test for the presence of live yeast cells. In oenology, the major spoilage yeast is *Brettanomyces bruxellensis*. *Candida blankii* has been detected in Iberian ham and meat.

18.8 SYMBIOSIS

An Indian study of seven bee species and nine plant species found 45 yeast species from 16 genera colonise the nectaries of flowers and honey stomachs of bees. Most were members of the genus *Candida*; the most common species in honey bee stomachs was *Dekkera intermedia*, while the most common species colonising flower nectaries was *Candida blankii*. Although the mechanism is not fully understood, it was found that *A. indica* flowers more if *Candida blankii* is present. In another example, *Spathaspora passalidarum*, found in the digestive tract of bess beetles, aids the digestion of plant cells by fermenting xylose. Many fruits produce different types of sugars that attract yeasts, which ferment the sugar and turns it into alcohol. Fruit eating mammals find the scent of alcohol attractive as it indicates a ripe, sugary fruit which provides more nutrition. In turn, the mammals helps disperse both the fruit's seeds and the yeast's spores.

Yeast and small hive beetle have mutualistic relationship. While small hive beetle is attracted by the pheromone released by the host honeybee, yeast can produce a similar pheromone which have the same attractive effect to the small hive beetle. Therefore, yeast facilitates SHB's infestation if the beehive contains yeast inside.

CHAPTER 19

PRODUCTION OF MICROBIAL INSECTICIDES

Microbial insecticides have emerged as an important tool in pest management, offering a more sustainable and environmentally friendly alternative to traditional chemical insecticides. These biological agents harness the power of microorganisms, such as bacteria, fungi, and viruses, to control pest populations effectively.

19.1 WHAT ARE MICROBIAL INSECTICIDES?

Microbial insecticides are biopesticides derived from naturally occurring microorganisms that can target and control insect pests. These agents may contain live microorganisms or the toxins produced by them. Common types of microbial insecticides include Bacillus thuringiensis (Bt), entomopathogenic fungi, and certain viruses like nucleopolyhedroviruses. The effectiveness of microbial insecticides lies in their specificity; they can effectively kill target pests while posing minimal risks to non-target organisms, including beneficial insects, humans, and the environment.

19.1.1 Types of Microbial Insecticides

- **Bacillus Thuringiensis (Bt):** One of the most widely used microbial insecticides, Bt is a soil-dwelling bacterium that produces crystalline proteins toxic to certain insect larvae, particularly lepidopterans (caterpillars). When ingested by the larvae, the proteins are activated by the alkaline conditions of the insect gut, leading to the formation of pores in the gut wall, causing paralysis and eventual death. Various formulations of Bt are available, targeting specific insect pests in crops like corn, cotton, and vegetables.
- **Entomopathogenic Fungi:** Fungi such as Beauveria bassiana and Metarhizium anisopliae are used as microbial insecticides that infect and kill insect pests. These fungi can penetrate the insect cuticle, proliferating within the host and ultimately leading to its death. They are effective against a wide range of insects, including aphids, whiteflies, and beetles, and can be applied as spores in powder or liquid formulations.

- **Nucleopolyhedroviruses (NPVs):** These are viruses that specifically infect insect hosts, particularly lepidopterans. NPVs are produced as polyhedra, which protect the viral particles from environmental degradation. When insects ingest these polyhedra, the virus infects their gut cells, causing death. NPVs can be effective in controlling various pests, including caterpillars and moths.

19.2 PRODUCTION OF MICROBIAL INSECTICIDES

The production of microbial insecticides involves several steps, including strain selection, fermentation, formulation, and quality control.

19.2.1 Strain Selection

The first step in producing microbial insecticides is the selection of effective microbial strains that exhibit insecticidal properties against target pests. Researchers screen natural isolates from the environment, such as soil, plant materials, and insect cadavers, to identify strains with promising insecticidal activity. This involves assessing their pathogenicity, virulence, and specificity against specific insect pests.

19.2.2 Fermentation

Once a suitable strain has been selected, the next step is to produce the microbial insecticide in bulk through fermentation. This process typically involves culturing the selected microorganisms in a controlled environment, where conditions such as temperature, pH, and nutrient availability can be optimized for maximum growth and production of insecticidal compounds. For bacteria like Bacillus thuringiensis, fermentation can be carried out in either solid-state or submerged fermentation systems. In solid-state fermentation, the bacterium is grown on solid substrates, such as grains or agricultural by-products, while submerged fermentation involves growing the bacteria in liquid media. The choice of fermentation method depends on the specific requirements of the microorganism and the intended formulation of the insecticide.

19.2.3 Harvesting and Formulation

After fermentation, the next step is to harvest the microbial biomass or the insecticidal compounds produced during fermentation. This may involve centrifugation, filtration, or other separation techniques to isolate the desired product. The harvested material is then formulated into a usable product. This could be in the form of a wettable powder, granule, or liquid formulation that can be easily applied in agricultural settings. Formulation may also include additives to enhance stability, shelf life, and efficacy, ensuring that the microbial insecticide remains effective under various environmental conditions.

19.2.4 Quality Control

Quality control is a critical step in the production of microbial insecticides. This involves testing the product for its viability, potency, and purity to ensure that it meets regulatory standards and performs effectively in the field. Quality control measures may include testing for the presence of contaminants, such as pathogens or other unwanted microorganisms, and ensuring that the microbial count remains within specified limits.

19.3 MECHANISMS OF ACTION

Microbial insecticides employ various mechanisms to control insect pests, making them effective in integrated pest management (IPM) programs.

19.3.1 Toxin Production

Bacillus thuringiensis produces specific insecticidal proteins that disrupt the gut function of target insects. Upon ingestion, the proteins are activated in the alkaline environment of the insect gut, leading to the formation of pores in the gut wall. This disrupts the gut integrity, causing the insect to cease feeding and eventually leading to death.

19.3.2 Infection and Pathogenicity

Entomopathogenic fungi infect insects by penetrating their cuticles. Once inside the host, these fungi proliferate and produce enzymes that break down the insect's internal tissues, ultimately leading to its death. This mode of action makes these fungi effective against a wide range of insect pests.

19.3.3 Viral Infection

Nucleopolyhedroviruses infect specific insect hosts, leading to disease. Upon ingestion, the virus replicates within the host's cells, causing cell death and ultimately leading to the host's demise. This targeted approach allows NPVs to effectively control specific pest populations.

19.4 ADVANTAGES OF MICROBIAL INSECTICIDES

19.4.1 Environmental Safety

Microbial insecticides are generally considered safer for the environment and non-target organisms. Their specificity allows them to target specific pests without harming beneficial insects, such as pollinators and natural predators. This reduces the risk of ecological imbalances often associated with chemical insecticides.

19.4.2 Reduced Chemical Residues

Microbial insecticides degrade more rapidly in the environment compared to synthetic chemicals, resulting in lower chemical residues on crops and in the surrounding ecosystem. This is particularly important for consumers concerned about food safety and the impacts of chemical residues on health.

19.4.3 Resistance Management

The use of microbial insecticides can help mitigate the development of resistance in pest populations. By incorporating microbial insecticides into pest management strategies, farmers can reduce reliance on chemical insecticides and disrupt pest resistance mechanisms.

19.4.4 Compatibility with Integrated Pest Management (IPM)

Microbial insecticides can be easily integrated into IPM programs, where multiple control strategies are employed to manage pest populations. Their use alongside cultural practices, biological control agents, and selective chemical applications can enhance overall pest management effectiveness.

19.5 CHALLENGES IN THE PRODUCTION AND USE OF MICROBIAL INSECTICIDES

19.5.1 Limited Shelf Life and Stability

Microbial insecticides can have a limited shelf life due to the viability of the microorganisms. Formulating products that remain stable and effective over time is crucial to ensure consistent performance in the field. Researchers continue to develop formulations that enhance the stability and viability of microbial products.

19.5.2 Application Techniques

The application of microbial insecticides can be more complex than traditional chemical insecticides. Factors such as the timing of application, environmental conditions, and the specific target pest must be carefully considered to achieve optimal efficacy. Education and training for farmers on proper application techniques are essential for successful implementation.

19.5.3 Regulatory Challenges

The registration and regulation of microbial insecticides can be more complex than for chemical pesticides. Regulatory agencies require extensive testing to assess the safety and efficacy of these products before they can be approved for use. This can lead to longer timelines for bringing new microbial insecticides to market.

19.6 FUTURE PROSPECTS

The future of microbial insecticides looks promising as research continues to uncover new microbial strains and their insecticidal properties. Advances in biotechnology, such as genetic engineering and synthetic biology, may enable the development of more effective microbial insecticides with enhanced specificity and efficacy. Moreover, the increasing demand for sustainable agricultural practices and the reduction of chemical pesticide use are likely to drive the growth of microbial insecticide production. As consumers become more environmentally conscious and regulations regarding chemical pesticide use become stricter, the adoption of microbial insecticides will likely increase.

19.7 CONCLUSION

Microbial insecticides represent a significant advancement in pest management, offering a sustainable and environmentally friendly alternative to conventional chemical insecticides. Through the utilization of naturally occurring microorganisms, these biopesticides provide effective control of insect pests while minimizing risks to non-target organisms and the environment. As research and production techniques continue to evolve, microbial insecticides are poised to play an increasingly important role in integrated pest management strategies, contributing to the future of sustainable agriculture.

CHAPTER 20

PRODUCTION OF RHIZOBIUM BIOFERTILIZER

Rhizobium biofertilizer is a vital agricultural input that enhances soil fertility and promotes sustainable farming practices. These nitrogen-fixing bacteria form symbiotic relationships with leguminous plants, facilitating the conversion of atmospheric nitrogen into a form usable by plants. The production of Rhizobium biofertilizer is a crucial step in harnessing the benefits of these microorganisms, as it involves cultivating, formulating, and applying them effectively to optimize crop yields.

20.1 WHAT IS RHIZOBIUM?

Rhizobium is a genus of Gram-negative soil bacteria that fix nitrogen. *Rhizobium* species form an endosymbiotic nitrogen-fixing association with roots of (primarily) legumes and other flowering plants. The bacteria colonize plant cells to form root nodules, where they convert atmospheric nitrogen into ammonia using the enzyme nitrogenase.

The ammonia is shared with the host plant in the form of organic nitrogenous compounds such as glutamine or ureides. The plant, in turn, provides the bacteria with organic compounds made by photosynthesis. This mutually beneficial relationship is true of all of the rhizobia, of which the genus *Rhizobium* is a typical example. *Rhizobium* is also capable of solubilizing phosphate.

Rhizobium is a genus of soil bacteria known for its ability to fix atmospheric nitrogen through symbiosis with legumes. These bacteria colonize the root nodules of leguminous plants, where they convert atmospheric nitrogen into ammonia, a form that plants can absorb and utilize.

This process is crucial for maintaining soil fertility and reducing the need for synthetic nitrogen fertilizers, which can have harmful environmental impacts. The relationship between Rhizobium and legumes is mutualistic; while Rhizobium gains carbohydrates and a protective environment from the plant, the plant benefits from the increased availability of nitrogen, promoting its growth and productivity. Common leguminous crops that benefit from Rhizobium symbiosis include peas, beans, lentils, clover, and alfalfa.

20.1.1 History

Martinus Beijerinck was the first to isolate and cultivate a microorganism from the nodules of legumes in 1888. He named it *Bacillus radicicola*, which is now placed in *Bergey's Manual of Determinative Bacteriology* under the genus *Rhizobium*.

20.1.2 Research

Rhizobium forms a symbiotic relationship with certain plants, such as legumes, fixing nitrogen from the air into ammonia, which acts as a natural fertilizer for the plants. The Agricultural Research Service is conducting research involving the genetic mapping of various rhizobial species with their respective symbiotic plant species, like alfalfa or soybean.

The goal of this research is to increase the plants' productivity without using fertilizers. In molecular biology, *Rhizobium* has been identified as a contaminant of DNA extraction kit reagents and ultrapure water systems, which may lead to its erroneous appearance in microbiota or metagenomic datasets. The presence of nitrogen-fixing bacteria as contaminants may be due to the use of nitrogen gas in ultra-pure water production to inhibit microbial growth in storage tanks.

20.2 SIGNIFICANCE OF RHIZOBIUM BIOFERTILIZER

20.2.1 Enhanced Soil Fertility

Rhizobium biofertilizer improves soil fertility by increasing the nitrogen content in the soil, which is essential for plant growth. This enhances the nutrient availability for crops and contributes to healthier soils.

20.2.2 Sustainable Agriculture

The use of biofertilizers like Rhizobium reduces the reliance on chemical fertilizers, which can lead to soil degradation and water pollution. By incorporating Rhizobium into agricultural practices, farmers can promote sustainable farming, improve biodiversity, and reduce environmental impacts.

20.2.3 Increased Crop Yields

The enhanced nitrogen availability from Rhizobium biofertilizer leads to increased crop yields. Leguminous plants that form symbiotic relationships with Rhizobium can grow faster, produce more biomass, and yield higher quantities of grain or forage.

20.2.4 Soil Structure Improvement

The application of Rhizobium biofertilizer can improve soil structure by enhancing organic matter content and promoting the activity of beneficial soil microorganisms. This contributes to better water retention, aeration, and overall soil health.

20.3 PRODUCTION OF RHIZOBIUM BIOFERTILIZER

The production of Rhizobium biofertilizer involves several steps, including strain selection, cultivation, formulation, and quality control. These processes are essential to ensure that the final product is effective and viable for agricultural use.

20.3.1 Strain Selection

The first step in producing Rhizobium biofertilizer is selecting the appropriate bacterial strain. Different Rhizobium species exhibit varying levels of nitrogen-fixing ability and compatibility with specific leguminous plants. Therefore, it is crucial to choose a strain that is well-suited for the target crop. Strains can be isolated from soil samples or obtained from culture collections.

20.3.2 Cultivation

Once a suitable strain has been selected, the next step is to cultivate the bacteria in a controlled environment. This is typically done through fermentation, where the bacteria are grown in nutrient-rich media that provide the necessary conditions for optimal growth. The fermentation process can be performed in either solid-state or liquid-state systems.

- **Liquid Fermentation:** In liquid fermentation, a nutrient broth is prepared, typically containing sources of carbon, nitrogen, and minerals essential for bacterial growth. The chosen Rhizobium strain is inoculated into the broth and incubated under controlled conditions, such as temperature, pH, and aeration. This method allows for rapid bacterial growth and biomass production.
- **Solid-State Fermentation:** In solid-state fermentation, the Rhizobium strain is grown on solid substrates, such as agricultural residues or other organic materials. This method can be more cost-effective and environmentally friendly, utilizing waste materials while providing the necessary nutrients for bacterial growth.

20.3.3 Harvesting

After fermentation, the next step is to harvest the biomass of the Rhizobium bacteria. This involves separating the bacterial cells from the fermentation medium. Various methods can be employed, including centrifugation, filtration, or sedimentation. The goal is to obtain a concentrated suspension of viable Rhizobium cells for formulation.

20.3.4 Formulation

The harvested Rhizobium biomass is then formulated into a suitable product for application. This may involve mixing the bacterial suspension with carriers or stabilizers to enhance shelf life and facilitate application. Common formulation types include:

- **Powdered Formulations:** The Rhizobium biomass can be dried and ground into a fine powder, which can be mixed with other organic materials or fertilizers. This form is easy to transport and apply in the field.
- **Liquid Formulations:** The Rhizobium suspension can be formulated into a liquid product for direct application to soil or seeds. Liquid formulations are often easier to use and can provide a quick inoculation of the target crops.

20.3.5 Quality Control

Quality control is essential to ensure the effectiveness and safety of the Rhizobium biofertilizer. This involves testing the product for viability, purity, and nitrogen-fixing efficiency. Various laboratory techniques, such as colony-forming unit (CFU) counts and biochemical assays, can be employed to assess the quality of the biofertilizer.

20.4 APPLICATION OF RHIZOBIUM BIOFERTILIZER

20.4.1 Seed Inoculation

One of the most effective ways to apply Rhizobium biofertilizer is through seed inoculation. Before planting, seeds can be coated with the Rhizobium biofertilizer, allowing the bacteria to establish themselves in the soil as the seeds germinate. This ensures that the plants have immediate access to the nitrogen-fixing bacteria.

20.4.2 Soil Application

Rhizobium biofertilizer can also be applied directly to the soil. This method involves broadcasting or incorporating the biofertilizer into the soil before planting. It is important to apply the biofertilizer at the right time to maximize its effectiveness, typically during land preparation.

20.4.3 Foliar Application

In some cases, foliar application of Rhizobium biofertilizer can be beneficial. This involves spraying the biofertilizer solution onto the leaves of the plants. While this method is less common, it can help establish a strong population of Rhizobium in the rhizosphere, especially during the early stages of plant growth.

20.5 BENEFITS OF USING RHIZOBIUM BIOFERTILIZER

20.5.1 Cost-Effectiveness

By reducing the need for synthetic nitrogen fertilizers, Rhizobium biofertilizer can lower production costs for farmers. The use of biofertilizers can also improve the overall profitability of leguminous crops, leading to increased income.

20.5.2 Environmental Sustainability

The use of Rhizobium biofertilizer promotes sustainable agricultural practices by enhancing soil health, reducing chemical inputs, and minimizing the environmental impacts associated with synthetic fertilizers.

This is particularly important in the face of global challenges such as soil degradation and water pollution.

20.5.3 Improved Crop Quality

Crops grown with Rhizobium biofertilizer often exhibit improved quality characteristics, such as higher protein content and better taste. This can enhance marketability and consumer satisfaction, providing additional economic benefits to farmers.

20.5.4 Enhanced Soil Microbial Diversity

The application of Rhizobium biofertilizer can enhance soil microbial diversity and activity. This leads to improved soil health, resilience, and overall fertility, creating a more sustainable agricultural ecosystem.

20.6 CHALLENGES AND FUTURE PROSPECTS

20.6.1 Strain Adaptability

Not all Rhizobium strains are equally effective for all leguminous crops. It is essential to identify and promote the use of locally adapted strains that can establish successfully in specific soil and climatic conditions.

Research efforts should focus on isolating and characterizing new, effective Rhizobium strains for different agricultural contexts.

20.6.2 Application Timing and Techniques

Optimizing application methods and timing is crucial for maximizing the effectiveness of Rhizobium biofertilizer. Farmers may require training and guidance on best practices to ensure successful application and establishment of the bacteria in the soil.

20.6.3 Awareness and Adoption

Raising awareness among farmers about the benefits of Rhizobium biofertilizer and promoting its use in agricultural practices is essential. Extension services and educational programs can play a significant role in increasing adoption rates and promoting sustainable farming practices.

20.7 CONCLUSION

The production of Rhizobium biofertilizer represents a significant advancement in sustainable agriculture. By harnessing the nitrogen-fixing capabilities of Rhizobium bacteria, farmers can enhance soil fertility, improve crop yields, and reduce their reliance on synthetic fertilizers.

Through effective strain selection, cultivation, and application methods, Rhizobium biofertilizer can be a valuable tool in promoting sustainable agricultural practices. As research continues to advance, the potential for Rhizobium biofertilizer to contribute to global food security and environmental sustainability remains promising.

CHAPTER 21

PRODUCTION OF FERMENTED FOODS

Fermented foods have been a fundamental part of human diets for thousands of years, providing not only preservation benefits but also enhanced flavors, textures, and nutritional profiles. The process of fermentation involves the metabolic activity of microorganisms such as bacteria, yeast, and molds, which transform raw ingredients into products that can be enjoyed for their unique characteristics.

21.1 UNDERSTANDING FERMENTATION

Fermentation is a biochemical process that converts sugars and carbohydrates into acids, gases, or alcohol through the action of microorganisms. This process is primarily anaerobic, meaning it occurs in the absence of oxygen. The most common forms of fermentation include lactic acid fermentation, alcoholic fermentation, and acetic acid fermentation.

21.1.1 Lactic Acid Fermentation

Lactic acid fermentation is a metabolic process by which glucose or other six-carbon sugars (also, disaccharides of six-carbon sugars, e.g. sucrose or lactose) are converted into cellular energy and the metabolite lactate, which is lactic acid in solution. It is an anaerobic fermentation reaction that occurs in some bacteria and animal cells, such as muscle cells. If oxygen is present in the cell, many organisms will bypass fermentation and undergo cellular respiration; however, facultative anaerobic organisms will both ferment and undergo respiration in the presence of oxygen. Sometimes even when oxygen is present and aerobic metabolism is happening in the mitochondria, if pyruvate is building up faster than it can be metabolized, the fermentation will happen anyway. Lactate dehydrogenase catalyzes the interconversion of pyruvate and lactate with concomitant interconversion of NADH and NAD^+. In *homolactic fermentation*, one molecule of glucose is ultimately converted to two molecules of lactic acid. *Heterolactic fermentation*, by contrast, yields carbon dioxide and ethanol in addition to lactic acid, in a process called the phosphoketolase pathway.

History

Several chemists discovered during the 19th century some fundamental concepts of the domain of organic chemistry. One of them for example was the French chemist

Joseph Louis Gay-Lussac, who was especially interested in fermentation processes, and he passed this fascination to one of his best students, Justus von Liebig. With a difference of some years, each of them described, together with colleagues, the chemical structure of the lactic acid molecule as we know it today. They had a purely chemical understanding of the fermentation process, which means that you can't see it using a microscope, and that it can only be optimized by chemical catalyzers.

In 1857, the French chemist Louis Pasteur first described lactic acid as the product of a microbial fermentation. During this time, he worked at the University of Lille, where a local distillery asked him for advice concerning some fermentation problems. Per chance and with the badly equipped laboratory he had at that time, he was able to discover that in this distillery, two fermentations were taking place, a lactic acid one and an alcoholic one, both induced by microorganisms. He then continued the research on these discoveries in Paris, where he also published his theories that presented a stable contradiction to the purely chemical version represented by Liebig and his followers.

Even though Pasteur described some concepts that are still accepted today, Liebig refused to accept them. But even Pasteur himself wrote that he was "driven" to a completely new understanding of this chemical phenomenon. Even if Pasteur didn't find every detail of this process, he still discovered the main mechanism of how the microbial lactic acid fermentation works. He was the first to describe fermentation as a "form of life without air".

Although this chemical process had not been properly described before Pasteur's work, people had been using microbial lactic acid fermentation for food production much earlier. Chemical analysis of archeological finds show that milk fermentation uses predate the historical period; its first applications were probably a part of the Neolithic Revolution. Since milk naturally contains lactic acid bacteria, the discovery of the fermentation process was quite evident, since it happens spontaneously at an adequate temperature.

The problem of these first farmers was that fresh milk is nearly indigestible by adults, so they had an interest to discover this mechanism. In fact, lactic acid bacteria contain the needed enzymes to digest lactose, and their populations multiply strongly during the fermentation. Therefore, milk fermented even a short time contains enough enzymes to digest the lactose molecules, after the milk is in the human body, which allows adults to consume it. Even safer was a longer fermentation, which was practiced for [cheesemaking]. This process was also discovered a very long time ago, which is proven by recipes for cheese production on Cuneiform scripts, the first written documents that exist, and a bit later in Babylonian and Egyptian texts. What is interesting is the theory of the competitive advantage of fermented milk products. The idea of this theory is that the women of these first settled farmer clans could shorten the time between two children thanks to the additional lactose uptake from milk consumption. This factor may have given them an important advantage to out-compete the hunter-gatherer societies.

With the increasing consumption of milk products these societies developed a lactase persistence by epigenetic inheritance, which means that the milk-digesting enzyme lactase was present in their bodies during the whole lifetime, so they could drink unfermented milk as adults too. This early habituation to lactose consumption in the first settler societies can still be observed today in regional differences of this mutation's concentration. It is estimated that about 65% of world population still lacks it. Since these first societies came from regions around eastern Turkey to central Europe, the gene appears more frequently there and in North America, as it was settled by Europeans. It is because of the dominance of this mutation that Western cultures believe it is unusual to have a lactose intolerance, when it is in fact more common than the mutation. On the contrary, lactose intolerance is much more present in Asian countries.

Milk products and their fermentation have had an important influence on some cultures' development. This is the case in Mongolia, where people often practice a pastoral form of agriculture. The milk that they produce and consume in these cultures is mainly mare milk and has a long tradition. But not every part or product of the fresh milk has the same meaning. For instance, the fattier part on the top, the "deež", is seen as the most valuable part and is therefore often used to honor guests. Very important with often a traditional meaning as well are fermentation products of mare milk, like for example the slightly-alcoholic yogurt kumis. Consumption of these peaks during cultural festivities such as the Mongolian lunar new year (in spring).

The time of this celebration is called the "white month", which indicates that milk products (called "white food" together with starchy vegetables, in comparison to meat products, called "black food") are a central part of this tradition. The purpose of these festivities is to "close" the past year – clean the house or the yurt, honor the animals for having provided their food, and prepare everything for the coming summer season – to be ready to "open" the new year. Consuming white food in this festive context is a way to connect to the past and to a national identity, which is the Mongolian empire personified by Genghis Khan.

During the time of this empire, the fermented mare milk was the drink to honor and thank warriors and leading persons, it was not meant for everybody. Although it eventually became a drink for normal people, it has kept its honorable meaning. Like many other traditions, this one feels the influence of globalization. Other products, like industrial yogurt, coming mainly from China and western countries, have tended to replace it more and more, mainly in urban areas. However, in rural and poorer regions it is still of great importance.

Biochemistry

Homofermentative Process

Homofermentative bacteria convert glucose to two molecules of lactate and use this reaction to perform substrate-level phosphorylation to make two molecules of ATP:

$$Glucose + 2\,ADP + 2\,P_i \rightarrow 2\,Lactate + 2\,ATP$$

Heterofermentative Process

Heterofermentative bacteria produce less lactate and less ATP, but produce several other end products:

$$Glucose + ADP + P_i \rightarrow Lactate + Ethanol + CO_2 + ATP$$

Examples include:

- Leuconostoc mesenteroides,
- Lactobacillus bifermentous, and
- Leuconostoc lactis.

Bifidum Pathway

Bifidobacterium bifidum utilizes a lactic acid fermentation pathway that produces more ATP than either homolactic fermentation or heterolactic fermentation:

$$2\,Glucose + 5\,ADP + 5\,P_i \rightarrow 3Acetate + 2\,Lactate + 5\,ATP$$

Major Genera of Lactose-Fermenting Bacteria

Some major bacterial strains identified as being able to ferment lactose are in the genera *Escherichia, Citrobacter, Enterobacter* and *Klebsiella* . All four of these groups fall underneath the family of *Enterobacteriaceae.* These four genera are able to be separated from each other by using biochemical testing, and simple biological tests are readily available. Apart from whole-sequence genomics, common tests include H2S production, motility and citrate use, indole, methyl red and Voges-Proskauer tests.

Applications

Lactic acid fermentation is used in many areas of the world to produce foods that cannot be produced through other methods. The most commercially important genus of lactic acid-fermenting bacteria is *Lactobacillus*, though other bacteria and even yeast are sometimes used. Two of the most common applications of lactic acid fermentation are in the production of yogurt and sauerkraut.

Fermented Fish

In some Asian cuisines, fish is traditionally fermented with rice to produce lactic acid that preserves the fish. Examples of these dishes include burong isda of the Philippines; narezushi of Japan; and pla ra of Thailand. The same process is also used for shrimp in the Philippines in the dish known as balao-balao.

Sauerkraut

Lactic acid fermentation is also used in the production of sauerkraut. The main type of bacteria used in the production of sauerkraut is of the genus *Leuconostoc.* As in yogurt, when the acidity rises due to lactic acid-fermenting organisms, many other pathogenic microorganisms are killed. The bacteria produce lactic acid, as well as simple alcohols and other hydrocarbons. These may then combine to form esters, contributing to the unique flavor of sauerkraut.

Yogurt

The main method of producing yogurt is through the lactic acid fermentation of milk with harmless bacteria. The primary bacteria used are typically *Lactobacillus bulgaricus* and *Streptococcus thermophilus*, and United States as well as European law requires all yogurts to contain these two cultures (though others may be added as probiotic cultures). These bacteria produce lactic acid in the milk culture, decreasing its pH and causing it to congeal. The bacteria also produce compounds that give yogurt its distinctive flavor. An additional effect of the lowered pH is the incompatibility of the acidic environment with many other types of harmful bacteria. For a probiotic yogurt, additional types of bacteria such as *Lactobacillus acidophilus* are also added to the culture.

In Vegetables

Lactic acid bacteria (LAB) already exists as part of the natural flora in most vegetables. Lettuce and cabbage were examined to determine the types of lactic acid bacteria that exist in the leaves. Different types of LAB will produce different types of silage fermentation, which is the fermentation of the leafy foliage. Silage fermentation is an anaerobic reaction that reduces sugars to fermentation byproducts like lactic acid.

Lactate Fermentation and Muscle Cramps

During the 1990s, the lactic acid hypothesis was created to explain why people experienced burning or muscle cramps that occurred during and after intense exercise. The hypothesis proposes that a lack of oxygen in muscle cells results in a switch from cellular respiration to fermentation. Lactic acid created as a byproduct of fermentation of pyruvate from glycolysis accumulates in muscles causing a burning sensation and cramps.

Research from 2006 has suggested that acidosis isn't the main cause of muscle cramps. Instead cramps may be due to a lack of potassium in muscles, leading to contractions under high stress. Animals, in fact, do not produce lactic acid during fermentation. Despite the common use of the term lactic acid in the literature, the byproduct of fermentation in animal cells is lactate. Another change to the lactic acid hypothesis is that when sodium lactate is inside of the body, there is a higher period of exhaustion in the host after a period of exercise.

Lactate fermentation is important to muscle cell physiology. When muscle cells are undergoing intense activity, like sprinting, they need energy quickly. There is only enough ATP stored in muscles cells to last a few seconds of sprinting. The cells then default to fermentation, since they are in an anaerobic environment. Through lactate fermentation, muscle cells are able to regenerate NAD+ to continue glycolysis, even under strenuous activity.

The vaginal environment is heavily influenced by lactic acid producing bacteria. *Lactobacilli* spp. that live in the vaginal canal assist in pH control. If the pH in the vagina becomes too basic, more lactic acid will be produced to lower the pH back to a more acidic level. Lactic acid producing bacteria also act as a protective barrier against possible pathogens such as bacterial vaginosis and vaginitis species, different fungi, and protozoa through the production of hydrogen peroxide, and antibacterial compounds. It is unclear if further use of lactic acid, through fermentation, in the vaginal canal is present

Benefits for the Lactose Tntolerant

In small amounts, lactic acid is good for the human body by providing energy and substrates while it moves through the cycle. In lactose intolerant people, the fermentation of lactose to lactic acid has been shown in small studies to help lactose intolerant people. The process of fermentation limits the amount of lactose available. With the amount of lactose lowered, there is less build up inside of the body, reducing bloating. Success of lactic fermentation was most evident in yogurt cultures. Further studies are being conducted on other milk products like acidophilus milk.

21.1.2 Alcoholic Fermentation

Ethanol fermentation, also called alcoholic fermentation, is a biological process which converts sugars such as glucose, fructose, and sucrose into cellular energy, producing ethanol and carbon dioxide as by-products. Because yeasts perform this conversion in the absence of oxygen, alcoholic fermentation is considered an anaerobic process. It also takes place in some species of fish (including goldfish and carp) where (along with lactic acid fermentation) it provides energy when oxygen is scarce. Ethanol fermentation is the basis for alcoholic beverages, ethanol fuel and bread dough rising.

21.1.3 Acetic Acid Fermentation

Acetic acid fermentation involves the conversion of alcohol into acetic acid by acetic acid bacteria (AAB), such as Acetobacter and Gluconobacter. This type of fermentation is essential in the production of vinegar, where the acetic acid gives the vinegar its characteristic sour flavor.

21.2 THE PRODUCTION PROCESS OF FERMENTED FOODS

The production of fermented foods generally involves several key steps: selection of raw materials, inoculation with starter cultures, fermentation, and post-fermentation processing. Each of these steps is critical to achieving the desired flavor, texture, and quality of the final product.

21.2.1 Selection of Raw Materials

The first step in producing fermented foods is selecting high-quality raw materials. The choice of ingredients significantly impacts the fermentation process and the flavor profile of the final product. For example, in yogurt production, fresh milk is preferred for its nutrient content, while in sauerkraut, fresh cabbage is essential for optimal fermentation.

21.2.2 Inoculation with Starter Cultures

Once the raw materials are prepared, the next step is to inoculate them with starter cultures. These are specific strains of microorganisms selected for their desirable fermentation characteristics. For instance, in yogurt production, Lactobacillus bulgaricus and Streptococcus thermophilus are commonly used. The addition of starter cultures ensures a consistent fermentation process and helps to achieve the desired taste and texture.

21.2.3 Fermentation

Fermentation is the core of the production process and can vary in duration and temperature depending on the type of food being produced. For yogurt, fermentation typically occurs at temperatures around 110°F (43°C) for 4 to 8 hours, allowing the bacteria to proliferate and thicken the milk. In sauerkraut production, cabbage is mixed with salt and left to ferment at room temperature for several weeks, during which lactic acid bacteria develop and acidify the mixture. During fermentation, several biochemical changes occur. Sugars are metabolized by microorganisms, leading to the production of organic acids, gases, and alcohol. The acidity generated during fermentation not only contributes to the flavor but also helps to preserve the food by creating an unfavorable environment for spoilage organisms.

21.2.4 Post-Fermentation Processing

After fermentation is complete, the final product may undergo additional processing steps, such as pasteurization, packaging, or aging, depending on the type of food. For example, yogurt may be cooled and flavored before being packaged, while cheese may be aged to develop its flavor and texture. Proper handling and storage are essential to maintain the quality and safety of fermented foods.

21.3 TYPES OF FERMENTED FOODS

21.3.1 Dairy Products

Dairy fermentation produces a range of products such as yogurt, cheese, and kefir. Yogurt is made by fermenting milk with specific bacterial cultures, resulting in a creamy, tangy product rich in probiotics. Cheese production involves curdling milk and aging it to develop flavor and texture. Kefir is a fermented milk drink that contains a diverse array of beneficial bacteria and yeasts, making it a potent probiotic food.

21.3.2 Vegetables

Vegetable fermentation is a time-honored method for preserving seasonal produce. Sauerkraut, kimchi, and pickles are examples of fermented vegetables. Sauerkraut is made from shredded cabbage fermented with salt, while kimchi is a spicy Korean dish made from fermented vegetables, often including napa cabbage and radishes. Pickles are cucumbers fermented in a brine solution, resulting in a tangy, crunchy snack.

21.3.3 Grains

Fermented grains contribute to a variety of foods, including sourdough bread, beer, and fermented porridges.

Sourdough bread is made by fermenting flour and water with wild yeast and lactic acid bacteria, giving the bread a distinctive flavor and texture. Beer production involves fermenting grains, primarily barley, with yeast to produce alcohol and carbonation.

21.3.4 Beverages

Fermented beverages include kombucha, kvass, and mead. Kombucha is a fermented tea beverage made by inoculating sweetened tea with a symbiotic culture of bacteria and yeast (SCOBY), resulting in a fizzy, tangy drink. Kvass is a traditional Slavic beverage made from fermented bread, while mead is an alcoholic drink made by fermenting honey with water and sometimes fruit.

21.4 HEALTH BENEFITS OF FERMENTED FOODS

21.4.1 Improved Digestive Health

Fermented foods are rich in probiotics, which are beneficial bacteria that promote gut health.

Regular consumption of probiotics can help balance the gut microbiota, improve digestion, and alleviate gastrointestinal issues such as bloating, constipation, and diarrhea.

21.4.2 Enhanced Nutrient Absorption

Fermentation can enhance the bioavailability of nutrients, making them easier for the body to absorb. For instance, the fermentation process breaks down anti-nutrients present in grains and legumes, allowing for better absorption of vitamins and minerals.

21.4.3 Strengthened Immune System

Probiotics found in fermented foods can support the immune system by enhancing the gut barrier function and modulating immune responses. A healthy gut microbiota is crucial for overall health and well-being, as it plays a significant role in immune function.

21.4.4 Potential Mental Health Benefits

Emerging research suggests a link between gut health and mental health, often referred to as the gut-brain axis. Consuming fermented foods may positively influence mood and cognitive function by promoting a healthy gut microbiome.

21.5 CHALLENGES IN THE PRODUCTION OF FERMENTED FOODS

21.5.1 Contamination Risks

The fermentation process can be sensitive to contamination from undesirable microorganisms. Maintaining hygienic practices and controlling fermentation conditions is essential to prevent spoilage and ensure product safety.

21.5.2 Standardization of Fermentation Processes

Achieving consistency in flavor, texture, and quality can be challenging due to variations in fermentation conditions, raw materials, and microbial strains. Developing standardized protocols and using specific starter cultures can help improve consistency.

21.5.3 Consumer Acceptance

While interest in fermented foods is growing, some consumers may be hesitant to embrace certain fermented products due to unfamiliar flavors or textures.

Educating consumers about the health benefits and culinary uses of fermented foods can help increase acceptance.

21.6 CONCLUSION

The production of fermented foods is a time-honored practice that continues to thrive in modern culinary traditions. Through the metabolic activities of microorganisms, raw ingredients are transformed into flavorful, nutritious products that offer numerous health benefits.

From dairy products to vegetables and beverages, the diversity of fermented foods reflects the rich cultural heritage and innovations in food processing. As the demand for healthy and sustainable food options increases, fermented foods are likely to play an even more prominent role in diets worldwide, promoting not only individual health but also sustainable food systems.

SECTION 6:

ALCOHOL-BASED FERMENTATION INDUSTRIES

CHAPTER 22

PRODUCTION OF BEER: RAW MATERIALS AND BREWING PROCESSES

Beer is one of the oldest and most widely consumed alcoholic beverages in the world, with a rich history that spans thousands of years. The production of beer involves a complex interplay of various raw materials and brewing processes, each contributing to the final product's flavor, aroma, and overall quality.

22.1 RAW MATERIALS FOR BEER PRODUCTION

The quality and characteristics of beer are significantly influenced by the choice of raw materials. The four primary ingredients in beer production are water, malted grains, hops, and yeast. Each ingredient plays a crucial role in the brewing process, contributing to the beer's flavor, aroma, and mouthfeel.

22.1.1 Water

Water is the most abundant ingredient in beer, typically comprising about 90-95% of the final product. The quality of water used in brewing can greatly affect the taste of the beer. Different water sources have varying mineral compositions, which can influence the brewing process and the flavor profile of the beer. For instance, water rich in minerals like calcium and magnesium can enhance the beer's body and mouthfeel. Conversely, water with high sulfate content can impart a crisp and dry finish, which is often desirable in certain beer styles, such as pale ales. Breweries often analyze their water to determine its mineral content and make adjustments as needed. This process, known as water treatment, can involve the addition of salts or the use of filtration methods to achieve the desired water profile for specific beer styles.

22.1.2 Malted Grains

Malted grains are the backbone of beer production, providing the sugars necessary for fermentation. Barley is the most commonly used grain, although other grains like wheat, corn, and rice can also be used. The malting process involves soaking the grains in water to allow them to germinate, then drying them in a kiln to halt germination. This process develops enzymes that convert starches into fermentable sugars during mashing. The choice of malt influences the beer's color, flavor, and aroma. Base malts, such as pale malt, provide the primary sugars for fermentation and contribute to the beer's body. Specialty malts, on the other hand, add unique flavors and colors. For example, roasted malts impart chocolate or coffee notes, while caramel malts add sweetness and a rich golden hue.

22.1.3 Hops

Hops are the flowers of the hop plant, Humulus lupulus, and they play a crucial role in beer production. Hops contribute bitterness to balance the sweetness of the malt, as well as aroma and flavor to the finished beer. They contain essential oils and alpha acids, which are responsible for the bitterness and aromatic qualities. Different hop varieties exhibit distinct flavor profiles, ranging from floral and citrusy to earthy and spicy. The timing of hop additions during the brewing process influences their contribution to the beer. Hops added early in the boil provide bitterness, while late additions or dry hopping introduce more aromatic qualities. The careful selection and timing of hops allow brewers to create a wide range of beer styles with diverse flavor characteristics.

22.1.4 Yeast

Yeast is a vital ingredient in beer production, responsible for fermentation, where it converts sugars from malted grains into alcohol and carbon dioxide. The choice of yeast strain significantly impacts the beer's flavor, aroma, and mouthfeel. There are two primary types of yeast used in brewing: ale yeast (Saccharomyces cerevisiae) and lager yeast (Saccharomyces pastorianus). Ale yeast ferments at warmer temperatures, producing fruity and complex flavors, making it suitable for styles like pale ales and IPAs. Lager yeast ferments at cooler temperatures, resulting in clean and crisp flavors, typical of lagers and pilsners. The fermentation process can vary depending on the yeast strain used and the desired characteristics of the final product.

22.2 BREWING PROCESSES

22.2.1 Mashing

Mashing is the first step in the brewing process, where malted grains are mixed with hot water in a mash tun. The heat activates enzymes in the malt, which convert starches into fermentable sugars. The temperature and duration of mashing can vary depending on the desired beer style. Generally, mashing occurs at temperatures between 150-160°F (65-71°C) for about an hour. During mashing, the mash forms a thick mixture, and the enzymes break down the starches into sugars. This process not only extracts sugars but also imparts flavors and colors to the beer. After mashing, the mixture is referred to as "mash," and the next step is lautering.

22.2.2 Lautering

Lautering is the process of separating the liquid wort from the solid grain husks. After mashing, the mash is transferred to a lauter tun, where it is allowed to settle. The liquid portion, known as wort, is drained off while the solid grains are retained. To extract additional sugars, hot water may be sprayed or poured over the grain bed in a process called sparging. This step ensures maximum sugar extraction from the grains, which is crucial for fermentation.

22.2.3 Boiling

Once the wort is collected, it is transferred to the kettle for boiling. Boiling serves several purposes: it sterilizes the wort, removes undesirable compounds, and allows for the addition of hops. Hops are added at various points during the boil, depending on the desired bitterness and aroma. Generally, hops added early in the boil contribute more bitterness, while those added later enhance aroma. The boiling process typically lasts between 60 to 90 minutes, depending on the recipe and style of beer being produced. After boiling, the wort is cooled rapidly to prevent the growth of unwanted microorganisms.

22.2.4 Fermentation

After cooling, the wort is transferred to a fermentation vessel, and yeast is added to initiate fermentation. The yeast consumes the fermentable sugars in the wort, producing alcohol and carbon dioxide as byproducts. The fermentation process can take anywhere from a few days to several weeks, depending on the type of beer being brewed and the yeast strain used. During fermentation, the temperature must be carefully controlled to ensure optimal yeast activity. For ales, fermentation typically occurs at temperatures between 65-75°F (18-24°C), while lagers ferment at cooler temperatures, around 45-55°F (7-13°C). As fermentation progresses, the beer develops its characteristic flavors and aromas.

22.2.5 Conditioning

After fermentation is complete, the beer undergoes conditioning, a maturation process that allows flavors to meld and any harsh or off-flavors to dissipate. Conditioning can take place in the fermentation vessel or in separate conditioning tanks. Some beers may also undergo secondary fermentation during conditioning, especially those that are bottle-conditioned. During conditioning, the beer may be clarified and filtered to remove any remaining yeast or particulate matter. Carbon dioxide produced during fermentation can naturally carbonate the beer, contributing to its effervescence.

22.2.6 Packaging

Once the beer has conditioned and reached the desired flavor profile, it is ready for packaging. This can involve bottling, canning, or kegging the beer for distribution.

Before packaging, the beer may be pasteurized or filtered to ensure stability and prevent spoilage. Labeling and marketing play a crucial role in the final steps of production, as breweries strive to create unique identities for their products. Packaging not only serves a functional purpose but also reflects the brewery's branding and the beer's style.

22.3 CONCLUSION

The production of beer is a fascinating interplay of raw materials and processes, each contributing to the final product's complexity and character. From the selection of water, malted grains, hops, and yeast to the intricate brewing processes of mashing, lautering, boiling, fermentation, conditioning, and packaging, every step is essential in crafting a quality beer. As craft brewing continues to grow in popularity, innovative techniques and experimentation with ingredients are pushing the boundaries of beer production, creating a diverse array of flavors and styles that delight beer enthusiasts worldwide. Understanding the raw materials and brewing processes involved in beer production allows consumers to appreciate the artistry and science behind this ancient beverage.

CHAPTER 23

PRODUCTION OF WINES AND SPIRITS

Wine and spirits are two of the most popular alcoholic beverages globally, each with its own distinct production processes, ingredients, and cultural significance. The production of wines involves the fermentation of grapes, while spirits are typically produced through fermentation followed by distillation.

23.1 RAW MATERIALS FOR WINE AND SPIRIT PRODUCTION

The primary raw material for both wines and spirits is fermentable sugars, which are most commonly derived from fruits and grains. The choice of raw materials significantly influences the flavor profile, aroma, and quality of the final products.

23.1.1 Grapes

In wine production, grapes are the key ingredient. Different grape varieties impart unique flavors, aromas, and characteristics to the wine. Grapes are classified into two main categories: red and white. Red grapes are used for red and rosé wines, while white grapes are used primarily for white wines. The cultivation of grapes, known as viticulture, is influenced by various factors, including climate, soil type, and vineyard management practices. The timing of grape harvest is crucial, as it determines the sugar, acid, and tannin levels in the fruit. Grapes are typically harvested when they have reached optimal ripeness, allowing for the best balance of flavors. After harvest, grapes are sorted and crushed to extract their juice, which will undergo fermentation.

23.1.2 Grains

In the production of spirits, grains such as barley, corn, wheat, and rye are commonly used. These grains are rich in starch, which can be converted into fermentable sugars through the malting and mashing processes. The choice of grain influences the flavor and texture of the spirit. For example, whiskey is often made from a combination of malted barley and other grains, while vodka can be produced from various sources, including grains, potatoes, and even fruits.

23.1.3 Other Fruits

While grapes are the primary fruit used in wine production, other fruits can also be fermented to create fruit wines or spirits. Examples include apple cider (from apples), pear wine (from pears), and cherry wine (from cherries). These fruit-based beverages offer a wide range of flavors and aromas, expanding the diversity of wines and spirits available.

23.1.4 Water

Water is an essential component in both wine and spirit production, affecting the final product's taste and quality. In wine production, water is used during the winemaking process, while in spirit production, it is often used for dilution and to adjust the final alcohol content. The mineral composition of the water can also impact the flavor profile, especially in the case of spirits.

23.2 THE PRODUCTION PROCESS OF WINES

23.2.1 Harvesting

Harvesting grapes is the first step in the winemaking process. Grapes are typically harvested during the late summer to early fall, depending on the climate and grape variety. Harvesting can be done by hand or by machine, with each method having its advantages. Hand-harvesting allows for greater selectivity and care, while machine harvesting is more efficient for larger vineyards.

23.2.2 Crushing and Pressing

After harvesting, the grapes are sorted and crushed to extract their juice. For white wines, the grapes are pressed to separate the juice from the skins immediately. In contrast, for red wines, the grapes are usually crushed, and the skins are left in contact with the juice during fermentation to impart color and tannins. The pressing process for white wines typically involves hydraulic presses, which gently extract the juice without damaging the delicate flavors. The resulting juice, known as must, is then ready for fermentation.

23.2.3 Fermentation

Fermentation is the process where yeast converts the sugars in the grape juice into alcohol and carbon dioxide. Yeast can be naturally occurring on grape skins or added as cultured strains.

The choice of yeast strain can influence the flavor and aroma profile of the wine. Fermentation temperatures and durations vary depending on the wine style, with white wines often fermented at cooler temperatures to preserve delicate aromas and red wines at warmer temperatures to enhance flavor extraction. During fermentation, the must is monitored for temperature, sugar levels, and other factors. For red wines, the skins are kept in contact with the juice to extract color and tannins, while for white wines, skin contact is minimal or absent.

23.2.4 Aging

After fermentation, the wine may be aged in various vessels, such as stainless steel tanks or oak barrels. Aging allows the wine to develop complexity, enhance flavors, and soften tannins. The choice of aging vessel can significantly impact the final product, as oak barrels impart flavors such as vanilla, spice, and toast. The aging process can last from a few months to several years, depending on the desired wine style. After aging, the wine is clarified and filtered to remove any sediment before bottling.

23.2.5 Bottling

Bottling is the final step in the winemaking process. The wine is carefully transferred to bottles, sealed with corks or screw caps, and labeled. Proper bottling techniques are essential to preserve the wine's quality and prevent spoilage.

23.3 THE PRODUCTION PROCESS OF SPIRITS

The production of spirits typically involves additional steps beyond fermentation, including distillation and aging. The primary steps in spirit production are fermentation, distillation, aging, and bottling.

23.3.1 Fermentation

Similar to wine production, the fermentation of grains or fruits is the first step in spirit production. Yeast is added to the mash or wash, converting sugars into alcohol. The fermentation process can take several days, depending on the raw materials and desired alcohol content.

23.3.2 Distillation

Distillation is the key process that differentiates spirits from wines. This process involves heating the fermented liquid to separate alcohol from water and other components based on their boiling points. The vaporized alcohol is then collected and cooled, condensing it back into liquid form. The distillation process can be performed using various types of stills, including pot stills and column stills. Pot stills are typically used for artisanal and small-batch spirits, allowing for greater flavor retention, while column stills are used for large-scale production, resulting in higher alcohol concentrations.

23.3.3 Aging

After distillation, many spirits are aged in barrels to develop their flavors and aromas. The aging process can vary significantly depending on the type of spirit. For example, whiskey is often aged in oak barrels for several years, allowing it to absorb flavors from the wood and develop complexity. In contrast, vodka is typically not aged, as its production focuses on achieving a clean and neutral flavor. The length of aging and the type of barrels used can significantly influence the final product. Some spirits, like rum, may be aged in a variety of barrel types, including those previously used for whiskey or sherry, imparting additional flavors.

23.3.4 Bottling

Once the aging process is complete, the spirit is filtered and diluted to achieve the desired alcohol content. It is then bottled, labeled, and prepared for distribution. The bottling process is critical for maintaining the quality and integrity of the spirit.

23.4 CONCLUSION

The production of wines and spirits is a fascinating blend of art and science, involving various raw materials and processes that ultimately shape the final product. From the careful selection of grapes and grains to the intricate processes of fermentation, distillation, and aging, each step plays a vital role in creating unique and enjoyable beverages.

As consumer preferences continue to evolve, winemakers and distillers alike are exploring innovative techniques and ingredients, pushing the boundaries of flavor and quality. Understanding the production processes behind wines and spirits enriches our appreciation of these beloved beverages, inviting us to savor not only their flavors but also the craftsmanship and tradition that go into each bottle.

CHAPTER 24
PRODUCTION OF VINEGAR

Vinegar is a sour liquid produced by the fermentation of ethanol or sugars, and it has been a staple in culinary traditions across the globe for centuries. Not only is vinegar a key ingredient in various dishes and condiments, but it also has applications in preservation, cleaning, and even medicinal practices.

24.1 RAW MATERIALS FOR VINEGAR PRODUCTION

The production of vinegar primarily relies on two main raw materials: ethanol and sugars. Different types of vinegar can be made depending on the source of these raw materials.

24.1.1 Ethanol Sources

Ethanol, also known as ethyl alcohol, is the key ingredient in vinegar production. It can be derived from various sources:

- **Wine:** Wine vinegar is produced from the fermentation of wine. The alcohol present in the wine undergoes a second fermentation to produce acetic acid, giving the vinegar its characteristic sour flavor.
- **Beer:** Beer vinegar is created from the fermentation of beer. The malted grains used in brewing beer provide sugars that are converted into alcohol before being fermented into vinegar.
- **Cider:** Cider vinegar is made from the fermentation of apple cider. The natural sugars in the apples are fermented into alcohol and then converted into acetic acid.
- **Grains:** Distilled vinegars can be produced from grain-based sources, such as corn or wheat. These grains are fermented to produce alcohol, which is then transformed into vinegar.

24.1.2 Sugars

Some vinegars can be produced from sugar-rich sources without the initial production of alcohol.

In such cases, the fermentation of sugar directly produces acetic acid. Common sources of sugar for vinegar production include:

- **Sugar Cane:** Cane sugar is often used to make vinegar in tropical regions, where sugar cane is abundant. The fermentation of cane sugar leads to the production of cane vinegar, which has a distinct flavor.

- **Molasses:** Molasses, a byproduct of sugar refining, can also be fermented to produce vinegar. This type of vinegar is commonly used in Caribbean cuisine.
- **Fruits:** Other fruits besides apples can also be fermented to produce vinegar. For example, grape vinegar can be made from crushed grapes or fruit juices, while other fruits like peaches or berries can yield unique flavored vinegars.

24.2 FERMENTATION PROCESSES

24.2.1 Alcoholic Fermentation

The first step in vinegar production is alcoholic fermentation, where sugars are converted into ethanol by the action of yeast. This process can occur naturally or be facilitated by the addition of cultured yeast strains. During this stage, yeasts consume the sugars present in the raw material and produce alcohol and carbon dioxide. The temperature and conditions of fermentation play a crucial role in determining the quality and flavor of the alcohol produced. For instance, the fermentation of grape juice to produce wine typically occurs at controlled temperatures to enhance flavor complexity. Once the alcoholic fermentation is complete, the resulting liquid contains a significant amount of ethanol and is ready for the next stage of vinegar production.

24.2.2 Acetic Acid Fermentation

The second fermentation process is acetic acid fermentation, where the ethanol produced in the first stage is converted into acetic acid by acetic acid bacteria (AAB). The most common bacteria involved in this process belong to the genera Acetobacter and Gluconobacter. During acetic acid fermentation, the bacteria oxidize ethanol in the presence of oxygen, producing acetic acid and water. This stage requires aeration, as the bacteria thrive in oxygen-rich environments. The process can take anywhere from a few days to several weeks, depending on the conditions and the desired acidity of the final product. To facilitate this process, vinegar producers often use specialized fermentation vessels that allow for aeration while preventing contamination. The temperature and pH levels of the fermentation environment must also be monitored to ensure optimal bacterial activity.

24.3 TYPES OF VINEGAR

24.3.1 Wine Vinegar

Wine vinegar is made from fermented wine and has a distinct flavor that varies depending on the type of wine used. Red wine vinegar, for example, has a rich, fruity flavor, while white wine vinegar is lighter and more acidic. Wine vinegar is commonly used in salad dressings, marinades, and sauces.

24.3.2 Apple Cider Vinegar

Apple cider vinegar is produced from the fermentation of apple cider. It is well-known for its tangy flavor and potential health benefits, such as aiding digestion and regulating blood sugar levels. Apple cider vinegar is often used in salad dressings, baking, and as a natural remedy.

24.3.3 Balsamic Vinegar

Balsamic vinegar is a traditional Italian vinegar made from cooked grape must. It undergoes a lengthy aging process in wooden barrels, resulting in a thick, sweet, and complex flavor. Balsamic vinegar is commonly used in dressings, marinades, and as a finishing touch for dishes.

24.3.4 Rice Vinegar

Rice vinegar is made from fermented rice and is a staple in Asian cuisines. It has a mild flavor and is often used in sushi rice, pickling, and salad dressings. There are various types of rice vinegar, including white rice vinegar and black rice vinegar, each with different flavor profiles.

24.3.5 Distilled Vinegar

Distilled vinegar, also known as white vinegar, is made from the fermentation of grain-based ethanol.

It has a sharp, clean flavor and is often used for pickling, cleaning, and cooking. Due to its high acidity, distilled vinegar is a popular choice for various culinary and household applications.

24.3.6 Coconut Vinegar

Coconut vinegar is produced from the sap of coconut trees or from fermented coconut water. It has a unique flavor profile, with a hint of sweetness and acidity. Coconut vinegar is commonly used in Southeast Asian cuisines and is often employed in salads, marinades, and dipping sauces.

24.4 QUALITY CONTROL IN VINEGAR PRODUCTION

24.4.1 Microbial Monitoring

Microbial monitoring is essential to ensure that the correct strains of yeast and acetic acid bacteria are present during fermentation. Regular testing can help identify any contamination or spoilage organisms that could affect the quality of the vinegar.

24.4.2 pH Testing

The pH level of vinegar is a critical factor in determining its acidity. Regular pH testing is conducted during production to ensure that the vinegar meets the desired acidity standards. Vinegar typically has a pH of around 2.5 to 3.5, making it acidic enough to inhibit the growth of harmful microorganisms.

24.4.3 Sensory Evaluation

Sensory evaluation involves tasting and assessing the flavor, aroma, and overall quality of the vinegar. Trained panels or sensory experts may be employed to evaluate the final product, ensuring it meets the desired flavor profile and quality standards.

24.5 CONCLUSION

The production of vinegar is a fascinating and intricate process that transforms various raw materials into a versatile and widely used product. From the initial fermentation of sugars to the subsequent conversion of ethanol into acetic acid, each step is essential in shaping the final vinegar. With a diverse range of types, including wine vinegar, apple cider vinegar, balsamic vinegar, and distilled vinegar, the culinary applications of vinegar are vast and varied. Understanding the production process and the factors that influence vinegar quality not only enhances our appreciation for this age-old condiment but also allows us to explore its many uses in cooking, preservation, and beyond. As consumer interest in artisanal and health-oriented products continues to grow, vinegar production is likely to evolve, offering new flavors and applications to enrich our culinary experiences.

SECTION 7:

PRODUCTION OF COMMODITIES OF MEDICAL IMPORTANCE

CHAPTER 25

PRODUCTION OF ANTIBIOTICS AND ANTI-TUMOR AGENTS

The discovery and production of antibiotics and anti-tumor agents have revolutionized medicine, providing effective treatments for infections and cancer. Both classes of compounds are produced through intricate biological processes, primarily involving microorganisms.

25.1 THE SIGNIFICANCE OF ANTIBIOTICS

Antibiotics are substances that inhibit the growth of or destroy bacteria. They are critical in treating bacterial infections, preventing complications, and saving lives. The emergence of antibiotic-resistant bacteria has highlighted the need for ongoing research and development in antibiotic production, ensuring that effective treatments remain available.

25.1.1 History of Antibiotic Production

The journey of antibiotic production began with the discovery of penicillin by Alexander Fleming in 1928. This groundbreaking discovery marked the beginning of the antibiotic era. Penicillin was initially produced through the fermentation of the Penicillium chrysogenum fungus. Following this, other antibiotics, such as streptomycin, tetracycline, and erythromycin, were discovered and developed from various microbial sources.

25.2 RAW MATERIALS FOR ANTIBIOTIC PRODUCTION

The raw materials used in antibiotic production primarily include microorganisms and substrates that provide the necessary nutrients for fermentation. Common sources of antibiotics include bacteria and fungi, which possess the ability to produce secondary metabolites with antimicrobial properties.

25.2.1 Microorganisms

Several species of bacteria and fungi are known for their antibiotic production capabilities. Some of the most notable include:

- **Fungi:** The Penicillium and Aspergillus genera are renowned for producing various antibiotics, including penicillin and cephalosporins.
- **Bacteria:** Actinomycetes, particularly Streptomyces species, are prolific

producers of antibiotics. Streptomyces griseus produces streptomycin, while Streptomyces aureofaciens produces tetracycline.

25.2.2 Substrates

The substrates used for growing microorganisms in fermentation processes are vital for antibiotic production. These substrates typically include:

- **Carbohydrates:** Sugars, such as glucose and sucrose, provide essential energy for microbial growth.
- **Nitrogen Sources:** Peptones, yeast extract, and ammonium salts serve as nitrogen sources that promote protein synthesis in microorganisms.
- **Minerals and Vitamins:** Essential minerals (such as calcium, magnesium, and potassium) and vitamins (such as B vitamins) are also added to the growth media to support microbial metabolism and growth.

25.3 THE PRODUCTION PROCESS OF ANTIBIOTICS

25.3.1 Fermentation

The fermentation process is the cornerstone of antibiotic production. It typically occurs in bioreactors under controlled conditions to optimize microbial growth and antibiotic synthesis. The steps involved in the fermentation process are as follows:

- **Inoculation:** The selected microorganism is inoculated into a nutrient-rich medium, which provides the necessary substrates for growth.
- **Cultivation:** The bioreactor is maintained at specific temperature, pH, and oxygen levels to promote optimal microbial growth. During this stage, the microorganisms proliferate and produce antibiotics as secondary metabolites.
- **Monitoring:** Throughout the fermentation process, various parameters are monitored, including biomass concentration, pH, and antibiotic levels. Samples are taken at regular intervals to assess the progress of antibiotic production.

25.3.2 Extraction

Once fermentation is complete, the next step is to extract the antibiotics from the fermentation broth. This process involves several methods:

- **Filtration:** Solid materials, such as microbial biomass and debris, are removed from the broth through filtration.
- **Solvent Extraction:** The antibiotic compounds are separated from the

aqueous medium using organic solvents. This method helps concentrate the antibiotics and remove impurities.

- **Adsorption:** Adsorbent materials may be used to bind the antibiotic compounds, allowing for their subsequent elution.

25.3.3 Purification

The extracted antibiotics undergo purification to achieve the desired level of purity and remove any remaining impurities. Common purification techniques include:

- **Chromatography:** Techniques such as high-performance liquid chromatography (HPLC) are used to separate and purify the antibiotic compounds based on their chemical properties.
- **Crystallization:** Antibiotics can be crystallized from the solution to achieve higher purity levels. This process involves cooling the solution to promote the formation of pure antibiotic crystals.

25.3.4 Formulation

After purification, the final antibiotics are formulated into suitable dosage forms for administration. This can include tablets, capsules, injectables, or liquid formulations. Quality control measures are implemented to ensure that the final product meets safety, efficacy, and regulatory standards.

25.4 THE SIGNIFICANCE OF ANTI-TUMOR AGENTS

Anti-tumor agents, also known as anticancer drugs, play a crucial role in the treatment of cancer. They work by inhibiting cancer cell proliferation, inducing apoptosis (programmed cell death), or targeting specific pathways involved in tumor growth. The production of anti-tumor agents has evolved significantly over the years, leading to a wide range of effective therapies.

25.4.1 History of Anti-Tumor Agent Production

The history of anti-tumor agent production dates back to the discovery of the first chemotherapy agents, such as nitrogen mustard compounds, derived from the study of chemical warfare agents. Subsequent discoveries led to the development of various classes of anticancer drugs, including alkylating agents, antimetabolites, and natural products.

25.5 RAW MATERIALS FOR ANTI-TUMOR AGENT PRODUCTION

The raw materials used in the production of anti-tumor agents vary depending on the type of drug being synthesized. Natural products, particularly those derived from plants and microorganisms, have historically played a significant role in anti-cancer drug discovery.

25.5.1 Natural Products

Several anti-tumor agents are derived from natural sources, including:

- **Plants:** Many well-known anti-cancer drugs, such as paclitaxel (Taxol) from the Pacific yew tree, are derived from plant extracts. Other examples include vincristine and vinblastine from the periwinkle plant.
- **Microorganisms:** Some antibiotics and anticancer compounds are produced by bacteria and fungi. For instance, the antibiotic doxorubicin is derived from the bacterium Streptomyces peucetius.

25.5.2 Synthetic Compounds

In addition to natural products, many anti-tumor agents are synthesized through chemical processes. Synthetic chemistry allows for the modification of existing compounds to enhance their efficacy, reduce toxicity, or improve bioavailability.

25.6 THE PRODUCTION PROCESS OF ANTI-TUMOR AGENTS

The production of anti-tumor agents involves several steps, including extraction (for natural products), synthesis (for synthetic compounds), purification, and formulation.

25.6.1 Extraction and Isolation

For anti-tumor agents derived from natural sources, extraction and isolation are crucial steps. The process generally involves:

- **Harvesting:** Plant or microbial materials are harvested, ensuring that the source is sustainable and ethically managed.
- **Extraction:** Various solvents (e.g., ethanol, methanol, water) are used to extract the active compounds from the raw materials. This process can involve maceration, percolation, or reflux extraction.
- **Isolation:** After extraction, the desired anti-tumor compounds are isolated through techniques such as chromatography, precipitation, or crystallization.

25.6.2 Synthesis

For synthetic anti-tumor agents, the production process involves chemical synthesis. This can include:

- **Multi-Step Synthesis:** The synthesis of complex molecules may require several steps, with intermediate compounds formed along the way. Each step is carefully monitored to ensure yield and purity.
- **Optimization:** Chemists often optimize reaction conditions, such as temperature and pH, to enhance the efficiency of the synthesis process.

25.6.3 Purification

Similar to antibiotics, anti-tumor agents undergo purification to achieve the required level of purity. Techniques such as chromatography, filtration, and crystallization are commonly employed.

25.6.4 Formulation

Once purified, anti-tumor agents are formulated into suitable dosage forms, including tablets, capsules, injectables, or infusions. Quality control measures are implemented to ensure that the final product meets safety and efficacy standards.

25.7 CONCLUSION

The production of antibiotics and anti-tumor agents represents a significant achievement in modern medicine, providing essential treatments for bacterial infections and cancer. The intricate processes involved in the production of these compounds highlight the importance of microorganisms, raw materials, and advanced techniques in pharmaceutical development.

As research continues to advance, the ongoing exploration of new antibiotics and anti-tumor agents will be vital in addressing the challenges posed by antibiotic resistance and cancer treatment. Understanding the production methods and the significance of these pharmaceuticals not only enriches our appreciation for these life-saving drugs but also underscores the importance of continued investment in research and innovation in the field of medicine.

CHAPTER 26

PRODUCTION OF ERGOT ALKALOIDS

Ergot alkaloids are a group of naturally occurring compounds produced by certain fungi of the genus *Claviceps*, particularly *Claviceps purpurea*. These compounds have a complex structure and significant biological activity, making them valuable in various medicinal applications.

26.1 UNDERSTANDING ERGOT ALKALOIDS

Ergot alkaloids are a diverse group of compounds that share a common structure derived from the ergot fungus. They are known for their pharmacological properties, including vasoconstriction, neuroactivity, and modulation of neurotransmitter release.

Historically, these compounds were associated with ergotism, a condition resulting from the consumption of contaminated grains, particularly rye, leading to symptoms such as hallucinations, convulsions, and gangrene.

However, modern medicine has harnessed the therapeutic potential of ergot alkaloids, making them crucial in treating various conditions, including migraines, Parkinson's disease, and certain gynecological disorders.

26.1.1 Historical Context

The history of ergot alkaloids dates back to ancient times when they were linked to outbreaks of ergotism in Europe. The identification of the causal agent, *Claviceps purpurea*, and the isolation of its active compounds began in the late 19th century. The first significant breakthrough was the extraction of ergotamine, a potent vasoconstrictor, which laid the foundation for the development of pharmaceutical applications.

Subsequent research led to the identification of other ergot alkaloids, such as ergonovine and bromocriptine, which have become integral to modern medicine.

26.2 RAW MATERIALS FOR ERGOT ALKALOID PRODUCTION

The production of ergot alkaloids primarily involves cultivating the ergot fungus and using suitable substrates for fermentation. The choice of raw materials is crucial in maximizing the yield and potency of the alkaloids.

26.2.1 The Ergot Fungus

The primary raw material for the production of ergot alkaloids is the fungus *Claviceps purpurea*. This fungus infects the ovaries of certain cereal crops, particularly rye, leading to the formation of sclerotia, which are hardened fungal tissues that contain the alkaloids. The cultivation of *C. purpurea* can occur naturally in contaminated fields or through controlled laboratory methods.

26.2.2 Substrates

In addition to the ergot fungus itself, suitable substrates are essential for promoting optimal growth and alkaloid production. Common substrates include:

- **Cereal Grains**: Rye is the most common host for *C. purpurea*, providing a natural environment for the fungus to grow and produce sclerotia.
- **Synthetic Media**: In laboratory settings, various synthetic media can be formulated to provide the necessary nutrients for fungal growth. These media often contain a source of carbohydrates, nitrogen, vitamins, and minerals that promote fungal metabolism and alkaloid synthesis.
- **Fermentation Conditions**: Factors such as pH, temperature, and oxygen levels must be carefully controlled during the fermentation process to optimize alkaloid production. The ideal conditions can vary depending on the specific strain of *C. purpurea* and the desired alkaloids.

26.3 THE PRODUCTION PROCESS OF ERGOT ALKALOIDS

26.3.1 Cultivation of Claviceps purpurea

The first step in the production of ergot alkaloids is cultivating the fungus. This can be done in natural settings, where *C. purpurea* infects rye crops, or through controlled laboratory methods.

- **Natural Cultivation:** In agricultural fields, rye is planted, and the *C. purpurea* fungus naturally infects the plants. Farmers can harvest the sclerotia after the rye crop matures, ensuring that the ergot sclerotia are collected in sufficient quantities.
- **Laboratory Cultivation**: In controlled environments, *C. purpurea* can be cultured on agar plates or in liquid media. The fungus is inoculated into sterile growth media, allowing it to proliferate under optimal conditions. This method allows for the production of high-purity alkaloids and facilitates research into different strains and their alkaloid profiles.

26.3.2 Fermentation

After cultivating the fungus, the next step is fermentation, where the sclerotia are harvested and processed to extract the ergot alkaloids.

- **Inoculation**: The harvested sclerotia can be inoculated into liquid culture media, providing the necessary nutrients for growth. This fermentation process typically occurs in bioreactors, allowing for the control of environmental conditions.
- **Monitoring Growth**: Throughout fermentation, parameters such as pH, temperature, and nutrient levels are monitored to ensure optimal growth. The fungus produces ergot alkaloids as secondary metabolites during this phase.
- **Harvesting Alkaloids**: After the fermentation period, which can last from several days to weeks, the culture broth is harvested, containing both the fungal biomass and the produced alkaloids.

26.3.3 Extraction

The extraction process is essential for isolating the ergot alkaloids from the fermentation broth. Various methods can be employed to achieve this:

- **Solvent Extraction**: The fermentation broth is treated with organic solvents, such as methanol, ethanol, or chloroform, which help dissolve the alkaloids. This process separates the alkaloids from the aqueous medium.
- **Filtration**: Solid materials, including fungal biomass and debris, are removed through filtration, leaving behind a liquid extract containing the dissolved alkaloids.
- **Concentration**: The extracted liquid may undergo evaporation to concentrate the alkaloids and remove excess solvent, resulting in a more potent extract.

26.3.4 Purification

Purification is a critical step in the production of ergot alkaloids to achieve high purity levels and remove impurities or byproducts. Several techniques are commonly used:

- **Chromatography**: Techniques such as high-performance liquid chromatography (HPLC) or column chromatography are employed to separate and purify individual alkaloids based on their chemical properties. This process allows for the isolation of specific compounds, such as ergotamine or ergonovine.
- **Crystallization**: The purified alkaloids can be crystallized from the solution to achieve even higher purity levels. This process involves cooling the solution to promote the formation of pure alkaloid crystals.

- **Quality Control**: Throughout the purification process, quality control measures are implemented to assess the potency and purity of the final products. Analytical techniques, such as mass spectrometry or nuclear magnetic resonance (NMR), are used to confirm the identity and concentration of the alkaloids.

26.4 APPLICATIONS OF ERGOT ALKALOIDS

26.4.1 Migraine Treatment

Ergotamine, one of the most well-known ergot alkaloids, is used to treat acute migraine attacks. It works by constricting blood vessels and alleviating the vasodilation associated with migraines. The drug is often formulated in various forms, including tablets, nasal sprays, and injectable forms.

26.4.2 Obstetric Applications

Ergonovine, another ergot alkaloid, is utilized to manage postpartum hemorrhage by promoting uterine contraction. It is administered after childbirth to help reduce bleeding and facilitate uterine recovery.

26.4.3 Parkinson's Disease Treatment

Bromocriptine, derived from ergot alkaloids, is used in treating Parkinson's disease and hyperprolactinemia. It acts as a dopamine agonist, helping to restore the balance of neurotransmitters in the brain.

26.4.4 Other Therapeutic Uses

Ergot alkaloids have also been investigated for their potential applications in treating various conditions, including psychiatric disorders, endometriosis, and obesity. Ongoing research continues to explore the pharmacological potential of these compounds, leading to the development of novel therapeutics.

26.5 CONCLUSION

The production of ergot alkaloids from the fungus *Claviceps purpurea* exemplifies the intricate processes involved in harnessing nature's resources for medicinal purposes. From the historical context of ergotism to the modern applications of these compounds in treating migraines, postpartum hemorrhage, and Parkinson's disease, ergot alkaloids have a significant impact on healthcare. The methods of cultivation, fermentation, extraction, and purification highlight the complexity of producing these valuable pharmaceuticals. As research advances, further understanding of ergot alkaloids may lead to the discovery of new therapeutic agents and improved treatment options for various medical conditions.

CHAPTER 27

MICROBIAL TRANSFORMATION OF STEROIDS AND STEROLS

The microbial transformation of steroids and sterols represents a fascinating area of study within microbiology and biochemistry. These processes involve the modification of steroidal compounds by microorganisms, leading to a range of potential applications in pharmaceuticals, agriculture, and industrial biotechnology.

27.1 UNDERSTANDING STEROIDS AND STEROLS

27.1.1 Steroids

A steroid is an organic compound with four fused rings (designated A, B, C, and D) arranged in a specific molecular configuration. Steroids have two principal biological functions: as important components of cell membranes that alter membrane fluidity; and as signaling molecules. Examples include the lipid cholesterol, sex hormones estradiol and testosterone, anabolic steroids, and the anti-inflammatory corticosteroid drug dexamethasone. Hundreds of steroids are found in fungi, plants, and animals. All steroids are manufactured in cells from the sterols lanosterol (opisthokonts) or cycloartenol (plants). Lanosterol and cycloartenol are derived from the cyclization of the triterpene squalene.

The steroid nucleus (core structure) is called gonane (cyclopentanoperhydrophenanthrene). It is typically composed of seventeen carbon atoms, bonded in four fused rings: three six-member cyclohexane rings (rings A, B and C in the first illustration) and one five-member cyclopentane ring (the D ring). Steroids vary by the functional groups attached to this four-ring core and by the oxidation state of the rings. Sterols are forms of steroids with a hydroxy group at position three and a skeleton derived from cholestane. Steroids can also be more radically modified, such as by changes to the ring structure, for example, cutting one of the rings. Cutting Ring B produces secosteroids one of which is vitamin D_3.

Nomenclature

Rings and Functional Groups

Gonane, also known as steran or cyclopentanoperhydrophenanthrene, the simplest steroid and the nucleus of all steroids and sterols, is composed of seventeen carbon atoms in carbon-carbon bonds forming four fused rings in a three-dimensional shape. The three cyclohexane rings (A, B, and C in the first illustration) form the skeleton of a perhydro derivative of phenanthrene.

The D ring has a cyclopentane structure. When the two methyl groups and eight carbon side chains (at C-17, as shown for cholesterol) are present, the steroid is said to have a cholestane framework.

The two common 5α and 5β stereoisomeric forms of steroids exist because of differences in the side of the largely planar ring system where the hydrogen (H) atom at carbon-5 is attached, which results in a change in steroid A-ring conformation. Isomerisation at the C-21 side chain produces a parallel series of compounds, referred to as isosteroids.

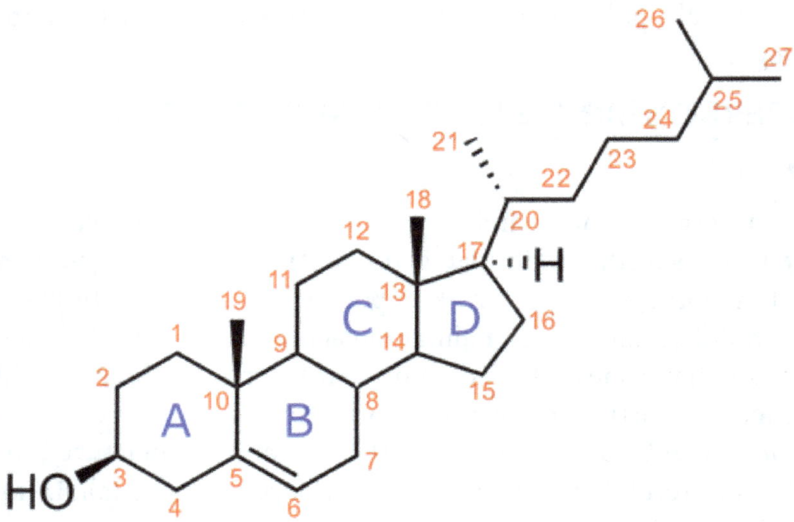

Fig 1: Cholesterol, a Prototypical Animal Sterol. This Structural Lipid and Key Steroid Biosynthetic

In addition to the ring scissions (cleavages), expansions and contractions (cleavage and reclosing to a larger or smaller rings)—all variations in the carbon-carbon bond framework—steroids can also vary:

- In the bond orders within the rings,
- In the number of methyl groups attached to the ring (and, when present, on the prominent side chain at C17),
- In the functional groups attached to the rings and side chain, and
- In the configuration of groups attached to the rings and chain.

For instance, sterols such as cholesterol and lanosterol have a hydroxyl group attached at position C-3, while testosterone and progesterone have a carbonyl (oxo substituent) at C-3. Among these compounds, only lanosterol has two methyl groups at C-4. Cholesterol which has a C-5 to C-6 double bond, differs from testosterone and progesterone which have a C-4 to C-5 double bond.

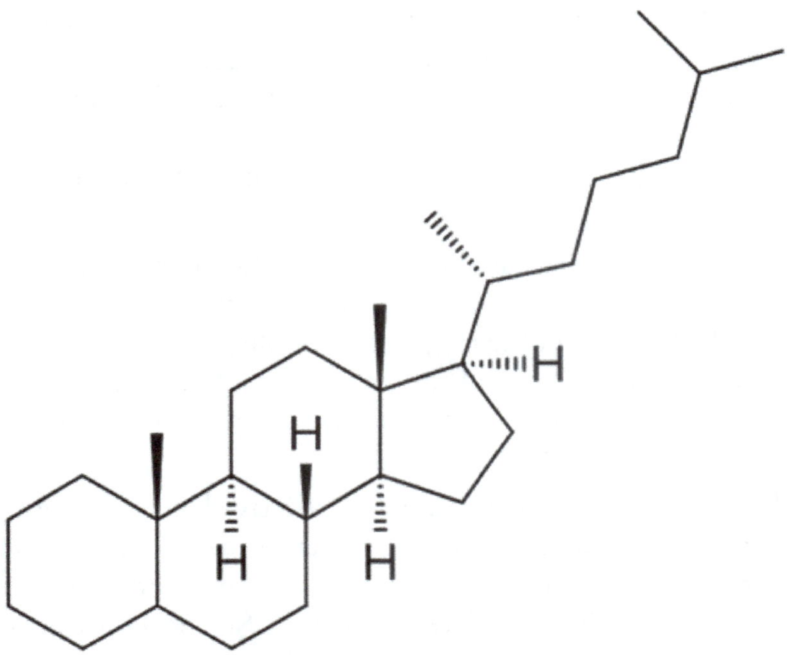

Fig 2: 5α-Cholestane, a Common Steroid Core

Naming Convention

Almost all biologically relevant steroids can be presented as a derivative of a parent cholesterol-like hydrocarbon structure that serves as a skeleton. These parent structures have specific names, such as pregnane, androstane, etc. The derivatives carry various functional groups called suffixes or prefixes after the respective numbers, indicating their position in the steroid nucleus.

There are widely used trivial steroid names of natural origin with significant biologic activity, such as progesterone, testosterone or cortisol. Some of these names are defined in The Nomenclature of Steroids. These trivial names can also be used as a base to derive new names, however, by adding prefixes only rather than suffixes, e.g., the steroid 17α-hydroxyprogesterone has a hydroxy group (-OH) at position 17 of the steroid nucleus comparing to progesterone.

The letters α and β denote absolute stereochemistry at chiral centers—a specific nomenclature distinct from the R/S convention of organic chemistry to denote absolute configuration of functional groups, known as Cahn–Ingold–Prelog priority rules. The R/S convention assigns priorities to substituents on a chiral center based on their atomic number. The highest priority group is assigned to the atom with the highest atomic number, and the lowest priority group is assigned to the atom with the lowest atomic number.

The molecule is then oriented so that the lowest priority group points away from the viewer, and the remaining three groups are arranged in order of decreasing priority around the chiral center. If this arrangement is clockwise, it is assigned an R configuration; if it is counterclockwise, it is assigned an S configuration. In contrast, steroid nomenclature uses α and β to denote stereochemistry at chiral centers.

The α and β designations are based on the orientation of substituents relative to each other in a specific ring system. In general, α refers to a substituent that is oriented towards the plane of the ring system, while β refers to a substituent that is oriented away from the plane of the ring system. In steroids drawn from the standard perspective used in this paper, α-bonds are depicted on figures as dashed wedges and β-bonds as solid wedges.

The name "11-deoxycortisol" is an example of a derived name that uses cortisol as a parent structure without an oxygen atom (hence "deoxy") attached to position 11 (as a part of a hydroxy group). The numbering of positions of carbon atoms in the steroid nucleus is set in a template found in the Nomenclature of Steroids that is used regardless of whether an atom is present in the steroid in question.

Unsaturated carbons (generally, ones that are part of a double bond) in the steroid nucleus are indicated by changing -ane to -ene. This change was traditionally done in the parent name, adding a prefix to denote the position, with or without Δ which designates unsaturation, for example, 4-pregnene-11β,17α-diol-3,20-dione (also Δ^4-pregnene-11β,17α-diol-3,20-dione) or 4-androstene-3,11,17-trione (also Δ^4-androstene-3,11,17-trione). However, the Nomenclature of Steroids recommends the locant of a double bond to be always adjacent to the syllable designating the unsaturation, therefore, having it as a suffix rather than a prefix, and without the use of the Δ character, i.e., pregn-4-ene-11β,17α-diol-3,20-dione or androst-4-ene-3,11,17-trione.

The double bond is designated by the lower-numbered carbon atom, i.e. "Δ^4-" or "4-ene" means the double bond between positions 4 and 5. The saturation of carbons of a parent steroid can be done by adding "dihydro-" prefix, i.e., a saturation of carbons 4 and 5 of testosterone with two hydrogen atoms is 4,5α-dihydrotestosterone or 4,5β-dihydrotestosterone. Generally, when there is no ambiguity, one number of a hydrogen position from a steroid with a saturated bond may be omitted, leaving only the position of the second hydrogen atom, e.g., 5α-dihydrotestosterone or 5β-dihydrotestosterone.

The Δ^5-steroids are those with a double bond between carbons 5 and 6 and the Δ^4 steroids are those with a double bond between carbons 4 and 5.

The abbreviations like "P4" for progesterone and "A4" for androstenedione for refer to Δ^4-steroids, while "P5" for pregnenolone and "A5" for androstenediol refer to Δ^5-steroids.

The suffix -ol denotes a hydroxy group, while the suffix -one denotes an oxo group. When two or three identical groups are attached to the base structure at different positions, the suffix is indicated as -diol or -triol for hydroxy, and -dione or -trione for oxo groups, respectively. For example, 5α-pregnane-3α,17α-diol-20-one has a hydrogen atom at the 5α position (hence the "5α-" prefix), two hydroxy groups (-OH) at the 3α and 17α positions (hence "3α,17α-diol" suffix) and an oxo group (=O) at the position 20 (hence the "20-one" suffix). However, erroneous use of suffixes can be found, e.g., "5α-pregnan-17α-diol-3,11,20-trione" [*sic*] — since it has just one hydroxy group (at 17α) rather than two, then the suffix should be -ol, rather than -diol, so that the correct name to be "5α-pregnan-17α-ol-3,11,20-trione".

According to the rule set in the Nomenclature of Steroids, the terminal "e" in the parent structure name should be elided before the vowel (the presence or absence of a number does not affect such elision). This means, for instance, that if the suffix immediately appended to the parent structure name begins with a vowel, the trailing "e" is removed from that name. An example of such removal is "5α-pregnan-17α-ol-3,20-dione", where the last "e" of "pregnane" is dropped due to the vowel ("o") at the beginning of the suffix -ol. Some authors incorrectly use this rule, eliding the terminal "e" where it should be kept, or vice versa.

The term "11-oxygenated" refers to the presence of an oxygen atom as an oxo (=O) or hydroxy (-OH) substituent at carbon 11. "Oxygenated" is consistently used within the chemistry of the steroids since the 1950s. Some studies use the term "11-oxyandrogens" as an abbreviation for 11-oxygenated androgens, to emphasize that they all have an oxygen atom attached to carbon at position 11. However, in chemical nomenclature, the prefix "oxy" is associated with ether functional groups, i.e., a compound with an oxygen atom connected to two alkyl or aryl groups (R-O-R), therefore, using "oxy" within the name of a steroid class may be misleading. One can find clear examples of "oxygenated" to refer to a broad class of organic molecules containing a variety of oxygen containing functional groups in other domains of organic chemistry, and it is appropriate to use this convention.

Even though "keto" is a standard prefix in organic chemistry, the 1989 recommendations of the Joint Commission on Biochemical Nomenclature discourage the application of the prefix "keto" for steroid names, and favor the prefix "oxo" (e.g., 11-oxo steroids rather than 11-keto steroids), because "keto" includes the carbon that is part of the steroid nucleus and the same carbon atom should not be specified twice.

Species Distribution

Steroids are found in all domains of life including bacteria, archaea, and eukaryotes. In eukaryotes, steroids are found in fungi, plants, and animals.

Eukaryotic and Prokaryotic

- Eukaryotic cells, which include animals, plants, fungi, and protists, have complex cellular structures with a true nucleus and membrane-bound organelles.
- Steroids are integral to eukaryotic cellular membranes, where they help maintain membrane integrity and function.
- During eukaryogenesis (the emergence of modern eukaryotic cells), steroids likely played a role in the acquisition of mitochondria via endocytosis.
- In prokaryotes, biosynthetic pathways exist for the tetracyclic steroid framework (e.g., in myxobacteria) - where its origin from eukaryotes is conjectured - and the more-common pentacyclic triterpinoid hopanoid framework.

Fungal

Fungal steroids include the ergosterols, which are involved in maintaining the integrity of the fungal cellular membrane. Various antifungal drugs, such as amphotericin B and azole antifungals, utilize this information to kill pathogenic fungi. Fungi can alter their ergosterol content (e.g. through loss of function mutations in the enzymes ERG3 or ERG6, inducing depletion of ergosterol, or mutations that decrease the ergosterol content) to develop resistance to drugs that target ergosterol.

Ergosterol is analogous to the cholesterol found in the cellular membranes of animals (including humans), or the phytosterols found in the cellular membranes of plants. All mushrooms contain large quantities of ergosterol, in the range of tens to hundreds of milligrams per 100 grams of dry weight. Oxygen is necessary for the synthesis of ergosterol in fungi.

Ergosterol is responsible for the vitamin D content found in mushrooms; ergosterol is chemically converted into provitamin D2 by exposure to ultraviolet light. Provitamin D2 spontaneously forms vitamin D2. However, not all fungi utilize ergosterol in their cellular membranes; for example, the pathogenic fungal species *Pneumocystis jirovecii* does not, which has important clinical implications (given the mechanism of action of many antifungal drugs).

Plant

Plant steroids include steroidal alkaloids found in Solanaceae and Melanthiaceae (specially the genus Veratrum), cardiac glycosides, the phytosterols and the brassinosteroids (which include several plant hormones).

Animal

Animal steroids include compounds of vertebrate and insect origin, the latter including ecdysteroids such as ecdysterone (controlling molting in some species).

Vertebrate examples include the steroid hormones and cholesterol; the latter is a structural component of cell membranes that helps determine the fluidity of cell membranes and is a principal constituent of plaque (implicated in atherosclerosis). Steroid hormones include:

- **Sex Hormones,** which influence sex differences and support reproduction. These include androgens, estrogens, and progestogens.
- **Corticosteroids,** including most synthetic steroid drugs, with natural product classes the glucocorticoids (which regulate many aspects of metabolism and immune function) and the mineralocorticoids (which help maintain blood volume and control renal excretion of electrolytes)
- **Anabolic Steroids,** natural and synthetic, which interact with androgen receptors to increase muscle and bone synthesis. In popular use, the term "steroids" often refers to anabolic steroids.

Types

By Function

The major classes of steroid hormones, with prominent members and examples of related functions, are:

- **Corticosteroids:**
  - *Glucocorticoids:* Cortisol, a glucocorticoid whose functions include immunosuppression.
  - *Mineralocorticoids:* Aldosterone, a mineralocorticoid that helps regulate blood pressure through water and electrolyte balance.
- **Sex Steroids:**
  - *Progestogens:* Progesterone, which regulates cyclical changes in the endometrium of the uterus and maintains a pregnancy.
  - *Androgens:* Testosterone, which contributes to the development and maintenance of male secondary sex characteristics.
  - *Estrogens:* Estradiol, which contributes to the development and maintenance of female secondary sex characteristics.

Additional classes of steroids include:

- Neurosteroids such as DHEATooltip dehydroepiandrosterone and allopregnanolone.
- Bile acids such as taurocholic acid.
- Aminosteroid neuromuscular blocking agents (mainly synthetic) such as pancuronium bromide.
- Steroidal antiandrogens (mainly synthetic) such as cyproterone acetate.
- Steroidogenesis inhibitors (mainly exogenous) such as alfatradiol.
- Membrane sterols such as cholesterol, ergosterol, and various phytosterols.
- Toxins such as steroidal saponins and cardenolides/cardiac glycosides.

As well as the following class of secosteroids (open-ring steroids):

- Vitamin D forms such as ergocalciferol, cholecalciferol, and calcitriol

Intact Ring System

Steroids can be classified based on their chemical composition. One example of how MeSH performs this classification is available at the Wikipedia MeSH catalog. Examples of this classification include:

Class	Example	Number of Carbon Atoms
Cholestanes	Cholesterol	27
Cholanes	Cholic acid	24
Pregnanes	Progesterone	21
Androstanes	Testosterone	19
Estranes	Estradiol	18

In biology, it is common to name the above steroid classes by the number of carbon atoms present when referring to hormones: C_{18}-steroids for the estranes (mostly estrogens), C_{19}-steroids for the androstanes (mostly androgens), and C_{21}-steroids for the pregnanes (mostly corticosteroids). The classification "17-ketosteroid" is also important in medicine.

The gonane (steroid nucleus) is the parent 17-carbon tetracyclic hydrocarbon molecule with no alkyl sidechains.

Cleaved, Contracted, and Expanded Rings

Secosteroids (Latin *seco*, "to cut") are a subclass of steroidal compounds resulting, biosynthetically or conceptually, from scission (cleavage) of parent steroid rings (generally one of the four). Major secosteroid subclasses are defined by the steroid carbon atoms where this scission has taken place. For instance, the prototypical secosteroid cholecalciferol, vitamin D_3 (shown), is in the 9,10-secosteroid subclass and derives from the cleavage of carbon atoms C-9 and C-10 of the steroid B-ring; 5,6-secosteroids and 13,14-steroids are similar.

Norsteroids (nor-, L. *norma*; "normal" in chemistry, indicating carbon removal) and homosteroids (homo-, Greek *homos*; "same", indicating carbon addition) are structural subclasses of steroids formed from biosynthetic steps. The former involves enzymic ring expansion-contraction reactions, and the latter is accomplished (biomimetically) or (more frequently) through ring closures of acyclic precursors with more (or fewer) ring atoms than the parent steroid framework.

Combinations of these ring alterations are known in nature. For instance, ewes who graze on corn lily ingest cyclopamine (shown) and veratramine, two of a sub-family of steroids where the C- and D-rings are contracted and expanded respectively via a biosynthetic migration of the original C-13 atom. Ingestion of these C-nor-D-homosteroids results in birth defects in lambs: cyclopia from cyclopamine and leg deformity from veratramine. A further C-nor-D-homosteroid (nakiterpiosin) is excreted by Okinawan cyanobacteriosponges. e.g., *Terpios hoshinota*, leading to coral mortality from black coral disease. Nakiterpiosin-type steroids are active against the signaling pathway involving the smoothened and hedgehog proteins, a pathway which is hyperactive in a number of cancers.

Biological Significance

Steroids and their metabolites often function as signalling molecules (the most notable examples are steroid hormones), and steroids and phospholipids are components of cell membranes. Steroids such as cholesterol decrease membrane fluidity. Similar to lipids, steroids are highly concentrated energy stores. However, they are not typically sources of energy; in mammals, they are normally metabolized and excreted. Steroids play critical roles in a number of disorders, including malignancies like prostate cancer, where steroid production inside and outside the tumour promotes cancer cell aggressiveness.

Biosynthesis and Metabolism

The hundreds of steroids found in animals, fungi, and plants are made from lanosterol (in animals and fungi) or cycloartenol (in other eukaryotes). Both lanosterol and cycloartenol derive from cyclization of the triterpenoid squalene. Lanosterol and cycloartenol are sometimes called protosterols because they serve as the starting compounds for all other steroids.

Steroid biosynthesis is an anabolic pathway which produces steroids from simple precursors. A unique biosynthetic pathway is followed in animals (compared to many other organisms), making the pathway a common target for antibiotics and other anti-infection drugs. Steroid metabolism in humans is also the target of cholesterol-lowering drugs, such as statins. In humans and other animals the biosynthesis of steroids follows the mevalonate pathway, which uses acetyl-CoA as building blocks for dimethylallyl diphosphate (DMAPP) and isopentenyl diphosphate (IPP).

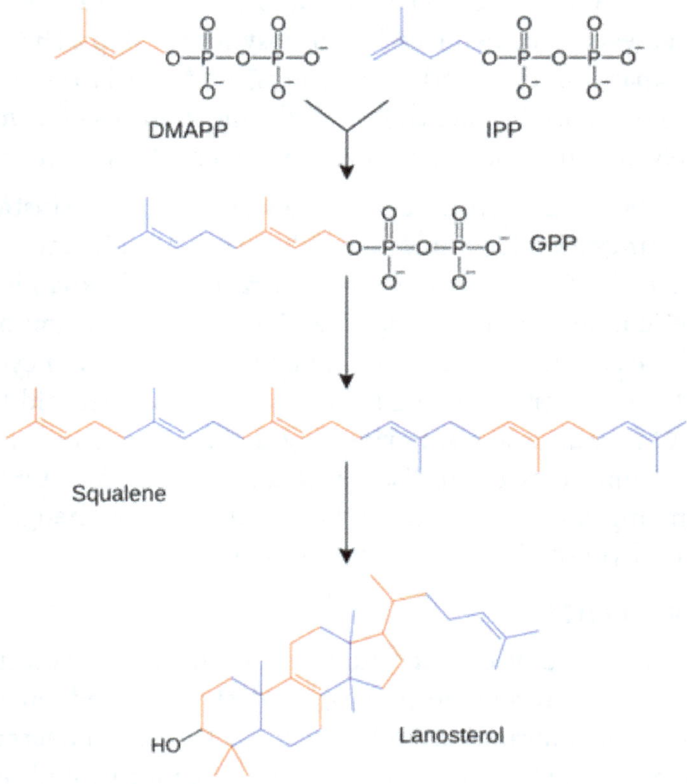

Fig 3: Simplification of the End of the Steroid Synthesis Pathway, where the Intermediates Isopentenyl Pyrophosphate (PP or IPP) and Dimethylallyl Pyrophosphate (DMAPP) form Geranyl Pyrophosphate (GPP), Squalene and Lanosterol (The First Steroid in the Pathway)

In subsequent steps DMAPP and IPP conjugate to form farnesyl diphosphate (FPP), which further conjugates with each other to form the linear triterpenoid squalene. Squalene biosynthesis is catalyzed by squalene synthase, which belongs to the squalene/phytoene synthase family. Subsequent epoxidation and cyclization of squalene generate lanosterol, which is the starting point for additional modifications into other steroids (steroidogenesis). In other eukaryotes, the cyclization product of epoxidized squalene (oxidosqualene) is cycloartenol.

Mevalonate Pathway

The mevalonate pathway (also called HMG-CoA reductase pathway) begins with acetyl-CoA and ends with dimethylallyl diphosphate (DMAPP) and isopentenyl diphosphate (IPP). DMAPP and IPP donate isoprene units, which are assembled and modified to form terpenes and isoprenoids (a large class of lipids, which include the carotenoids and form the largest class of plant natural products). Here, the isoprene units are joined to make squalene and folded into a set of rings to make lanosterol. Lanosterol can then be converted into other steroids, such as cholesterol and ergosterol. Two classes of drugs target the mevalonate pathway: statins (like rosuvastatin), which are used to reduce elevated cholesterol levels, and bisphosphonates (like zoledronate), which are used to treat a number of bone-degenerative diseases.

Steroidogenesis

Steroidogenesis is the biological process by which steroids are generated from cholesterol and changed into other steroids. The pathways of steroidogenesis differ among species. The major classes of steroid hormones, as noted above (with their prominent members and functions), are the progestogens, corticosteroids (corticoids), androgens, and estrogens. Human steroidogenesis of these classes occurs in a number of locations:

- Progestogens are the precursors of all other human steroids, and all human tissues which produce steroids must first convert cholesterol to pregnenolone. This conversion is the rate-limiting step of steroid synthesis, which occurs inside the mitochondrion of the respective tissue.
- Cortisol, corticosterone, aldosterone are produced in the adrenal cortex.
- Estradiol, estrone and progesterone are made primarily in the ovary, estriol in placenta during pregnancy, and testosterone primarily in the testes (some testosterone may also be produced in the adrenal cortex).
- Estradiol is converted from testosterone directly (in males), or via the primary pathway DHEA – androstenedione – estrone and secondarily via testosterone (in females).
- Stromal cells have been shown to produce steroids in response to signaling produced by androgen-starved prostate cancer cells.
- Some neurons and glia in the central nervous system (CNS) express the enzymes required for the local synthesis of pregnenolone, progesterone, DHEA and DHEAS, *de novo* or from peripheral sources.

Alternative Pathways

In plants and bacteria, the non-mevalonate pathway (MEP pathway) uses pyruvate and glyceraldehyde 3-phosphate as substrates to produce IPP and DMAPP. During diseases pathways otherwise not significant in healthy humans can become utilized. For example, in one form of congenital adrenal hyperplasia a deficiency in the 21-hydroxylase enzymatic pathway leads to an excess of 17α-Hydroxyprogesterone (17-OHP) - this pathological excess of 17-OHP in turn may be converted to dihydrotestosterone (DHT, a potent androgen) through among others 17,20 Lyase (a member of the cytochrome P450 family of enzymes), 5α-Reductase and 3α-Hydroxysteroid dehydrogenase.

Catabolism and Excretion

Steroids are primarily oxidized by cytochrome P450 oxidase enzymes, such as CYP3A4. These reactions introduce oxygen into the steroid ring, allowing the cholesterol to be broken up by other enzymes into bile acids. These acids can then be eliminated by secretion from the liver in bile. The expression of the oxidase gene can be upregulated by the steroid sensor PXR when there is a high blood concentration of steroids. Steroid hormones, lacking the side chain of cholesterol and bile acids, are typically hydroxylated at various ring positions or oxidized at the 17 position, conjugated with sulfate or glucuronic acid and excreted in the urine.

Isolation, Structure Determination, and Methods of Analysis

Steroid *isolation*, depending on context, is the isolation of chemical matter required for chemical structure elucidation, derivitzation or degradation chemistry, biological testing, and other research needs (generally milligrams to grams, but often more or the isolation of "analytical quantities" of the substance of interest (where the focus is on identifying and quantifying the substance (for example, in biological tissue or fluid). The amount isolated depends on the analytical method, but is generally less than one microgram.

The methods of isolation to achieve the two scales of product are distinct, but include extraction, precipitation, adsorption, chromatography, and crystallization.

In both cases, the isolated substance is purified to chemical homogeneity; combined separation and analytical methods, such as LC-MS, are chosen to be "orthogonal"—achieving their separations based on distinct modes of interaction between substance and isolating matrix—to detect a single species in the pure sample.

Structure determination refers to the methods to determine the chemical structure of an isolated pure steroid, using an evolving array of chemical and physical methods which have included NMR and small-molecule crystallography.

Methods of analysis overlap both of the above areas, emphasizing analytical methods to determining if a steroid is present in a mixture and determining its quantity.

Chemical Synthesis

Microbial catabolism of phytosterol side chains yields C-19 steroids, C-22 steroids, and 17-ketosteroids (i.e. precursors to adrenocortical hormones and contraceptives). The addition and modification of functional groups is key when producing the wide variety of medications available within this chemical classification. These modifications are performed using conventional organic synthesis and/or biotransformation techniques.

Precursors

- **Semisynthesis:** The semisynthesis of steroids often begins from precursors such as cholesterol, phytosterols, or sapogenins. The efforts of Syntex, a company involved in the Mexican barbasco trade, used *Dioscorea mexicana* to produce the sapogenin diosgenin in the early days of the synthetic steroid pharmaceutical industry.
- **Total Synthesis:** Some steroidal hormones are economically obtained only by total synthesis from petrochemicals (e.g. 13-alkyl steroids). For example, the pharmaceutical Norgestrel begins from methoxy-1-tetralone, a petrochemical derived from phenol.

27.1.2 Sterols

Sterol is an organic compound with formula $C_{17}H_{28O}$, whose molecule is derived from that of gonane by replacement of a hydrogen atom on C3 position by a hydroxyl group. It is therefore an alcohol of gonane. More generally, any compounds that contain the gonane structure, additional functional groups, and/or modified ring systems derived from gonane are called steroids. Therefore, sterols are a subgroup of the steroids. They occur naturally in most eukaryotes, including plants, animals, and fungi, and can also be produced by some bacteria (however likely with different functions). The most familiar type of animal sterol is cholesterol, which is vital to cell membrane structure, and functions as a precursor to fat-soluble vitamins and steroid hormones.

While technically alcohols, sterols are classified by biochemists as lipids (fats in the broader sense of the term).

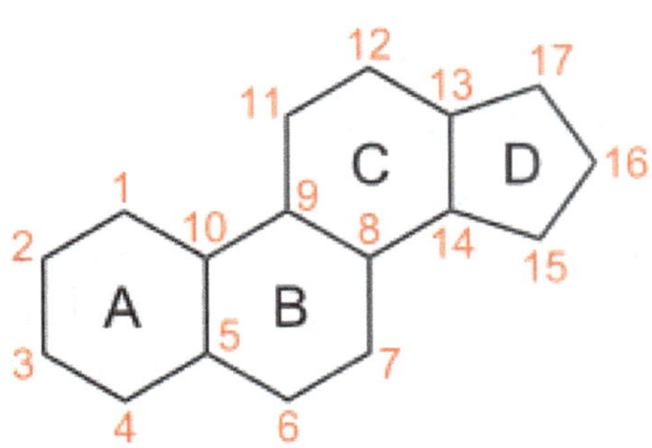

Fig 4: The Gonane skeleton, with the IUPAC recommended numbering of the carbon atoms

Types

- **Phytosterols:** Phytosterols are sterols naturally found in plants. Notable examples of phytosterols include campesterol, sitosterol, and stigmasterol.
- **Zoosterols:** Zoosterols are sterols found in animals. The most significant zoosterol is cholesterol.
- **Mycosterols:** Sterols found in fungi are called mycosterols. A common example is ergosterol, a mycosterol present in the cell membrane of fungi, where it serves a role similar to cholesterol in animal cells.

Role in Biochemistry

Sterols and related compounds play essential roles in the physiology of eukaryotic organisms, and are essential for normal physiology of plants, animals, and fungi. For example, cholesterol forms part of the cellular membrane in animals, where it affects the cell membrane's fluidity and serves as secondary messenger in developmental signaling.

In humans and other animals, corticosteroids such as cortisol act as signaling compounds in cellular communication and general metabolism. Sterols are common components of human skin oils.

Phytosterols as a Nutritional Supplement

Phytosterols, more commonly known as plant sterols, have been shown in clinical trials to block cholesterol absorption sites in the human intestine, thus helping to reduce cholesterol absorption in humans. They are currently approved by the U.S. Food and Drug Administration for use as a food supplement; however, there is some concern that they may block absorption not only of cholesterol, but of other important nutrients as well. At present, the American Heart Association has recommended that supplemental plant sterols be taken only by those diagnosed with elevated cholesterol, and has particularly recommended that they not be taken by pregnant women or nursing mothers. Preliminary research has shown that phytosterols may have anticancer effects.

Chemical Classification and Structure

Sterols are a subgroup of steroids with a hydroxyl group at the 3-position of the A-ring. They are amphipathic lipids synthesized from acetyl-coenzyme A via the HMG-CoA reductase pathway. The overall molecule is quite flat. The hydroxyl group on the A ring is polar. The rest of the aliphatic chain is non-polar.

27.2 TYPES OF MICROBIAL TRANSFORMATION

Microbial transformations of steroids and sterols can occur through various biochemical pathways, including hydroxylation, oxidation, reduction, and side-chain cleavage. Each of these processes can lead to significant changes in the chemical structure and biological activity of the steroid or sterol.

27.2.1 Hydroxylation

Hydroxylation is the addition of hydroxyl groups (-OH) to steroid or sterol structures. This process is critical for the activation or inactivation of steroid hormones, influencing their biological effects. Certain microorganisms, such as fungi and bacteria, possess enzymes called hydroxylases that facilitate this transformation. For example, the fungus *Aspergillus niger* has been shown to hydroxylate steroids, such as progesterone, resulting in the production of hydroxyprogesterone. This modified compound can exhibit enhanced biological activity compared to its parent steroid.

27.2.2 Oxidation and Reduction

Oxidation and reduction reactions play a significant role in the microbial transformation of steroids. These processes involve the gain or loss of electrons, leading to changes in the oxidation state of specific carbon atoms in the steroid structure.

Microorganisms can employ a variety of enzymes, such as oxidoreductases, to facilitate these reactions. For instance, the bacterium *Corynebacterium* can reduce cortisone to hydrocortisone, a compound with higher anti-inflammatory activity. This transformation is particularly relevant in the pharmaceutical industry, where specific steroid derivatives are sought for therapeutic applications.

27.2.3 Side-Chain Cleavage

Side-chain cleavage refers to the removal or modification of alkyl side chains attached to the steroid core structure. This process can lead to the formation of smaller molecules with distinct biological activities. Certain microorganisms, particularly actinomycetes and fungi, are known to perform side-chain cleavage on steroid compounds.

The cleavage of the side chain of cholesterol can yield significant intermediates, such as pregnenolone, which serves as a precursor for various steroid hormones. This conversion is of considerable interest for the synthesis of steroid-based pharmaceuticals.

27.3 MICROORGANISMS INVOLVED IN STEROID TRANSFORMATION

Several groups of microorganisms are known to participate in the transformation of steroids and sterols. These include fungi, bacteria, and yeast, each exhibiting unique capabilities for steroid modification.

27.3.1 Fungi

Fungi have long been recognized for their ability to transform steroid compounds. Many species possess enzymes that facilitate hydroxylation, oxidation, and side-chain cleavage. Fungal strains such as *Aspergillus*, *Rhizopus*, and *Penicillium* have been extensively studied for their steroid-transforming capabilities. The transformation pathways utilized by fungi are often complex and can lead to the production of multiple metabolites. For instance, *Aspergillus niger* can convert testosterone into various hydroxylated derivatives, demonstrating the potential for producing diverse steroidal compounds through microbial fermentation.

27.3.2 Bacteria

Bacterial species, particularly actinomycetes, play a crucial role in the microbial transformation of steroids. *Corynebacterium*, *Nocardia*, and *Rhodococcus* are among the genera known for their steroid-transforming abilities. These bacteria can carry out a wide range of reactions, including hydroxylation, reduction, and side-chain cleavage. For example, *Nocardia* species have been reported to hydroxylate steroids such as progesterone and cortisol, leading to the production of biologically active derivatives. The enzymatic machinery present in these bacteria offers significant potential for biotechnological applications in steroid modification.

27.3.3 Yeast

Yeast, particularly species from the genus *Candida*, are also known for their ability to transform sterols and steroids.

Yeasts possess unique enzymatic pathways that enable the hydroxylation and oxidation of sterols, leading to the production of valuable metabolites. *Candida tropicalis* is one yeast species that has been extensively studied for its ability to convert cholesterol into various hydroxylated derivatives, showcasing its potential for industrial applications in steroid transformation.

27.4 SIGNIFICANCE OF MICROBIAL TRANSFORMATION

27.4.1 Pharmaceutical Applications

Microbial transformation processes are of particular importance in the pharmaceutical industry, where they provide an eco-friendly alternative to traditional chemical synthesis. The ability of microorganisms to produce specific steroid derivatives with enhanced pharmacological properties makes them valuable in drug development. For instance, the microbial conversion of cortisone to hydrocortisone has led to the production of a potent anti-inflammatory medication. Additionally, the synthesis of steroid hormones, such as progesterone and estrogen, through microbial transformation has opened new avenues for hormone replacement therapy.

27.4.2 Agricultural Applications

In agriculture, microbial transformations can lead to the development of bioactive compounds with potential pesticide or herbicide properties. The conversion of plant sterols into active metabolites may enhance the growth and resistance of crops, contributing to sustainable agricultural practices. Moreover, microbial transformation processes can be employed to develop biopesticides that target specific pests without harming beneficial organisms. This approach aligns with the growing demand for environmentally friendly agricultural solutions.

27.4.3 Biotechnology and Industrial Applications

The application of microbial transformation in biotechnology has gained increasing interest for producing fine chemicals, flavorings, and biofuels. The ability of microorganisms to modify steroids and sterols can lead to the development of novel products with enhanced functionalities. For example, the transformation of cholesterol into steroid-based biofuels through microbial fermentation can offer a sustainable energy source. The use of microorganisms in bioprocessing also reduces the need for hazardous chemicals, making the production process more environmentally friendly.

27.5 FUTURE PERSPECTIVES

The microbial transformation of steroids and sterols holds great promise for future research and development. Advances in genetic engineering and metabolic engineering techniques can enhance the capabilities of microorganisms to produce specific steroid derivatives with desired properties. Additionally, the exploration of microbial consortia—combinations of different microorganisms—may lead to synergistic effects in steroid transformation, allowing for more efficient bioprocesses. Furthermore, the integration of systems biology and omics technologies can provide insights into the metabolic pathways involved in steroid transformation, facilitating the optimization of production processes.

27.6 CONCLUSION

Microbial transformation of steroids and sterols represents a dynamic and promising area of study with significant implications for pharmaceuticals, agriculture, and biotechnology. Through the use of various microorganisms, these processes allow for the modification of steroidal compounds, leading to the production of novel metabolites with enhanced properties. As research continues to advance in this field, the potential for developing sustainable and eco-friendly solutions for steroid transformation is increasingly apparent, paving the way for innovative applications in various industries.

CHAPTER 28
VACCINES

Vaccines have emerged as one of the most effective public health tools for preventing infectious diseases. They work by stimulating the immune system to recognize and combat pathogens, thereby providing immunity without causing the disease itself. The development and administration of vaccines have played a crucial role in reducing morbidity and mortality from infectious diseases, leading to significant advancements in global health.

28.1 A BRIEF HISTORY OF VACCINATION

The practice of vaccination dates back centuries. The earliest form of immunization can be traced to ancient China, where people would inhale powdered smallpox scabs to protect themselves from the disease. However, the modern concept of vaccination began in the late 18th century with Edward Jenner's discovery in 1796. Jenner inoculated a young boy with cowpox, a milder virus, and observed that he did not develop smallpox. This pioneering work laid the foundation for the development of vaccines as a preventive measure against infectious diseases. The 20th century witnessed significant advancements in vaccine development, particularly with the advent of more sophisticated technologies and a better understanding of immunology. Vaccines for diseases such as polio, measles, mumps, rubella, and hepatitis B were developed and widely distributed, resulting in dramatic declines in morbidity and mortality associated with these illnesses.

28.2 TYPES OF VACCINES

Vaccines can be classified into several categories based on their composition and method of action. Each type has unique characteristics and is suited for specific pathogens.

28.2.1 Live Attenuated Vaccines

Live attenuated vaccines contain weakened forms of the pathogen that cause the disease. These vaccines stimulate a strong immune response and often provide lifelong immunity after just one or two doses. Examples include the measles, mumps, and rubella (MMR) vaccine and the oral polio vaccine. However, because they contain live viruses, these vaccines are not suitable for individuals with weakened immune systems.

28.2.2 Inactivated or Killed Vaccines

Inactivated or killed vaccines are made from pathogens that have been killed or inactivated so that they cannot cause disease. While these vaccines are generally safe and stable, they may require multiple doses to achieve adequate immunity. Examples include the influenza vaccine and the hepatitis A vaccine.

28.2.3 Subunit, Recombinant, and Conjugate Vaccines

These vaccines consist of specific pieces of the pathogen, such as proteins or sugars, rather than the entire organism. Subunit vaccines contain purified components of the pathogen, while recombinant vaccines are produced using genetic engineering techniques. Conjugate vaccines link polysaccharide antigens to protein carriers to enhance the immune response. The human papillomavirus (HPV) vaccine and the Haemophilus influenzae type b (Hib) vaccine are examples of this type.

28.2.4 mRNA Vaccines

A newer class of vaccines, mRNA vaccines, use a small piece of genetic material from the virus to instruct cells to produce a harmless protein that mimics a part of the virus. This prompts an immune response without introducing live virus into the body. The Pfizer-BioNTech and Moderna COVID-19 vaccines are notable examples of mRNA technology and have been crucial in the fight against the COVID-19 pandemic.

28.2.5 Viral Vector Vaccines

Viral vector vaccines use a harmless virus to deliver genetic material from the target pathogen into human cells. This approach allows the body to produce an immune response against the target pathogen. The Johnson & Johnson COVID-19 vaccine is an example of a viral vector vaccine.

28.3 MECHANISMS OF ACTION

Vaccines work by training the immune system to recognize and respond to specific pathogens. When a vaccine is administered, it introduces antigens—substances that stimulate an immune response—into the body. The immune system recognizes these antigens as foreign and mounts a response, which includes the production of antibodies and the activation of T cells.

28.3.1 Antibody Production

B cells, a type of white blood cell, are responsible for producing antibodies that bind to specific antigens. When the body is exposed to the actual pathogen in the future, these antibodies can quickly neutralize the pathogen and prevent infection.

28.3.2 Cell-Mediated Immunity

In addition to antibody production, vaccines also stimulate T cells, which play a crucial role in cell-mediated immunity. T cells can recognize and destroy infected cells or help other immune cells respond more effectively to the pathogen.

28.4 THE VACCINE DEVELOPMENT PROCESS

28.4.1 Preclinical Research

The vaccine development process begins with preclinical research, which involves laboratory studies and animal testing to evaluate the vaccine's safety and immune response. Researchers assess the vaccine's potential effectiveness and gather preliminary data on dosing.

28.4.2 Clinical Trials

If preclinical studies are promising, the vaccine moves to clinical trials, which are conducted in three phases:

- **Phase I Trials:** A small group of healthy volunteers receives the vaccine to evaluate its safety and determine the appropriate dosage.
- **Phase II Trials:** The vaccine is administered to a larger group, often consisting of individuals who are at risk for the disease, to further assess safety and immune response.
- **Phase III Trials:** The vaccine is given to thousands of participants in a controlled environment to evaluate its effectiveness and monitor for any adverse effects.

28.4.3 Regulatory Approval

Following successful clinical trials, vaccine manufacturers submit data to regulatory agencies, such as the U.S. Food and Drug Administration (FDA) or the European Medicines Agency (EMA), for review.

These agencies assess the vaccine's safety, efficacy, and manufacturing processes before granting approval for public use.

28.4.4 Post-Marketing Surveillance

Once a vaccine is approved and distributed, it continues to be monitored for safety and effectiveness through post-marketing surveillance. This phase involves collecting data on adverse events and long-term effects to ensure the ongoing safety of the vaccine.

28.5 THE IMPACT OF VACCINES ON PUBLIC HEALTH

Vaccines have had a profound impact on public health, leading to significant reductions in the incidence of infectious diseases worldwide. Vaccination campaigns have eradicated smallpox and significantly reduced the prevalence of diseases such as polio, measles, and diphtheria.

28.5.1 Herd Immunity

Widespread vaccination contributes to herd immunity, which occurs when a significant portion of the population is immunized, thereby protecting those who cannot be vaccinated, such as individuals with certain medical conditions or weakened immune systems. This collective immunity helps to prevent outbreaks and the spread of diseases within communities.

28.5.2 Economic Benefits

The economic impact of vaccines is substantial. Preventing diseases reduces healthcare costs associated with treatment and hospitalizations. Additionally, vaccination programs contribute to increased productivity and economic stability by preventing illness-related absenteeism.

28.5.3 Global Health Initiatives

Vaccines play a crucial role in global health initiatives aimed at eradicating or controlling infectious diseases, particularly in low- and middle-income countries. Organizations like the World Health Organization (WHO) and Gavi, the Vaccine Alliance, work to ensure equitable access to vaccines, especially for vulnerable populations.

28.6 CHALLENGES AND CONTROVERSIES

Despite the successes of vaccination programs, several challenges and controversies persist. Misinformation and vaccine hesitancy have emerged as significant barriers to achieving high vaccination rates. Concerns about vaccine safety, fueled by misinformation, can lead to decreased public trust and lower vaccination uptake. Additionally, the global distribution of vaccines, particularly during the COVID-19 pandemic, has highlighted disparities in access to immunization. Ensuring that vaccines reach underserved populations remains a critical challenge for public health authorities.

28.7 CONCLUSION

Vaccines are a cornerstone of modern medicine, providing a safe and effective means of preventing infectious diseases. Through the stimulation of the immune system, vaccines protect individuals and communities from potentially devastating illnesses. The history of vaccination is a testament to human ingenuity and scientific progress, leading to significant advancements in public health. As research continues to evolve, addressing challenges such as vaccine hesitancy and equitable access to immunization will be essential for maintaining and expanding the successes of vaccination programs worldwide. Vaccines not only save lives but also represent a crucial investment in the health and well-being of future generations.

CHAPTER 29

DRUG DISCOVERY IN MICROBIAL METABOLITES: THE SEARCH FOR MICROBIAL PRODUCTS WITH BIOACTIVE PROPERTIES

Microbial metabolites have long been recognized as a rich source of bioactive compounds with significant therapeutic potential. The discovery of antibiotics, anticancer agents, and immunosuppressive drugs from microorganisms has paved the way for innovative approaches in drug development.

29.1 THE IMPORTANCE OF MICROBIAL METABOLITES

Microbial metabolites are small organic compounds produced by microorganisms during their metabolic processes. These metabolites can be classified into primary metabolites, which are essential for growth and reproduction, and secondary metabolites, which often have ecological functions and are not directly involved in the organism's growth. Secondary metabolites, including antibiotics, alkaloids, and polyketides, have gained significant attention in drug discovery due to their diverse biological activities.

The unique chemical structures and mechanisms of action of microbial metabolites make them valuable candidates for drug development. Historically, more than half of the drugs currently in use have been derived from natural products, with a significant proportion originating from microbial sources. The exploration of microbial metabolites continues to be a promising avenue for identifying novel therapeutics in the face of emerging infectious diseases and drug-resistant pathogens.

29.2 HISTORICAL CONTEXT: DISCOVERIES THAT CHANGED MEDICINE

The journey of drug discovery from microbial metabolites began in earnest with the serendipitous discovery of penicillin by Alexander Fleming in 1928. This antibiotic, derived from the fungus *Penicillium notatum*, revolutionized medicine and set the stage for the development of a plethora of antibiotic compounds. Following this breakthrough, a surge of interest in microbial metabolites led to the discovery of other important drugs, such as streptomycin from *Streptomyces griseus*, which became a cornerstone in the treatment of tuberculosis.

The mid-20th century marked a golden age in the discovery of microbial-derived pharmaceuticals, with researchers identifying numerous antibiotics, antifungals, and antiparasitic agents. The intricate relationships between microbes and their environments, along with their ability to produce diverse bioactive compounds, have driven ongoing research in this field.

29.3 MECHANISMS OF BIOACTIVITY

Microbial metabolites exhibit a wide range of bioactive properties, often targeting specific biological pathways in humans or pathogens. Understanding the mechanisms of action is crucial for the development and optimization of these compounds for therapeutic use.

29.3.1 Antibiotic Action

Many microbial metabolites, particularly antibiotics, function by disrupting essential cellular processes in bacteria. For example, penicillin inhibits cell wall synthesis, leading to bacterial lysis and death. Other antibiotics, such as tetracycline and macrolides, interfere with protein synthesis by binding to bacterial ribosomes. This targeted approach minimizes harm to human cells while effectively combating bacterial infections.

29.3.2 Anticancer Activity

Some microbial metabolites have shown promising anticancer properties by targeting specific cancer cell pathways. For instance, the alkaloid vincristine, derived from the *Catharanthus roseus* plant but initially discovered in microbial cultures, inhibits cell division by disrupting the mitotic spindle. Other microbial-derived compounds, such as doxorubicin, have been extensively studied for their efficacy in treating various cancer types.

29.3.3 Immunosuppressive Properties

Microbial metabolites also play a crucial role in immunosuppressive therapies. Cyclosporine, produced by the fungus *Tolypocladium inflatum*, is a well-known immunosuppressant used in organ transplantation to prevent rejection. Its ability to inhibit T-cell activation has made it a cornerstone in transplant medicine, highlighting the therapeutic potential of microbial metabolites in modulating the immune response.

29.4 THE PROCESS OF DRUG DISCOVERY FROM MICROBIAL METABOLITES

The discovery of bioactive compounds from microbial sources involves a series of well-defined steps, from initial screening to clinical trials. This process is critical for identifying and developing novel therapeutics.

29.4.1 Isolation and Characterization

The first step in drug discovery involves isolating microorganisms from diverse environments, including soil, marine habitats, and extreme environments. Once isolated, these microbes are cultured to induce the production of secondary metabolites. This process may involve optimizing growth conditions, such as temperature, pH, and nutrient availability. Once the microorganisms are cultured, the next step is to extract and characterize the metabolites produced. Various extraction techniques, such as solvent extraction and chromatography, are employed to isolate the bioactive compounds. Characterization techniques, including mass spectrometry and nuclear magnetic resonance (NMR) spectroscopy, help elucidate the chemical structures of these metabolites.

29.4.2 Screening for Bioactivity

After isolating and characterizing microbial metabolites, the next step involves screening these compounds for bioactivity. High-throughput screening techniques are often employed to assess the efficacy of these compounds against specific biological targets, such as pathogenic bacteria or cancer cell lines. This phase is crucial for identifying lead compounds with promising therapeutic potential.

29.4.3 Mechanistic Studies

Once bioactive compounds are identified, further studies are conducted to elucidate their mechanisms of action. Understanding how these compounds interact with biological systems helps in optimizing their structure for enhanced efficacy and reduced toxicity. Mechanistic studies may involve cellular assays, animal models, and molecular biology techniques to determine the specific pathways targeted by the metabolites.

29.4.4 Preclinical and Clinical Development

Successful candidates undergo rigorous preclinical and clinical development processes. Preclinical studies assess the safety and efficacy of the compounds in animal models, while clinical trials evaluate their safety, tolerability, and effectiveness in human subjects. The path from laboratory discovery to clinical application is complex and often requires significant time and resources.

29.5 CHALLENGES IN DRUG DISCOVERY FROM MICROBIAL METABOLITES

29.5.1 Rediscovery of Known Compounds

One of the primary challenges in the field is the rediscovery of known compounds. Many microbial metabolites have already been identified and characterized, leading to a saturated market for certain drugs. To overcome this challenge, researchers are exploring novel microbial sources and employing advanced techniques, such as metagenomics and synthetic biology, to discover new metabolites with unique properties.

29.5.2 Variability in Production

Microbial production of metabolites can be highly variable, influenced by factors such as growth conditions, strain differences, and nutrient availability. Standardizing production processes and optimizing fermentation conditions are essential for ensuring consistent yields of bioactive compounds.

29.5.3 Regulatory Hurdles

The path from discovery to clinical use involves navigating complex regulatory frameworks. Obtaining approval from regulatory agencies, such as the U.S. Food and Drug Administration (FDA), requires extensive documentation and data on safety, efficacy, and manufacturing processes. These regulatory hurdles can delay the introduction of promising microbial-derived drugs to the market.

29.5.4 Antimicrobial Resistance

The emergence of antimicrobial resistance presents a significant challenge for drug discovery. As pathogens develop resistance to existing antibiotics, the need for new microbial metabolites with novel mechanisms of action becomes increasingly urgent. Focusing on unexplored microbial niches and employing innovative screening methods can aid in the discovery of compounds effective against resistant strains.

29.6 FUTURE DIRECTIONS IN MICROBIAL DRUG DISCOVERY

29.6.1 Metagenomics and Genomic Mining

Metagenomics, the study of genetic material recovered directly from environmental samples, has opened new avenues for discovering microbial metabolites. By analyzing the genetic potential of unculturable microorganisms, researchers can identify novel biosynthetic pathways and predict the production of new compounds. Genomic mining techniques allow scientists to exploit the genetic blueprints of microbes to engineer strains capable of producing valuable metabolites.

29.6.2 Synthetic Biology

Synthetic biology combines biology and engineering principles to design and construct new biological parts and systems. By manipulating microbial pathways, scientists can optimize the production of bioactive metabolites and even create entirely new compounds with desired properties. This approach holds significant promise for accelerating drug discovery and addressing the challenges posed by antimicrobial resistance.

29.6.3 Collaboration and Interdisciplinary Research

Collaborative efforts among microbiologists, chemists, pharmacologists, and clinicians are essential for successful drug discovery. Interdisciplinary research fosters knowledge exchange and facilitates the translation of findings from the laboratory to clinical applications. Collaborative platforms that integrate diverse expertise can enhance the efficiency and effectiveness of drug discovery efforts.

29.7 CONCLUSION

The search for microbial products with bioactive properties is a dynamic and evolving field that holds great promise for drug discovery. Microbial metabolites have demonstrated remarkable therapeutic potential, leading to significant advancements in medicine. Despite the challenges faced in the discovery and development of these compounds, ongoing research, technological advancements, and collaborative efforts are poised to unlock the full potential of microbial metabolites. As we continue to explore the vast diversity of microbial life, the potential for discovering novel drugs that can combat emerging diseases and address global health challenges remains immense. The future of drug discovery from microbial metabolites is bright, offering hope for innovative therapies that can improve human health and well-being.

SECTION 8:
WASTE DISPOSAL

CHAPTER 30

TREATMENT OF WASTES IN INDUSTRY

The industrial sector plays a vital role in modern economies, contributing to economic growth and job creation. However, industrial activities often generate various types of waste that can pose significant environmental and public health risks if not managed properly. Effective waste treatment is essential to minimize these risks, comply with regulations, and promote sustainability.

30.1 TYPES OF INDUSTRIAL WASTES

Industrial wastes can be classified into several categories based on their origin, composition, and hazardous nature. Understanding these classifications is crucial for developing effective treatment strategies.

30.1.1 Solid Wastes

Solid wastes are generated from various industrial processes and can include scrap materials, packaging waste, and by-products. Common examples include metal shavings from machining, discarded packaging materials, and unused raw materials. The treatment of solid wastes typically involves recycling, incineration, or landfilling, depending on the composition and potential for recovery.

30.1.2 Liquid Wastes

Liquid wastes, often referred to as effluents, are generated during manufacturing processes, cleaning operations, and chemical reactions. These wastes can contain a variety of pollutants, including heavy metals, organic compounds, and toxic chemicals. Treatment methods for liquid wastes vary widely and may include physical, chemical, or biological processes designed to remove contaminants before discharge.

30.1.3 Gaseous Wastes

Gaseous wastes result from combustion processes, chemical reactions, and other industrial activities. Common pollutants in gaseous emissions include particulate matter, volatile organic compounds (VOCs), and greenhouse gases. Treatment of gaseous wastes is critical to reducing air pollution and may involve filtration, scrubbers, or catalytic converters to capture harmful substances.

30.1.4 Hazardous Wastes

Hazardous wastes are those that pose substantial risks to human health or the environment due to their toxicity, reactivity, or infectious nature. Examples include solvents, pesticides, and heavy metals. The treatment of hazardous wastes requires specialized handling and disposal methods to mitigate their harmful effects.

30.2 IMPORTANCE OF WASTE TREATMENT IN INDUSTRY

30.2.1 Environmental Protection

Proper treatment of industrial wastes helps prevent pollution of air, water, and soil, thereby protecting ecosystems and public health. Untreated wastes can contaminate natural resources, harm wildlife, and contribute to climate change. By implementing effective waste treatment practices, industries can minimize their ecological footprint and contribute to environmental sustainability.

30.2.2 Regulatory Compliance

Governments worldwide have established regulations and guidelines governing waste management in industries. Compliance with these regulations is crucial for avoiding legal penalties and maintaining a good reputation. Treatment methods must align with local, national, and international regulations to ensure safe and responsible waste disposal.

30.2.3 Economic Efficiency

Implementing efficient waste treatment practices can lead to cost savings for industries. By recycling and reusing materials, companies can reduce raw material costs and minimize waste disposal fees. Investing in advanced treatment technologies can enhance operational efficiency and create new revenue streams from recovered resources.

30.2.4 Corporate Social Responsibility

In an era where consumers are increasingly conscious of environmental issues, companies are expected to demonstrate corporate social responsibility (CSR) in their operations. Effective waste management practices enhance a company's image and reputation, fostering trust among consumers, stakeholders, and the community.

30.3 TREATMENT METHODS FOR INDUSTRIAL WASTES

30.3.1 Physical Treatment Methods

Physical treatment methods involve the removal of contaminants from waste without altering their chemical composition. Common techniques include:

- **Filtration:** Used to separate solid particles from liquids or gases. It is commonly employed in wastewater treatment to remove suspended solids.

- **Sedimentation:** This process relies on gravity to separate solid particles from liquids. It is often used in the treatment of wastewater to settle out sludge.
- **Incineration:** A thermal treatment process that involves burning waste materials at high temperatures, converting them into ash, gas, and heat. Incineration reduces the volume of waste and can generate energy.

30.3.2 Chemical Treatment Methods

Chemical treatment methods involve altering the chemical composition of waste to neutralize or remove contaminants. These methods include:

- **Neutralization:** This process involves adding acids or bases to neutralize hazardous waste, such as acidic or alkaline effluents, making them safer for disposal.
- **Oxidation and Reduction:** Chemical reactions that transform hazardous substances into less harmful compounds. For example, chlorine can be used to oxidize organic pollutants in wastewater.
- **Precipitation:** A chemical process that involves adding reagents to a liquid waste stream to precipitate dissolved contaminants, such as heavy metals, into solid forms that can be removed.

30.3.3 Biological Treatment Methods

Biological treatment methods utilize microorganisms to break down organic waste and pollutants. These methods are often employed in wastewater treatment and include:

- **Aerobic Treatment:** This process involves the use of oxygen-rich environments to promote the growth of aerobic microorganisms that decompose organic matter. Common examples include activated sludge systems and trickling filters.
- **Anaerobic Treatment:** In the absence of oxygen, anaerobic microorganisms decompose organic waste, producing biogas as a by-product. Anaerobic digestion is often used for treating organic waste from agricultural and food processing industries.
- **Bioremediation:** This technique employs microorganisms to clean up contaminated soil and water by breaking down pollutants. Bioremediation can be applied to treat oil spills, heavy metal contamination, and hazardous waste sites.

30.4 REGULATORY FRAMEWORKS FOR WASTE MANAGEMENT

Effective waste treatment in industries is guided by a complex framework of regulations and standards. Governments at various levels have established guidelines to ensure safe waste management practices.

30.4.1 Environmental Protection Agency (EPA)

In the United States, the Environmental Protection Agency (EPA) plays a critical role in regulating waste management through legislation such as the Resource Conservation and Recovery Act (RCRA) and the Clean Water Act (CWA). These laws establish standards for waste treatment, storage, and disposal, ensuring that industries comply with environmental protection measures.

30.4.2 International Standards

International organizations, such as the World Health Organization (WHO) and the International Organization for Standardization (ISO), provide guidelines and standards for waste management practices globally. For example, ISO 14001 outlines environmental management systems that help organizations minimize their environmental impact.

30.4.3 Local Regulations

In addition to national and international regulations, many regions and municipalities have their own waste management regulations. Industries must comply with these local laws to avoid fines and maintain their operating licenses.

30.5 CHALLENGES IN WASTE TREATMENT

30.5.1 Technological Limitations

While many treatment technologies exist, not all industries have access to advanced waste treatment systems. Smaller companies, in particular, may struggle to afford the necessary infrastructure, leading to inadequate waste management practices.

30.5.2 Complexity of Waste Streams

Industrial waste streams can be complex and variable, containing a mixture of hazardous and non-hazardous materials. This complexity complicates the selection of appropriate treatment methods and requires careful analysis to ensure compliance with regulations.

30.5.3 Public Perception and Awareness

Public awareness and perception of industrial waste management can influence regulatory decisions and community acceptance. Industries must actively engage with communities to promote transparency and address concerns about waste treatment practices.

30.6 FUTURE TRENDS IN INDUSTRIAL WASTE TREATMENT

30.6.1 Circular Economy

The concept of a circular economy emphasizes reducing waste and maximizing resource recovery. Industries are increasingly adopting circular economy principles by implementing practices such as recycling, upcycling, and waste-to-energy conversion. This approach promotes sustainability and reduces reliance on finite resources.

30.6.2 Advanced Treatment Technologies

Innovations in waste treatment technologies, such as membrane filtration, advanced oxidation processes, and bioreactors, are enhancing the efficiency and effectiveness of waste management. These technologies offer new solutions for treating complex waste streams and reducing environmental impact.

30.6.3 Digitalization and Data Analytics

The integration of digital technologies and data analytics is transforming waste management practices. Industries can utilize data-driven insights to optimize waste treatment processes, monitor compliance, and improve decision-making. Smart technologies enable real-time monitoring of waste streams, facilitating more efficient treatment and management.

30.7 CONCLUSION

The treatment of wastes in industry is a critical aspect of sustainable development and environmental protection. As industries continue to grow and evolve, the effective management of waste will remain a significant challenge. By employing appropriate treatment methods, complying with regulations, and embracing innovative technologies, industries can minimize their environmental impact and contribute to a healthier planet. The commitment to responsible waste management not only safeguards natural resources but also enhances corporate reputation and fosters a sustainable future for generations to come.

GLOSSARY

Aerobic Fermentation: A metabolic process that occurs in the presence of oxygen, where microorganisms convert sugars into carbon dioxide and water, producing energy.

Anaerobic Fermentation: A metabolic process that occurs in the absence of oxygen, where microorganisms convert sugars into alcohol or organic acids, producing energy.

Biocatalysis: The use of natural catalysts, such as enzymes, to accelerate chemical reactions in industrial processes.

Biopharmaceuticals: Medicinal products produced using biotechnological methods, including vaccines, antibodies, and gene therapies.

Bioreactor: A vessel or container used for the growth of microorganisms or cells under controlled conditions to produce biological products.

Bioremediation: The use of microorganisms to degrade or remove contaminants from soil, water, or other environments.

Biotechnology: A field of science that uses living organisms, cells, or biological systems to develop products and processes for various applications, including medicine, agriculture, and industry.

Cell Culture: The process of growing cells in a controlled environment, typically outside their natural habitat, for research, production, or therapeutic purposes.

Downstream Processing: The series of processes involved in recovering and purifying biological products after fermentation or cell growth, including separation, concentration, and purification.

Enzyme Engineering: The modification of enzymes to enhance their activity, stability, or specificity for industrial applications.

Fermentation Technology: The application of fermentation processes to produce various products, including food, beverages, and pharmaceuticals.

Fermentation: A metabolic process that converts carbohydrates into alcohol or organic acids using microorganisms under anaerobic or aerobic conditions.

Genetic Engineering: The direct manipulation of an organism's DNA to alter its characteristics, often involving the introduction of new genes.

Industrial Microbiology: The application of microbiological principles and techniques to produce products such as enzymes, antibiotics, and biofuels on an industrial scale.

Microbial Fuel Cells: Devices that use microorganisms to convert organic substrates into electricity through electrochemical reactions.

Microbial Pathogen: A microorganism that can cause disease in humans, animals, or plants.

Monoclonal Antibodies: Laboratory-produced molecules that can bind to specific targets in the body, used for diagnosis and treatment in medicine.

Probiotics: Live microorganisms that provide health benefits when consumed in adequate amounts, often used in food products.

Quality Control (QC): The procedures and tests conducted to ensure that products meet specific quality standards during production.

Recombinant DNA Technology: A set of techniques used to combine DNA from different sources, enabling the production of genetically modified organisms.

Scale-Up: The process of increasing the size or volume of a bioprocess from laboratory or pilot scale to full-scale production.

Screening: The process of testing large numbers of microorganisms or enzymes to identify those with desired traits for industrial applications.

Synthetic Biology: An interdisciplinary field that combines biology and engineering to design and construct new biological parts, devices, and systems.

Upstream Processing: The initial stages of bioprocessing, involving the preparation and cultivation of microbial cells or tissues before product recovery.

Waste Treatment: Processes and methods used to manage and dispose of waste materials, minimizing environmental impact and recovering valuable resources.

BIBLIOGRAPHY

Advances in Industrial Biotechnology: Bioprocessing and Bioreactor Design Publication: Springer Year: 2015

Applied Industrial Microbiology Publication: Academic Press Year: 2014

Bioprocess Engineering: Systems, Equipment, and Facilities Publication: Springer Year: 2010

Bioprocessing of Renewable Resources to Value-Added Products Publication: Wiley Year: 2016

Bioremediation: Principles and Practice Publication: Wiley Year: 2015

Biotechnology and Bioengineering Publication: John Wiley & Sons Year: 2011

Biotechnology and Biopharmaceuticals Publication: CRC Press Year: 2016

Biotechnology and the Biopharmaceutical Industry: A Handbook Publication: Wiley Year: 2016

Biotechnology for Beginners Publication: Greenleaf Book Group Press Year: 2017

Biotechnology for Environmental Sustainability Publication: Springer Year: 2012

Biotechnology: A Very Short Introduction Publication: Oxford University Press Year: 2015

Biotechnology: Academic Cell Update Edition Publication: Academic Press Year: 2015

Fermentation and Biochemical Engineering Handbook Publication: William Andrew Year: 2012

Fermentation Technology: A Practical Approach Publication: Springer Year: 2014

Fundamentals of Industrial Biotechnology Publication: Springer Year: 2016

Industrial Biocatalysis Publication: Wiley Year: 2015

Industrial Biotechnology: A Global Perspective Publication: Wiley Year: 2016

Industrial Biotechnology: Sustainable Production and Use of Biofuels and Biochemicals Publication: Wiley Year: 2017

Industrial Biotechnology: Techniques and Applications Publication: Springer Year: 2016

Industrial Microbiology and Biotechnology Publication: Academic Press Year: 2014

Industrial Microbiology: A Practical Approach Publication: Springer Year: 2016

Industrial Microbiology: A Practical Approach Publication: Taylor & Francis Year: 2008

Industrial Microbiology: A Practical Guide Publication: Wiley Year: 2015

Industrial Microbiology: An Introduction Publication: Wiley Year: 2006

Industrial Microbiology: Practical Applications Publication: Springer Year: 2013

Industrial Microbiology: Principles and Applications Publication: Springer Year: 2013

Industrial Microbiology: The Application of Microbiology to Industrial Processes Publication: Wiley Year: 2015

Microbial Biotechnology in Food and Health Publication: Wiley Year: 2016

Microbial Biotechnology: A Comprehensive Approach Publication: Wiley Year: 2015

Microbial Biotechnology: Fundamentals and Applications Publication: CRC Press Year: 2015

Microbial Biotechnology: Fundamentals of Applied Microbiology Publication: Wiley-Blackwell Year: 2009

Microbial Biotechnology: Principles and Applications Publication: Springer Year: 2016

Microbial Biotechnology: Theory and Applications Publication: Springer Year: 2015

Microbial Fuel Cells: From Fundamentals to Applications Publication: Wiley Year: 2013

Microbial Processes in Industrial Biotechnology Publication: Springer Year: 2014

Principles of Fermentation Technology Publication: Elsevier Year: 2015

Principles of Industrial Microbiology Publication: Jones & Bartlett Learning Year: 2015

INDEX

www.ingramcontent.com/pod-product-compliance
Lightning Source LLC
LaVergne TN
LVHW062217200926
848484LV00006B/1415